Natural Dyes and Pigments of Nepal Himalaya

Netra Lal Bhandari · Ganesh Bhandari ·
Maan Bahadur Rokaya

Natural Dyes and Pigments of Nepal Himalaya

Sources, Extraction, Characterization and Applications

Springer

Netra Lal Bhandari
Central Department of Chemistry
Tribhuvan University
Kirtipur, Kathmandu, Nepal

Maan Bahadur Rokaya
Department of Population Ecology
Institute of Botany
Czech Academy of Sciences
Pruhonice, Czech Republic

Department of Biodiversity Research,
Global Change Research Institute
Czech Academy of Sciences
Brno, Czech Republic

Ganesh Bhandari
Department of Nano Convergence
Engineering
Jeonbuk National University
Jeonju, Republic of Korea

ISBN 978-3-032-23228-1 ISBN 978-3-032-23229-8 (eBook)
https://doi.org/10.1007/978-3-032-23229-8

This Springer imprint is published by the registered company Springer Nature Switzerland AG
The registered company address is: Gewerbestrasse 11, 6330 Cham, Switzerland

Preface

Natural dyes and pigments have been in use in our civilisation since prehistoric times (Mollon 2003; Feeser et al. 2012). Before the discovery of synthetic dyes and pigments, the entire world was dependent on natural dyes and pigments for colouring textiles, decorating ceremonial and cultural items. These colourants were obtained from plant species, animal species, and minerals found across diverse ecological zones in Nepal (Press et al. 2000; Miehe et al. 2015). Natural dyes and pigments are important in expressing identity, social values, artistic creativity, and spiritual significance (Manandhar 2002).

The use of natural dyes in Nepal dates back to between 400 and 650 CE, and was recorded through archaeological discoveries in Samdzong, Upper Mustang, Nepal (Gleba et al. 2016). The use of colour is expressed in a unique and distinctive way. The country is rich in biodiversity, and the unique cultural heritage is reflected in the presence of 125 castes and ethnic groups speaking 123 languages (CBS 2023). This diversity can be well observed in Nepali art, architecture, rituals, and especially in traditional clothing (Malla 1993). Thus, natural dyes and pigments are an integral part of daily and ceremonial life in Nepal and provide the vibrant colours that are used during weddings, festivals, and religious ceremonies. Natural dyes and pigments are of significance because they are biodegradable, non-toxic, hypoallergenic, and capable of producing fine and beautiful colour variations that are impossible to replicate synthetically (Alegbe and Uthman 2024). Thus, natural dyes and pigments are still preferred, and it is important to have an in-depth study that not only systematically documents traditional knowledge regarding the use of dyes and pigments but also explores their sources, chemistry, extraction methods, characterisation, and applications.

Natural dyeing traditions have been maintained for generations in Nepal, particularly among indigenous and ethnic groups. However, this knowledge is not yet systematically documented, and there is an increasing risk of its disappearance (Malla 1993; Rajbhandari 2001; Manandhar 2002; Baral and Kurmi 2006). The use of dyes and pigments is also largely dominated by synthetic dyes. This book has been prepared to document, protect, and revitalise this important long-lasting cultural heritage by connecting it with scientific research and modern technological insights, ensuring the cultural and ecological significance of natural dyes and pigments for future generations.

This book, *Natural Dyes and Pigments of Nepal: Types, Extraction, Characterisation and Application*, presents comprehensive knowledge on natural dyes and pigments in seven chapters that guide readers from historical and natural prospects to scientific approaches, extraction techniques, analytical approaches, mordants and their chemistry, and applications. Chapter 1 introduces the history and origins of natural dyes and pigments that are obtained from plant species, animal species, minerals, and other sources. It presents the chemical structures of different compounds of dyes and pigments. It shows the significance of the use of natural dyes and pigments in our daily lives, along with their uses in food, medicine, crafts, and cultural items. Chapter 2 explains the chemistry of dyes and pigments by distinguishing natural dyes and pigments from synthetic ones. Chapter 3 presents a wide range of extraction techniques, including traditional aqueous and alkaline processes as well as advanced solvent-based, ultrasonic, microwave-assisted, enzymatic, and supercritical fluid methods, comparing them in terms of efficiency, suitability, and practical limitations. Chapter 4 focuses on the characterisation of natural dyes and describes physical, spectroscopic, chromatographic, microscopic, and thermal methods used to assess stability, safety, and quality. Chapter 5 covers mordants used in dyeing in depth, including natural, metallic, and emerging nano-mordants. This chapter explains the chemistry, bonding mechanisms, and environmental implications of different mordants. Chapter 6 deals with the use of natural dyes and pigments across sectors in textiles, cosmetics, leather processing, food products, pharmaceuticals, polymer colouring, and hair colouring. It also features advanced applications in dye-sensitised solar cells and corrosion inhibition in metals. Last but not least, Chap. 7 covers the feasibility of natural dyeing in Nepal by looking into traditional practices, market potential, sustainability challenges, and environmental impacts. It also explores with how the use of natural dyes and pigments can be made more practically on an industrial scale in Nepal.

This book, in seven chapters, integrates different aspects of natural dyes and pigments in detail. At present, there is widespread concern about environmental pollution and human health issues that are associated with synthetic dyes. This book argues that the use of natural dyes and pigments is an important source of sustainable, biodegradable, and culturally significant alternatives. Due to its rich biological resources and deeply rooted traditional knowledge, the Nepal Himalayas are well-positioned to contribute meaningfully to the global transition towards environmentally friendly and responsible colour technologies.

This book is expected to serve the interests of students, researchers, artisans, conservationists, environmentalists, industrialists, and policymakers who are not only keen to understand natural dyes and pigments but also interested in their application at a personal or industrial scale. This book will support the conservation of natural dyes and pigments and also inspire scientific innovation and sustainable industrial development. We express sincere gratitude to the scholars, artisans, communities, and institutions whose knowledge and support have contributed to the writing of this book. We hope that this book will serve as a valuable piece of information for future generations and contribute to safeguarding indigenous knowledge, strengthening scientific understanding, and providing opportunities for indigenous and rural communities to enhance their livelihoods through the industrialisation of natural dyes and pigments.

Kathmandu, Nepal
Netra Lal Bhandari
netra.bhandari@cdc.tu.edu.np

Jeonju, South Korea
Ganesh Bhandari
ganesh@jbnu.ac.kr

Prague, Czech Republic
November 2025
Maan Bahadur Rokaya
rokaya.m@czechglobe.cz

References

Alegbe EO, Uthman TO (2024) A review of history, properties, classification, applications and challenges of natural and synthetic dyes. Heliyon 10:e33646. https://doi.org/10.1016/j.heliyon.2024.e33646

Baral SR, Kurmi PP (2006) A compendium of medicinal plants in Nepal. Rachana Sharma, Kathmandu, Nepal

CBS (2023) National Census 2021 (Preliminary Report). Central Bureau of Statistics, Government of Nepal, Kathmandu, Nepal

Feeser A, Goggin MD, Tobin BF (2012) The materiality of color: the production, circulation, and application of dyes and pigments, 1400–1800. Ashgate Publishing, Ltd

Gleba M, Vanden Berghe I, Aldenderfer M (2016) Textile technology in Nepal in the 5th–7th centuries CE: the case of Samdzong. STAR: Science and Technology of Archaeological Research 2:25–35. https://doi.org/10.1080/20548923.2015.1110421

Malla L (1993) The natural dyes of Nepal. Research centre for applied science and technology, Tribhuvan University, Kirtipur, Nepal

Manandhar NP (2002) Plants and people of Nepal. Timber Press, Portland, Oregon, USA

Miehe G, Pendry CA, Chaudhary R (eds) (2015) Nepal: An introduction to the natural history, ecology and human environment in the himalayas. Edinburgh, UK: Royal Botanic Garden Edinburgh

Mollon JD (2003) The origins of modern color science. In: The science of color. Elsevier, pp 1–39

Press JR, Shrestha KK, Sutton DA (2000) Annotated checklist of the flowering plants of Nepal. Natural History Museum, London

Rajbhandari KR (2001) Ethnobotany of Nepal. Ethnobotanical Society of Nepal, Kathmandu, Nepal

Acknowledgments We gratefully acknowledge the various publishers and authors who granted permission to reproduce figures throughout this book. Permissions were granted for the following: Fig. 2.2 (Saikhao et al., 2018, *J. Clean Prod.* 197:106–113, Elsevier); Fig. 2.3 (Habib et al., 2024, *Sci. Rep.* 14:1–11, Nature); Fig. 2.4 (Pramod, 2023, *Peer Rev Ref J* 10:66–71); Fig. 2.5 (Barnett et al., 2006, *Opt. Laser Technol.* 38:445–453, Elsevier); Fig. 3.9 (Abou Elmatty et al., 2022, *Processes* 10:2111, MDPI); Figs. 4.1a, 4.3c, and 4.6 (Bhandari et al., 2021a, *Asian J. Chem.* 33:404–410); Fig. 4.1b (Al4.1b 2024, *ACS Omega* 9:18273–18277, American Chemical Society); Figs. 4.1c, 4.1d and 4.3c (Chang et al., 2013, *Int. J. Photoenergy* 2013:1–8, Hindawi); Figs. 4.3d and 6.7 (Bhandari et al., 2021b, *Egypt J. Chem.* 64:6681–6690); Fig. 4.4 (Ahn & Obendorf, 2006, *Fibers Polym.* 7:158–163, Springer Nature); Fig. 4.5 (Jamadar & Sannapapamma, 2018, *The Pharma Innovation J.* 7:474–479); Fig. 4.7 (Chajii et al., 2025, *Fibers Polym.* 26:2427–2448, Springer Nature); Fig. 4.8 (Nguyen & Bechtold, 2021, *J. Clean Prod.* 315:128195, Elsevier); Fig. 4.9 (Fajraoui et al., 2019, *Polym. Test.* 78:105988, Elsevier); Figs. 5.1 and 5.2 (Banc et al., 2023, *Foods* 12:270, MDPI; Zhang et al., 2022, *Arab J. Chem.* 15:103624, Elsevier; Lee et al., 2010, *J. Ethnopharmacol.* 131:567–574, Elsevier; Uemori et al., 2025, *Molecules* 30:2216, MDPI); Fig. 5.4 (El-Azab et al., 2010, *J. Dispersion Sci. Technol.* 31:1150–1156, Taylor & Francis); Fig. 5.5 (Gedik et al., 2014, *Fibers Polym.* 15:261–272, Springer Nature); Fig. 5.6 (Bukhari et al., 2017, *Text Cloth Sustain* 3:3, Springer Nature); Fig. 5.8 (Islam et al., 2025, *Environ. Sci. Pollut. Res.* 32:25511–25540, Springer Nature); Fig. 5.9 (Bhandari et al., 2020, *J. Inst. Sci. Tech.* 25:30–36); Fig. 5.10 (Khan et al., 2015, *J. Nat. Fibers* 12:243–255, Taylor & Francis); Figs. 5.11 and 5.12 (Sadannavar et al., 2024, *Sustainability* 16:6832, MDPI); Fig. 5.13 (Taheri et al., 2015, *Orient J. Chem.* 31:85–96); Fig. 5.14 (Singh et al., 2019, *Sustain. Chem. Pharm.* 14:100184, Elsevier); Fig. 5.15 (Geelani et al., 2017, *Text Cloth Sustain* 2:8, Springer Nature); Fig. 5.16 (İşmal & Yıldırım, 2019, Woodhead Publishing, pp. 57–82); Fig. 5.17 (Jha et al., 2015, *Der Pharmacia Lettre* 7:185–195); Figs. 5.18 and 5.19 (Assefi Pour et al., 2020, *Heliyon* 6:e04911, Elsevier); Fig. 5.20 (Negi, 2025, *Polymers* 17:871, MDPI); Figs. 5.22–5.25 (Dulo et al., 2021, *Waste Biomass Valor.* 12:6339–6374, Springer Nature); Fig. 6.1 (Semwal et al., 2012, *Trends Appl. Sci. Res.* 7:392–399); Fig. 6.6 (Sivakumar et al., 2009, *Ultrason. Sonochem.* 16:782–789, Elsevier); and Fig. 6.7 (Uddin et al., 2017, *ASRJETS* 34:1–7). All figures and tables used in this book are properly cited in the text and referenced.

We are grateful to Binu Timsina, Kamala Khanal, Trishna Bhandari and Asbin Rokaya for preparing the picture plates, editing the English-language texts, and proofreading. We are indebted to Rameshwar Adhikari, Sirisha G.C., Kabita Bist, Num Kant Parajuli, Santosh Chaudhary, Jyoti Ghimire, Basant Pokhrel, and Janak Bahadur Shahi, who contributed, either directly or indirectly, to the writing of this book.

The first author wishes to thank the University Grants Commission (UGC), Nepal, Bhaktapur, Nepal; the Nepal Academy of Science and Technology (NAST), Lalitpur, Nepal; the National Youth Council, Sanothimi, Bhaktapur, and the Provincial Research Project 2020, Lumbini Province, Nepal, for providing partial support for this research on natural dyes. Several results from his previous work are clearly presented in a number of chapters.

Competing Interests The authors have no competing interests to declare that are relevant to the content of this manuscript.

Introduction

Natural dyes and pigments are an important part of culture, art, and traditions in Nepal. Their use dates back to 400–650 CE. Despite the growing dominance of synthetic dyes, natural dyes and pigments are still in use for their biodegradability, safety, and unique aesthetic qualities. However, the traditional knowledge related to the use of natural dyes and pigments is under threat and is at risk of disappearing. This book documents and aims at revitalising the heritage by combining indigenous practices with modern scientific insights that are mentioned in the seven chapters. These chapters cover historical context, chemistry, extraction methods, characterization techniques, mordants, applications across various industries, and the feasibility of large-scale natural dyeing in Nepal. This book aims to support researchers, artisans, industrialists, and policymakers in promoting sustainable dyeing technologies and help in safeguarding cultural knowledge and enhancing livelihood opportunities for local communities.

Contents

About the Authors

Netra Lal Bhandari is currently employed as a researcher and faculty member of organic chemistry at the Central Department of Chemistry, Tribhuvan University (TU), Kathmandu, Nepal. Bhandari received his M.Sc. in Chemistry (Organic and Natural Products) in 2003 and his Ph.D. in natural fibre-based polymer composites under a fellowship from the University Grants Commission (UGC), Nepal, in 2014 (in collaboration with research institutes in India, Germany, and France) at TU, Nepal. His research primarily focuses on plant fibres, fibre modifications, fibre composites, the extraction and use of essential oils in cosmetics, and the development of green methods for extracting plant dyes for use in dyeing fabrics, textiles, hair, and other materials. He has published over 30 scientific research papers, including reviews. He is an active reviewer and supervisor of M.Sc. and Ph.D. research students.

Ganesh Bhandari is a Ph.D. scholar working on energy conversion and storage at the Department of Nano Convergence Engineering, Jeonbuk National University, Jeonju, South Korea. Bhandari received his M.Sc. in chemistry (organic chemistry) in 2019 from Tri-Chandra Multiple Campus, Tribhuvan University (TU), Kathmandu, Nepal. He completed his MSc dissertation on natural dye extraction from plants and their use for fibre/fabric dyeing and composite preparation under the fellowship of the Province Planning Commission, National Youth Council, Nepal, and Nepal Academy of Science and Technology, Kathmandu, Nepal. His research mainly focuses on plant-natural dye extraction, fibre and fabric dyeing, composite preparation, and preparation of electrocatalyst for water splitting. He has published over 17 scientific research papers, including reviews.

Maan Bahadur Rokaya is currently working as a research scientist at the Institute of Botany (Pruhonice), and Global Change Research Institute (Brno), both under Czech Academy of Sciences. Rokaya obtained his Master of Science (M.Sc.) in Botany (Plant Taxonomy and Systematics) from Tribhuvan University, Kathmandu, Nepal, in 2001, and his Ph.D. in Plant Ecology from Charles University in Prague, Czech Republic, in 2011. His research mainly focuses on plant population dynamics of important species, herbivory patterns along elevational gradients, the effects of climate change on plant species, and the uses of plants for human benefit, primarily in the Nepal Himalayas.

Chapter 1
History, Origin, and Uses of Natural Dyes and Pigments

Abstract History, sources, and uses of dyes and pigments derived from plants, insects, animals, and minerals are discussed in this chapter. The earliest use of natural dyes and pigments dates back to the Neolithic period. Due to their non-toxicity, biodegradability, and environmental sustainability, natural dyes and pigments are preferred over synthetic dyes and pigments. Thus, textile, food, and pharmaceutical industries widely use natural dyes and pigments, which also have medicinal properties having several health benefits.

1.1 Importance of Natural Dyes and Pigments

Dyes are substances that are soluble in water and used to colour clothes, paper, leather, and other materials commonly used daily in life. Dyes generally are permanent, and they do not change their colour or fade away when subjected to washing, heat, and light (Nambela 2023). The word 'dye' is derived from the Middle English word 'deie' or the Old English words 'dag' or 'dah', meaning 'colour or hue' (Anonymous 2025).

Pigments are substances that are insoluble in nature and are also used to colour different items. Pigments need to be suspended in a binder or medium such as oil, acrylic, or resin to attach to a surface. The word 'pigment' is derived from the Latin word 'pigmentum', meaning 'colouring substance'.

Dyes and pigments come from both natural and artificial sources (Kumar et al. 2021). Natural sources of dyes or pigments are plants, minerals, insects, and fungi. These sources are often preferred because they are environmentally sustainable, as they are biodegradable and cause less pollution. Natural dyes and pigments are traditionally used, and they are culturally intertwined with rituals, social identity, and unique craftsmanship passed down through generations (Iqbal and Ansari 2021). For example, the Tharu tribe in Nepal uses traditional tattoos, and knowledge of Japan's indigo dyeing has been passed down through generations (Manandhar 2002). Natural dyes or pigments have economic as well as social value. Thus, the cultivation of plant species for yielding dye or pigment can help in generating income for farmers or indigenous people who grow or collect them for trade.

N. L. Bhandari et al., *Natural Dyes and Pigments of Nepal Himalaya*,
https://doi.org/10.1007/978-3-032-23229-8_1

Pigments are obtained from both organic and inorganic sources. Examples of pigments obtained from organic sources are chlorophyll from plants and carmine from insects. Likewise, titanium dioxide (white) are obtained from inorganic sources by processing titanium-rich ores and ultramarine blue was originally obtained from lapis lazuli (Gürses et al. 2016).

1.2 History of Use of Dyes and Pigments in the World, the South Asian Continent and the Nepal Himalayas

The use of natural dyes and pigments is as old as human civilisation, with its origin dating back to the Neolithic period, which was around 8,000–3,000 B.C. (Mollon 2003; Feeser et al. 2012). Early use of dyes is found in cave paintings using ochre, a substance obtained from earth, charcoal, and plant parts such as fruits, leaves, and roots. The history of weaving dates back to around 5000 B.C., and this weaving slowly led to the use of dyeing clothes. Popular examples include the use of indigo colour obtained from the leaves of the indigo plant (*Indigofera tinctoria* L., Fabaceae) and woad herb (*Isatis tinctoria* L., Brassicaceae). These are some of the oldest known dyes used by human beings (Mullen 2008).

Nowadays, dyeing has become a much more sophisticated art than it was in ancient civilisations. By 3000 B.C., the first dye made using complex processes was known as Egyptian blue, copper calcium silicate ($CaCuSi_4O_{10}$). It was made by heating quartz sand, lime, a copper compound, and a flux like natron. After the Egyptians, the Chinese and other cultures developed more complicated methods for making dyes and using them to colour various items in their daily lives. The use of rare Tyrian purple in the Roman Empire and the madder-dyed textiles are 5,000 years old, and this shows the use of advanced methods for dyeing that slowly spread around the world, ultimately becoming part of people's cultures (Susmann 2020). In the dyeing process, the ancient methods used different mordants that helped produce a range of colours from a single dye or pigment source (Gaboriaud-Kolar et al. 2014). However, the use of natural sources suddenly reached an end after the Industrial Revolution (1760–1840 A.D.). William Henry Perkin, the 18-year-old British chemist, in 1856 accidentally discovered the first synthetic dye known as mauveine. It is also known as aniline purple or Perkin's mauve, and it was produced by the oxidation of impure aniline while trying to synthesise quinine. The purple dye that Perkin made was used to colour silk and many textiles, which then led to the establishment of artificial dye industries around the world (Johnston 2008).

The Indus Valley Civilisation (c. 3300–1300 BCE) has the earliest record of dyeing in the Indian subcontinent. Studies have confirmed that there was use of indigo (*Indigofera tinctoria* L., Fabaceae) and madder (*Rubia cordifolia* L. and *Rubia tinctorum* L., Rubiaceae), alongside saffron (*Crocus sativus* L., Iridaceae) and lac dye obtained from the lac resin secreted by the female lac insect, primarily the species *Kerria lacca* that belongs to the family Kerriidae (Balfour-Paul 2013). The dyeing methods became more advanced during the mediaeval and Mughal periods and are well exemplified by the *Kalamkari*, an ancient textile printing art (Sengupta 2019).

The first recorded use of dyes in Nepal was observed between 400 and 650 CE, and it was recorded in archaeological discoveries in Samdzong, Upper Mustang, Nepal (Gleba et al. 2016). Traditional painting on cotton or silk, usually depicting a Buddhist deity, scene, or mandala, is known as a *thangka* painting and is famous in Nepal. These *thanka* paintings use dyes or pigments obtained from plants, animals, and minerals. The famous traditionally woven cotton fabrics known as *Dhaka* in Nepal are also coloured using dyes or pigments obtained from local resources, as in *Thanka* (Pal 1974).

1.3 Origin of Dyes and Pigments

Although the use of artificial dyes and pigments is widespread, natural sources are still being used as they were in the past. The main sources of natural dyes and pigments are plant species.

1.3.1 Plant Species

There are about 450 plant species that are used as sources of dyes or pigments in India (Siva 2007), and 182 plant species in Nepal (Malla 1993) have been identified. However, the list given by Malla (1993) includes plant species that were not grown in Nepal and were imported from other countries, such as India and China. At present, handmade clothes often use natural dyes.

Different plant parts, such as roots, leaves, nuts, bark, outer shells, flowers, are the sources of natural dyes and pigments (Manandhar 2002; Bhandari et al. 2020b). These plant parts contain different compounds that give a range of colours. For example, yellow, one of the earliest used colours, was first obtained from the roots of turmeric (*Curcuma longa* L., Zingiberaceae), and a compound known as curcumin gives the yellow colour (Alemayehu and Teklemariam 2014; Gupta 2019). Pomegranate rind (*Punica granatum* L., Lythraceae) gives a range of colours from buttery and golden yellows to greens, greys, and blacks, and the compound granatonine is responsible for these colours (Gupta 2019). The yellow colour is obtained from oak bark (*Quercus velutina* Lam., Fagaceae), and the compound known as quercitron is responsible for giving its colour (Hansen 2011). Likewise, for blues, the classic dye indigo is extracted from the leaves of the *Indigofera* plant (*Indigofera tinctoria* L., Fabaceae), and it contains a compound known as indigotin. A similar shade comes from the woad plant (*Isatis tinctoria* L., Brassicaceae), and it contains a compound known as indirubin (Abdel-Kareem 2012). The spectrum of red and brown colours is obtained from henna leaves (*Lowsonia inermis* L., Lytheraceae) with a compound called lawone (Alem et al., 2020). The reddish-brown colour of safflowers (*Carthamus tinctorius* L., Asteraceae) contains the compound carthamine. Colours such as yellow to orange for yellow/brown onions and pink to maroon or even green for red onions contain the important compound pelargonidin (Gupta 2019). Chutro (*Berberis asiatica* Roxb. Ex DC., Berberidaceae) produces a range of colours due to different colouring compounds that are found in different plant parts. The different compounds are karachine, berberine, palmitine, oxyberberine, aromoline, berbamine, and oxyacanthine. The bark and roots of *B. asiatica* contain several alkaloids that impart a yellow colour (Gilbert (nee Stoker) and Cooke 2001).

1.3.1.1 List of Dye-Yielding Plant Species and Compounds

Table 1.1 enumerates 200 plant species, belonging to 166 genera and 83 families. All these plant species are found in Nepal and are used as a main source of natural dyes and pigments. Table 1.2 lists plant species that are imported from other countries for the purpose of obtaining colourants. It enumerates 32 species, belonging to 29 genera and 21 families. Similarly, Table 1.3 lists plant species used as mordants during the dyeing process, and it enumerates 16 species, 13 genera, and 13 families.

There was a total of 316 compounds present in plant species listed in Tables 1.1, 1.2 and 1.3. The compounds are categorised into ten categories (Figs. 1.1, 1.2, 1.3, 1.4, 1.5, 1.6, 1.7, 1.8, 1.9 and 1.10), and they are alkaloids (1–36), glycosides (37–58), lipids and fatty acids (59–64), organic acids (65–69), phenolics (70–160), polysaccharides (161–163), quinones (164–188), terpenoids (189–257), vitamins and pigments (258–274), and other (275–316). The 'other' category represents a range of natural compounds, such as sulphur compounds; iridoid and phenolic glycosides; xanthones and benzophenones; coumarins; polyacetylenes; sterols; fatty acids and vitamin E derivatives; amines (e.g., dopamine, serotonin, histamine); lichen acids; pyrethrins; cinerins; gingerols; shogaols; and usnic acid.

1.3.2 Insect Species and Animal Species

Beyond plant species, some of the most desired dyes and pigments are obtained from insects and marine animals (Degano et al. 2009). These animal-derived colours are intense and difficult to obtain from plants. Thus, they are used in luxury and valuable commodities.

The cochineal insect contains carminic acid that is responsible for giving a red colour when alum as a mordant is used. Likewise, kermesic acid from the Kermes insect, when mixed with an alum mordant, produces a prestigious crimson colour. Laccaic acids, derived from the secretion of lac insects on plants, also require a mordant like alum to give brilliant red and violet colours. However, the legendary Tyrian purple obtained from the Murex snail contains the 6,6'-dibromoindigo compounds, which does not require any mordant to give a good colour. Tekhelet blue, obtained from the related *Hexaplex trunculus* snail, contains the indigotin compound,

Table 1.1 List of plant species from which dye or pigments are obtained and that are found in Nepal. The plant species are arranged alphabetically with any synonyms in parentheses (), followed by the family in square brackets []. The available vernacular names are given for each plant and are arranged in alphabetical order: Dolpali (Dn), Chepang (Che), Danuwar (Dan), Darai (Dar), English (Eng), Gurung (Gur), Hindi (Hin), Limbu (Lim), Lepcha (Lep), Magar (Mag), Moonsoor (Moos), Nepali (Nep), Newari (Newa), Rai (Rai), Sherpa (She), Tamang (Tam), Thakali (Tha), and Tibetan (Tib). The plant parts used to obtain dye or pigment are followed by active compounds found in each plant species, their medicinal uses, and altitudinal ranges in Nepal (E as East, C as Centre, and W, as West Nepal) and distribution ranges in Nepal and the world. In the end, references are given

Latin names [Family]	Local names	Plant parts	Dye or pigment	Active compounds	Medicinal Uses	Altitudinal range, distribution in Nepal and in the world	References
Abies spectabilis (D. Don) Mirb. [Pinaceae]	Himalayan Silver Fir (Eng), Talish patra (Nep); Talisa, Talish Patra (San); Buk (Dn), Ba-lu (Tib); Kye (Manang)	Young cones	Voilet	Volatile oils (α-Pinene, β-Pinene, Limonene), Flavonoids, Lignans	Resin used for wounds, bronchitis, and as an incense	2400–4400 m (WC); Himalaya (Chitral to C. Nepal)	Malla (1993), Bhandari et al. (2020c)
Acacia catechu (L. f.) Willd. [Leguminosae]	Catechu Tree, Cutch tree (Eng), Khayer (Nep); Khair (Dan); Khaira (Dar); Khayar (Tha.); Khadira (San)	Woods, nuts, pods	Black, Brown, dark red	Flavonoids (Catechin, Epicatechin), Tannins, Phlobatannins	Used for sore throat, diarrhoea, and skin conditions	200–1400 m (WCE); Tropical Himalaya, India, Burma, Thailand, S. China	Malla (1993), Dangol (2002, 2008), Manandhar (2002), Nautiyal and Kaul (2003), Sigdel and Rokaya (2011), Bhandari et al. (2020c)

(continued)

Table 1.1 (continued)

Latin names [Family]	Local names	Plant parts	Dye or pigment	Active compounds	Medicinal Uses	Altitudinal range, distribution in Nepal and in the world	References
Acacia farnesiana Wild	Ganhaaune Khayer, Syaanimaini (Nep)	Bark	Black	Tannins, Flavonoids, Volatile oils (Farnesol), Alkaloids	Used for diarrhoea, sore throat, and skin inflammation	200–1100 m (WCE Nepal); Pantropic, a native of tropical America	Nautiyal and Kaul (2003)
Acacia nilotica (L.) Willd. ex Delile (*Acacia arabica* wild) [Leguminosae]	Gum arabic tree, Babul, Black babul (Eng), Babul, Babur (Nep); Babboola (San); Babul (Hin)	Woods, nuts, pods	Black	Tannins (Gallic acid, Ellagic acid), Flavonoids, Polysaccharides (Gum)	Used for diarrhoea, bleeding disorders, and dental hygiene	Up to 1500 m (?), Angola, Botswana, Tanzania, Malawi, Mozambique, Zambia, Zimbabwe and Limpopo, Mpumalanga, KwaZulu-Natal, South Africa, and east through the Arabian Peninsula to the Indian subcontinent and Burma	Malla (1993), Nautiyal and Kaul (2003)

(continued)

Table 1.1 (continued)

Latin names [Family]	Local names	Plant parts	Dye or pigment	Active compounds	Medicinal Uses	Altitudinal range, distribution in Nepal and in the world	References
Acacia rugata (Lam.) Voigt (*Acacia concinna* DC) [Leguminosae]	Soap pod (Eng), Sikakaee, Araare (Nep), Rangaso (Che), Aila (Tha)	Bark	Black	Saponins, Alkaloids, Tannins	Pods used as a hair cleanser and for skin diseases	400–800 m (WC); India, S. E. Asia, S. China, Malaysia	Malla (1993)
Aegle marmelos (L.) Correa [Rutaceae]	Bengal Quince (Eng), Bel (Nep)	Fruit rind, bark	Yellow	Alkaloids (Aegeline, Marmeline), Coumarins (Marmelosin, Scopoletin), Tannins	Unripe fruit is a well-known remedy for diarrhoea and dysentery; leaves for diabetes	600–1100 m (WCE Nepal); Himalaya (Kashmir to Nepal), India, Burma, Indo-China, Malaysia. Widely cultivated	Malla (1993), Nautiyal and Kaul (2003), Bhandari et al. (2020c)
Ageratina adenophora (Spreng.) R. M. King and H. Rob. (*Eupatorium adenophorum* Spreng.) [Asteraceae]	Ban mara, Kal jhar (Nep)	Leaves	Yellow	Diterpenes (Euptox A), Essential oils	Used to stop bleeding and for wound healing. Considered an invasive weed and can be toxic to livestock	850–2200 m (CE Nepal); A pantropic weed, native of Mexico	Bhandari et al. (2020c)

(continued)

Table 1.1 (continued)

Latin names [Family]	Local names	Plant parts	Dye or pigment	Active compounds	Medicinal Uses	Altitudinal range, distribution in Nepal and in the world	References
Ageratum conyzoides L. [Asteraceae]	Goat Weed (Eng), Ganmane ghans, Gane (Nep)	Flower	Varies with mordants (yellow, purple, and red)	Chromenes (Precocene I, II), Essential oils, Flavonoids, Alkaloids	Used to stop bleeding, for wounds, and as an anti-inflammatory; for pneumonia and fever	200–2000 m (WCE); Pantropical, widespread weed	Malla (1993)
Albizia odoratissima Benth. [Fabaceae]	Kalo Siris (Nep)	Bark	Brown	Tannins, Flavonoids, Saponins, Alkaloids	Bark used for asthma, allergies, and as an anti-anxiety agent	100–500 m (CE Nepal); Himalaya (Kumaun to Bhutan), India, Ceylon	Nautiyal and Kaul (2003)
Allium cepa L. [Liliaceae]	Onion (Eng), Pyaj (Nep)	Skin	Gold, yellow	Organosulfur compounds (Allicin, Quercetin), Flavonoids, Saponins	Used for colds, coughs, and as an antimicrobial; for cardiovascular health	Cultivated as a vegetable	Malla (1993), Bhandari et al. (2020c)

(continued)

Table 1.1 (continued)

Latin names [Family]	Local names	Plant parts	Dye or pigment	Active compounds	Medicinal Uses	Altitudinal range, distribution in Nepal and in the world	References
Alnus nepalensis D. Don [Betulaceae]	Alder, Nepal Black Cedar (Eng), Utis, Boshi swa (Nep), Bhomsing (Grur.), Bisuwaa (Rai), Koplong (Sher)	Bark	Brownish black Dyeing and tanning	Tannins, Flavonoids (Diarylheptanoids), Triterpenoids	Bark used for diarrhoea, dysentery, and internal haemorrhage	500–2600 m (WCE Nepal); Himalaya (Garhwal to Bhutan), Assam, Tibet, Burma, Indo-China, W. China	Malla (1993), Manandhar (2002)
Alnus nitida (Spach) Endl. [Betulaceae]	Utis (Nep)	Bark	Brownish black Dyeing and tanning	Tannins, Flavonoids, Triterpenoids	Used for sore throat, burns, and skin inflammation	2100–2200 m (WC Nepal); Himalaya (Kashmir to Nepal)	Manandhar (2002)
Aloe vera (L.) Burm. f. (*Aloe barbadensis* Mill.) [Asphodelaceae]	Ghiu kumari (Nep)	Whole plant	Red	Anthraquinones (Aloin, Aloe-emodin), Polysaccharides (Acemannan), Mannans	Gel is used topically for burns, wounds, and skin inflammation; juice is used internally for constipation and digestion	1200–1400 m (WC Nepal); Mediterranean and Canary Is., naturalised in Florida, West Indies, Central America and Asia	Bhandari et al. (2020c)

(continued)

Table 1.1 (continued)

Latin names [Family]	Local names	Plant parts	Dye or pigment	Active compounds	Medicinal Uses	Altitudinal range, distribution in Nepal and in the world	References
Amaranthus spinosus L. [Amaranthaceae]	Spiny Pigweed (Eng), Guti ghans, Kande lundo (Nep), Rangnan (Che), Katgeni (Dan), Kataya (Gur)	Leaves	Red	Flavonoids, Triterpenoids, Tannins, Saponins, Betalains	Used to treat fever, boils, eczema, and diarrhoea; also used as a galactagogue (promotes milk flow) and diuretic	150–1200 m (WCE Nepal); Cosmopolitan warm temperate and tropical weed	Malla (1993)
Annona reticulata L. [Annonaceae]	Ramphal (Nep)	Fruit	Red	Acetogenins (Annonacin, Reticulatacin), Alkaloids (Anonaine)	The bark is used as an astringent for diarrhoea and dysentery; unripe fruit is used for worms	E Nepal; West Indies. Cultivated throughout the tropics (Custard Apple)	Nautiyal and Kaul (2003)
Ardisia solanacea Roxb. (*Ardisia humilis* Vahl) [Myrsinaceae]	Damaai Phal, Tola para (Nep)	Red juice of the berries	Yellow	Saponins, Tannins, Flavonoids	Used for fever, rheumatism, and skin diseases	200–1100 m (WCE Nepal); Tropical Himalaya (Kumaun to Sikkim), India, Ceylon, Burma, W. China, Malaya	Malla (1993)

(continued)

Table 1.1 (continued)

Latin names [Family]	Local names	Plant parts	Dye or pigment	Active compounds	Medicinal Uses	Altitudinal range, distribution in Nepal and in the world	References
Arnebia benthamii (Wall. ex G. Don) I.M. Johnst	Dimok (Am); Koma, Muktsi (Km); Maharangi (Dn, Nep)	Root stock	Purple, Red	Naphthoquinones (Shikonin, Acetylshikonin), Phenolic acids	Root used for burns, wounds, and skin diseases; known for its red dye and antimicrobial properties	2800–3800 m (W Nepal); Himalaya (Kashmir to Nepal)	Ghimire et al. (2001), Lama et al. (2001), Rokaya et al. (2010)
Artocarpus heterophyllus Lam. [Moraceae]	Rukh katahar (Nep)	Bark	Yellow	Flavonoids (Artocarpin, Morin), Stilbenoids	Root used for diarrhoea and asthma; fruit is nutritious	800 m (C Nepal); Himalaya (Nepal), Assam, India, Ceylon, Burma. Known as the Jack Fruit, this tree is cultivated throughout the Tropics and is often found as an escape. It is thought to be indigenous along the Western Ghats of India	Manandhar (2002), Nautiyal and Kaul (2003)

(continued)

Table 1.1 (continued)

Latin names [Family]	Local names	Plant parts	Dye or pigment	Active compounds	Medicinal Uses	Altitudinal range, distribution in Nepal and in the world	References
Artocarpus integra (Thunb.) Merr. (*Artocarpus integrifolia* L. f.) [Moraceae]	Rukh katahar (Nep)	Wood and Bark	Yellow	Flavonoids (Artocarpin, Cycloartocarpin), Stilbenoids	Similar to jackfruit; wood used for treating skin diseases	200–1700 m (E Nepal); This native of Malaysia is widely cultivated; the flavour of its fruit being considered superior to that of the Jack Fruit (A. heterophyllus)	Malla (1993)
Artocarpus lakoocha Roxb. [Moraceae]	Badahar, Dahua (Nep)	Root	Yellow	Flavonoids (Artocarpin), Stilbenoids (Oxyresveratrol), Tannins	Wood used for skin diseases and as an astringent	Nepal; Himalaya (Kumaun to Bhutan), India, Ceylon, Burma, Malaysia	Nautiyal and Kaul (2003)
Azadirachta indica A. Juss. (*Melia azadirachta* L.) [Meliaceae]	Margosa Tree (Eng), Neem (Nep)	Bark	Brown; Seeds deep yellow colour (Manandhar 2002)	Triterpenoids/Limonoids (Azadirachtin, Nimbin, Salannin), Tannins	A powerful antiseptic; used for skin diseases, malaria, diabetes, and as an insect repellent. Fruits are toxic	700–1100 m (WE Nepal); Iran, Himalaya, east to China. Cultivated	Malla (1993), Manandhar (2002), Bhandari et al. (2020c)

(continued)

Table 1.1 (continued)

Latin names [Family]	Local names	Plant parts	Dye or pigment	Active compounds	Medicinal Uses	Altitudinal range, distribution in Nepal and in the world	References
Baccaurea ramiflora Lour. (*Baccaurea sapida* (Roxb.) Müll. Arg.) [Euphorbiaceae]	Burmese grape (Eng), Kala bogti (Hindi)	Leaves	Green	Tannins, Flavonoids, Alkaloids	Bark used for skin infections; fruit is edible and astringent	250–460 m (CE Nepal); Nepal, Bhutan, Assam, Burma, S. China, Indo-China, Malay Peninsula	Malla (1993)
Barleria prionitis L. [Acanthaceae]	Katsaraiya, Bhende kurro (Nep.)	Flower	Yellow	Iridoid glycosides (Barlerin, Shanzhiside), Phenylethanoid glycosides (Acteoside), Alkaloids, Triterpenoids	Used for rheumatism, joint pain, toothache, glandular swellings, and as a diuretic and anti-inflammatory agent	ca. 200 m (W Nepal); China (Yunnan), India	Bhandari et al. (2020c)
Bauhinia purpurea L. [Leguminosae]	Rato koiralo, Tanki, Koiralo (Nep); Koilara (Tha)	Flower	Rose purple dyeing and tanning	Flavonoids (Quercetin, Rutin), Tannins, Sterols	Bark used for diarrhoea and skin diseases; root as an antidote to snake venom	300–1600 m (WCE Nepal); Tropical Himalaya (Kashmir to Bhutan), India, S. E. Asia, S. and W. China, and cultivated widely in tropics	Malla (1993), Manandhar (2002)

(continued)

Table 1.1 (continued)

Latin names [Family]	Local names	Plant parts	Dye or pigment	Active compounds	Medicinal Uses	Altitudinal range, distribution in Nepal and in the world	References
Bauhinia variegata L. [Leguminosae]	Mountain ebony (Eng), Koiralo, Seto koiralo (Nep), Kunabhun (New), Kancanara, Kovindara (San.)	Bark	Bark is used for dyeing and tanning for brown shades; yellow (Nautiyal and Kaul 2003)	Flavonoids (Quercetin, Kaempferol), Tannins, Sterols	Bark and buds used for diarrhoea, worms, and skin diseases; for thyroid disorders	150–1900 m (WCE Nepal); Himalaya (Swat to Bhutan), India, Burma, China	Malla (1993), Manandhar (2002), Nautiyal and Kaul (2003)
Berberis aristata DC. [Balsaminaceae]	Ban Chutro, Braha mase chutro (Nep), Chutro (Nep), Komme (Gur), Pichyar (Tam), Chotro (Thar)	Root and bark	Yellow	Isoquinoline alkaloids (Berberine, Palmatine, Jatrorrhizine)	Used for eye infections, skin diseases, diarrhoea, and as an antipyretic (lowers fever). A source of the alkaloid berberine	1800–2500 m (WCE Nepal); India, Bhutan	Pohle (1990), Malla (1993, 1994), Manandhar (2002), Bhandari et al. (2020c)

(continued)

Table 1.1 (continued)

Latin names [Family]	Local names	Plant parts	Dye or pigment	Active compounds	Medicinal Uses	Altitudinal range, distribution in Nepal and in the world	References
Berberis asiatica Roxb. ex DC. [Balsaminaceae]	Chutro (Np), Jurshi (Rai), Kerpaak (She.), Pichyar (Tam.)	Roots and stem	Yellow	Isoquinoline alkaloids (Berberine, Oxyacanthine, Palmatine)	Similar to *B. aristata*; used for jaundice, diarrhoea, and eye ailments	1200–2500 m (WCE Nepal); Himalaya (Garhwal to Bhutan), Assam, China (Yunnan)	Pohle (1990), Manandhar (2002)
Berberis chitria Lindl. [Berberidaceae]	Chutro (Nep), Kelpa (Gur), Kerpaak (She)	Roots and stem	Yellow	Isoquinoline alkaloids (Berberine, Palmatine, Jatrorrhizine)	Used for fever, diarrhoea, and skin diseases	2000–3700 m (CE Nepal); Himalaya (Kashmir to Bhutan)	Pohle (1990), Manandhar (2002)
Berberis edgeworthiana C. K. Schneid. (*Berberis vulgaris* L.) [Berberidaceae]	Chutro (Nep)	Stem and roots	Yellow	Isoquinoline alkaloids (Berberine, Oxyacanthine)	Used for liver and gallbladder ailments, digestive issues, and fever	2700–3600 m (E Nepal); Himalaya (Punjab to Sikkim)	Pohle (1990), Malla (1993), Nautiyal and Kaul (2003)
Betula utilis D. Don [Betulaceae]	Bhoj patra (Nep), Taakpaa (She)	Bark, Gum	Brown	Triterpenoids (Betulin), Flavonoids, Tannins	Bark used for wound healing; sap as a tonic; leaves for gout and rheumatism	2700–4300 m (WCE Nepal); Bhutan, Tibet, W. China	Malla (1993), Bhandari et al. (2020c)

(continued)

Table 1.1 (continued)

Latin names [Family]	Local names	Plant parts	Dye or pigment	Active compounds	Medicinal Uses	Altitudinal range, distribution in Nepal and in the world	References
Bischofia javanica Blume [Euphorbiaceae]	Kainjal (Nep)	Bark	Red	Tannins, Flavonoids, Triterpenoids, Alkaloids	Used for wounds, ulcers, and as an astringent	150–1200 m (WCE Nepal); Himalaya, India east to C. China, Taiwan and Ryukyu, Malaysia, Polynesia	Manandhar (2002)
Bixa orellana L. [Bixaceae]	Annatto (Eng), Sindure (Nep)	Bulp of seeds	Brown, red, orange	Carotenoids (Bixin, Norbixin), Triterpenoids, Flavonoids	Seeds used as a digestive stimulant and antipyretic; as a colorant	~300 m (CE Nepal; India, Tropical America, Africa	Manandhar (2002), Nautiyal and Kaul (2003)
Brassica oleracea L. [Cruciferae]	Bandakovi (Nep)	Whole plant	Redish colour	Glucosinolates (Glucoraphanin, Sinigrin), Flavonoids, Vitamins (C, K)	Nutritional support; used for detoxification, stomach ulcers, and anti-inflammatory purposes	Native to W Europe; cultivated worldwide including in Nepal as a vegetable	Malla (1993)
Bridelia retusa (L.) A. Juss. (*Bridelia retusa* (L.) Spreng.) [Euphorbiaceae]	Gudi or Kaja, Gayo (Nep)	Bark	Green	Tannins, Flavonoids, Triterpenoids	Used for diarrhoea, diabetes, and joint pain	150–1200 m (WCE Nepal); India, Ceylon, Burma, Indo-China, Malay Peninsula, Sumatra	Malla (1993)

(continued)

Table 1.1 (continued)

Latin names [Family]	Local names	Plant parts	Dye or pigment	Active compounds	Medicinal Uses	Altitudinal range, distribution in Nepal and in the world	References
Buchanania latifolia Roxb. [Anacardiaceae]	Chiraunjee, Piyar (Nep)	Bark	Light yellow	Flavonoids, Tannins, Phenolic acids (Gallic acid, Ellagic acid), Phytosterols (β-sitosterol)	Used in treating skin diseases, burns, and coughs; the seeds are considered a brain tonic	150–2000 m (WC Nepal); Himalaya (Kumaun to Nepal), India, Burma, Thailand	Malla (1993)
Butea monosperma (Lam.) Kuntze [Leguminosae]	Palans (Nep), Plydhup (Dar); Palas (Moo); Terti (Rai); Marup (Sat); Dhak (Tha)	Flowers	Yellow and orange	Flavonoids (Butrin, Isobutrin), Glucosides (Butin), Tannins	Flowers used for diarrhoea; seeds as an anthelmintic; gum for diarrhoea and dysentery	150–1200 m (WCE Nepal); Tropical Himalaya, India, Ceylon, S. E. Asia, Malaysia	Malla (1993), Manandhar (2002), Nautiyal and Kaul (2003)
Calendula officinalis L. [Asteraceae]	Asarfi Phool (Nep)	Flowers	Yellow	Flavonoids (Quercetin, Isorhamnetin), Carotenoids (Lutein, Lycopene), Triterpenoid saponins, Essential oils	Topically for skin inflammation, wounds, burns, and rashes; internally for digestive ulcers	2600–4400 m (C Nepal); A cultivated plant, native of the Mediterranean region	Malla (1993), Bhandari et al. (2020c)

(continued)

Table 1.1 (continued)

Latin names [Family]	Local names	Plant parts	Dye or pigment	Active compounds	Medicinal Uses	Altitudinal range, distribution in Nepal and in the world	References
Calotropis gigantea (L.) Dryand. [Asclepiadaceae]	Aank (Nep), Aanak (Che), Ankon (Dan.), Akhua (Tha.)	Seed	Purple	Cardiac glycosides (Calotropin, Uscharin), Pregnane glycosides, Flavonoids	Used with extreme caution for skin diseases, leprosy, and worms. Toxic and can be fatal if misused	100–1000 m (WCE Nepal); Tropical Himalaya, India, east to W. and C. China, Malaysia	Malla (1993)
Caltha palustris L. [Ranuculaceae]	Eka aankhle phul (Nep)	Flowerhead	Golden colour	Toxic glycosides (Protoanemonin), Alkaloids, Flavonoids	Used for warts and fits. All parts are toxic and can irritate the skin	600–4000 m (Nepal), most temperate regions of N hemisphere	Malla (1993)
Canna indica L. [Cannaceae]	Sarbada phool (Nep)	Flowers	Green	Flavonoids, Alkaloids, Tannins	Rhizome used as a diuretic and diaphoretic (induces sweating); seeds used for infected wounds	900 m (CE Nepal); Himalaya, India, Ceylon, Malacca, Malay Is	Malla (1993)

(continued)

Table 1.1 (continued)

Latin names [Family]	Local names	Plant parts	Dye or pigment	Active compounds	Medicinal Uses	Altitudinal range, distribution in Nepal and in the world	References
Capsicum annuum L. [Solanaceae]	Khursani (Nep)	Fruit	Red	Capsaicinoids (Capsaicin, Dihydrocapsaicin), Carotenoids (Capsanthin, Beta-carotene), Flavonoids	Used as a stimulant, for circulatory problems, and topically for pain (in liniments)	E Nepal; Tropical America, widely cultivated as a vegetable elsewhere	Bhandari et al. (2020c)
Careya arborea Roxb. [Licythidaceae]	Kumbhi (Nep), Kumh (Tha)	Bark	Dark grey, black	Tannins, Flavonoids, Triterpenoids	Bark used for colds, coughs, and diarrhoea; leaves for headaches	200–600 m (CE Nepal); Afghanistan to N. Malaya	Malla (1993), Nautiyal and Kaul (2003)
Carissa carandas L. [Apocynaceae]	Karaunda (Nep)	Ripe fruits	Fully ripe violet cranberries, ripe red-violet ones, and slightly-ripe red cranberries	Triterpenoids (Ursolic acid, Oleanolic acid), Flavonoids, Volatile oils, Cardiac glycosides	Fruit used for anaemia, digestive issues, and skin diseases; root used for stomach pain and fever	100–170 m (C Nepal); India, Sri Lanka, Burma, Malaysia	Malla (1993)

(continued)

Table 1.1 (continued)

Latin names [Family]	Local names	Plant parts	Dye or pigment	Active compounds	Medicinal Uses	Altitudinal range, distribution in Nepal and in the world	References
Carpesium abrotanoides L. [Asteraceae]	Pig's Head (Eng)	Flowers	Yellow	Sesquiterpene lactones (Carpesiolin), Flavonoids, Volatile oils	Used for malaria, fever, and as an anthelmintic to expel worms	1400–2200 m (WCE Nepal); Europe, Caucasus, Asia Minor, Iran, Himalaya, China, Indo-China, Korea, Japan	Malla (1993)
Carthamus tinctorius L. [Asteraceae]	Kusum (Nep)	Flower	Red, yellow	Quinone-chalcones (Carthamin, Carthamone), Flavonoids (Safflor yellow), Serotonin derivatives	Used to improve blood circulation, alleviate menstrual pain, and for cardiovascular health	3000 m (C Nepal); A cultivated plant from W. Asia	Bhandari et al. (2020c)

(continued)

Table 1.1 (continued)

Latin names [Family]	Local names	Plant parts	Dye or pigment	Active compounds	Medicinal Uses	Altitudinal range, distribution in Nepal and in the world	References
Cassia fistula L. [Leguminosae]	Raj brichhya, Amaltas (Nep)	Bark and Pulps Barks	Yellow	Anthraquinones (Sennosides A & B, Rhein), Flavonoids, Tannins	Pulp of the fruit is a well-known mild laxative	150–1400 m (WCE Nepal); Widely cultivated in Africa, W. Asia, Himalaya, India, S. E. Asia, Malaysia, China, Polynesia; probably a native of E. India, Burma, Malay Isl	Malla (1993)
Cassia tora L. [Fabaceae]	Chakramandi, Sano tapre (Nep)	Seed	Blue or Red	Anthraquinones (Chrysophanol, Emodin), Flavonoids, Naphthopyrone glycosides (Cassiaside)	Seeds and leaves used for skin diseases like ringworm and for vision health	450–1300 m (WCE Nepal); Tropics, probably of S. American origin	Nautiyal and Kaul (2003)
Castanopsis indica (Roxb.) Miq. [Fabaceae]	Katus, Dhalne katus (Nep)	Bark	Brown	Tannins, Flavonoids, Triterpenoids	Bark used for diarrhoea and dysentery	1200–2900 m (CE Nepal); Himalaya (Nepal to NEFA), Khasia, Burma, W. China, Indo-China	Bhandari et al. (2020c)

(continued)

Table 1.1 (continued)

Latin names [Family]	Local names	Plant parts	Dye or pigment	Active compounds	Medicinal Uses	Altitudinal range, distribution in Nepal and in the world	References
Casuarina equisetifolia L. [Casurinaceae]	Jangali saru, Jungali jhau (Nep)	Bark	Brown	Tannins, Flavonoids, Triterpenoids	Bark and leaves used for diarrhoea, dysentery, and sore throats	Native to Australia, Polynesia and Malaysia. Introduced in Pakistan, India, Sri Lanka, Tropical Africa, Mascarene Isles, Europe. N, and S. America	Malla (1993)
Chenopodium album L. [Chenopodiaceae]	Bethe (Nep), Bathuwa (Dar), Sina (Lim), Chinwaam (Rai), Bethu (Tam), Bethuwa (Tha.)	Young shoots	Green	Saponins, Flavonoids, Alkaloids (Chenopodine)	Used as a digestive aid, laxative, and for liver disorders. Seeds are nutritious	Fields, waste places, roadsides, a difficult weed to control. Throughout temperate and tropical regions of the world	Malla (1993)

(continued)

Table 1.1 (continued)

Latin names [Family]	Local names	Plant parts	Dye or pigment	Active compounds	Medicinal Uses	Altitudinal range, distribution in Nepal and in the world	References
Cicer arietinum L. [Leguminosae]	Chana (Nep)	Leaves	Indigo	Flavonoids (Biochanin A), Saponins, Proteins	Nutritional support; used as an energy tonic and for blood sugar control	150–1300 m (WCE Nepal); Cultivated widely in the tropics and in warm regions in S. Europe, Africa, Asia (except E. Asia), Australia, America; probably a native of S. W. Asia	Malla (1993)
Cinnamomum tamala (Buch.-Ham.) Nees & Eberm. [Lauraceae]	Tejpat (Nep), Pinge (Gur), Pataarangkhi (Rai), Shishi (She)	Leaves	Gray-brown	Volatile oils (Linalool, Cinnamaldehyde, Eugenol), Tannins	Used for diabetes, colds, and indigestion	450–2000 m (WCE Nepal); Himalaya (Kashmir to Bhutan), India (Assam, Khasi)	Malla (1993)

(continued)

Table 1.1 (continued)

Latin names [Family]	Local names	Plant parts	Dye or pigment	Active compounds	Medicinal Uses	Altitudinal range, distribution in Nepal and in the world	References
Cinnamomum zeylanicum Breyn. [Lauraceae]	Dalchini (Nep)	Bark	Brownish-reddish	Volatile oils (Cinnamaldehyde, Eugenol, Linalool), Tannins	Used for diabetes, digestive issues, and as an antimicrobial	Native to Sri Lanka and tropical Asia, also grows in Malabar, Cochin-China, Sumatra and in Eastern Islands. Cultivated in different countries such as Brazil, Mauritius, India, Jamaica, etc	Malla (1993)
Clitoria ternatea L. [Leguminosae]	Aparajita (Nep)	Seeds	Blue	Triterpenoids (Taraxerol), Flavonoids (Kaempferol, Quercetin), Anthocyanins (Ternatins)	Root used for asthma and bronchitis; seeds as a laxative; known as a brain tonic	200 m (E Nepal); Widely cultivated in tropics; probably a native of S. America	Malla (1993)

(continued)

Table 1.1 (continued)

Latin names [Family]	Local names	Plant parts	Dye or pigment	Active compounds	Medicinal Uses	Altitudinal range, distribution in Nepal and in the world	References
Coffea arabica L. [Rubiaceae]	Coffee (Eng)	Bean (fruits)	Brown	Purine alkaloids (Caffeine, Theobromine), Chlorogenic acids, Diterpenes (Cafestol)	Used as a stimulant for fatigue and headaches; also as a diuretic	200–700 m (CE Nepal); native to E Africa and cultivated in moist, usually cool tropical regions	Malla (1993)
Coffea benghalensis B. Heyne ex Roth [Rubiaceae]	Wild Indian coffee (Eng)	Seeds	Brown	Purine alkaloids (Caffeine), Chlorogenic acids	Similar stimulant uses to common coffee	Tropical region	Malla (1993)
Coriaria napalensis Wall. [Coriariaceae]	Macchaino (Nep)	Leaves, bark	Black	Triterpenoids, Tannins, Glycosides, Neurotoxins (Tutin)	Used for skin diseases and as an insecticide. Extremely toxic and can cause convulsions and death	1200–2400 m (WCE Nepal); Himalaya (Kashmir to Bhutan), Assam, N. Burma, W. China	Malla (1993)
Cotinus coggygria Scop. (*Rhus cotinus* L.) [Anacardiaceae]	Smoke tree (Eng)	Leaves, bark, wood	Yellow	Flavonoids (Fustin, Sulfuretin), Tannins (Gallotannins), Volatile oils (Pinene, Myrcene)	Used as an antiseptic and anthelmintic; also for heart and skin diseases	1100–2400 m (WC Nepal); S. Europe, W. Asia, Himalaya, India, China	Malla (1993)

(continued)

Table 1.1 (continued)

Latin names [Family]	Local names	Plant parts	Dye or pigment	Active compounds	Medicinal Uses	Altitudinal range, distribution in Nepal and in the world	References
Curcuma aromatica Salisb. [Zingiberaceae]	Ban haledo, Jangali haldi (Nep)	Rhizome	Abir powder (red)	Curcuminoids (Curcumin), Volatile oils (Curcumene, Camphor, Borneol)	Used for skin care, wounds, and as an anti-inflammatory	700–1100 m (CE Nepal); Himalaya (Nepal, Sikkim), India, Ceylon	Malla (1993)
Curcuma longa L. [Zingiberaceae]	Besar (Nep)	Rhizome	Yellow	Curcuminoids (Curcumin, Demethoxycurcumin, Bisdemethoxycurcumin), Volatile oils (Turmerone, α-Atlantone)	A powerful anti-inflammatory; used for arthritis, wounds, and as a digestive	Cultivated throughout tropical Asia	Malla (1993), Nautiyal and Kaul (2003), Bhandari et al. (2020c)
Cuscuta reflexa Roxb. [Convolvulaceae]	Aakasbeli (Nep), Amarlati (Dar), Melan mili (Gur), Taarghey (Tam), Amabel (Tha)	Whole plant	Yellow	Flavonoids (Kaempferol, Quercetin), Alkaloids, Phenolic acids	Used for liver disorders, jaundice, and muscle pain	900–2800 m (WCE Nepal). Afghanistan, China, India, Indonesia, Malaysia, Myanmar, Pakistan, Sri Lanka, Thailand	Malla (1993)

(continued)

Table 1.1 (continued)

Latin names [Family]	Local names	Plant parts	Dye or pigment	Active compounds	Medicinal Uses	Altitudinal range, distribution in Nepal and in the world	References
Cynoglossum lanceolatum Forssk. (*Cynoglossum micranthum* Desf.) [Boraginaceae]	Nilkarai (Nep)	Whole plant	Green	Pyrrolizidine alkaloids, Flavonoids, Tannins	Used for coughs and bronchitis. Contains pyrrolizidine alkaloids which are liver toxic	150–3200 m (WCE Nepal); Africa, W. Asia, Himalaya, India, Ceylon, Burma, east to China, Malaysia	Malla (1993)
Dahlia pinnata Cav. [Asteraceae]	Lahure Phool (Nep)	Flowers	Orange	Flavonoids (Anthocyanins), Polyacetylenes, Polysaccharides (Inulin)	Tubers are used as a source of inulin for diabetics; also for skin rashes and inflammation	2600 m (C Nepal); A cultivated plant, native of Mexico	Malla (1993)
Daucus carota L. [Apiaceae]	Gajar (Nep)	Root	Ruby Red	Carotenoids (Beta-carotene, Lutein), Polyacetylenes (Falcarinol), Flavonoids, Volatile oils	Seeds are used as a diuretic and for kidney stones; the root is a source of Vitamin A for eye health	Cultivated and adventive worldwide in subtropical and temperate regions	Malla (1993)
Diospyros malabarica (Desr.) Kostel. (*Diospyros embryopteris* Pers.) [Ebenaceae]	Allo, Haluwabed (Nep.)	Fruit	Brown	Tannins, Naphthoquinones, Triterpenoids	Used for diarrhoea, bleeding wounds, and fever; unripe fruit for worms	500–1500 m (W Nepal); India, Ceylon, Malaysia	Malla (1993)

(continued)

Table 1.1 (continued)

Latin names [Family]	Local names	Plant parts	Dye or pigment	Active compounds	Medicinal Uses	Altitudinal range, distribution in Nepal and in the world	References
Drosera peltata Sm. [Droseraceae]	Pamga, Kitas (Nep)	Plant	Red	Naphthoquinones (Plumbagin), Flavonoids	Used for respiratory ailments like whooping cough, asthma, and bronchitis	1000–2500 m (Nepal); China, E and SE Asia; Australia	Malla (1993)
Eclipta prostrata (L.) L (*Eclipta alba* (L.) Hassk.) [Asteraceae]	Bhringaraj, Bhangeree Samo ghans (Nep)	Plant	Black, bluish	Coumarins (Wedelolactone, Demethylwedelolactone), Thiophene derivatives, Flavonoids	A key herb in Ayurveda for liver health (jaundice), hair growth, and as a black hair dye	200–1200 m (WCE Nepal); A pantropic weed	Malla (1993), Dangol (2002, 2008), Manandhar (2002)
Elsholtzia fruticosa (D. Don) Rehder (*Elsholtzia polystachya* Benth.) [Lamiaceae]	Saano simal (Nep)	Leaves	Brown or orange	Volatile oils (Elsholtzia ketone, Limonene), Flavonoids	Used for fever, headache, and as a diuretic	1800–4200 m (WCE Nepal); Punjab, Himalaya (Kashmir to Bhutan), India, Tibet, Burma, China	Malla (1993)

(continued)

Table 1.1 (continued)

Latin names [Family]	Local names	Plant parts	Dye or pigment	Active compounds	Medicinal Uses	Altitudinal range, distribution in Nepal and in the world	References
Embelia ribes Burm. f. [Myrsinaceae]	Bhyabhidanga (Nep)	Fruit	Red	Benzoquinones (Embelin), Tannins, Alkaloids	A primary anthelmintic for tapeworms; also used for obesity and skin diseases	100–2000 m (Nepal); China, India, Indonesia, Laos, Malaysia, Myanmar, New Guinea, Philippines, Sri Lanka, Thailand, Vietnam	Bhandari et al. (2020c)
Equisetum ramosissimum var. *huegelii* (Milde) Christenh. & Husby (*Equisetum debile* Roxb. ex Vaucher) [Equisetaceae]	Kukure ghans (Nep)	Whole plant	Yellow, grey pink, Yellow color depenind on mordant (Svanberg 1997)	Silica, Flavonoids (Equisetrin), Alkaloids (Nicotine, Palustrine), Saponins	Used as a diuretic for kidney and bladder problems; for brittle nails and hair	Native range of this variety is Tropical and Subtropical Asia to SW. Pacific	Malla (1993)
Euonymus tingens Wall. [Celastraceae]	Kasuree (Nep), Pheisign (Tam)	Inner Bark	Yellow	Alkaloids, Triterpenoids, Cardiac glycosides	Used for joint pain and digestive disorders	2300–3300 m (WCE Nepal); Himalaya (Simla to Bhutan), Assam, S. E. Tibet, N. Burma, W. China	Malla (1993)

(continued)

Table 1.1 (continued)

Latin names [Family]	Local names	Plant parts	Dye or pigment	Active compounds	Medicinal Uses	Altitudinal range, distribution in Nepal and in the world	References
Euphorbia pulcherrima Willd. ex Klotzsch [Euphorbiaceae]	Lalupate (Nep)	Flower petals	Reddish, brown	Diterpenes, Triterpenoids, Flavonoids, Latex	Latex used topically for skin infections. Can be a skin irritant and is toxic if ingested	1100–1200 m (C Nepal); Native of central America; widely cultivated in tropical areas	Bhandari et al. (2020a)
Everniastrum (*Parmelia*) *nepalense* (Taylor) Hale ex Sipman (Synonym: *Hypotrachyna nepalensis* (Taylor) Divakar; *Parmelia nepalensis* Taylor) [Parmeliaceae]	Jyau (Nep)	Whole plant	Brown	Lichen acids (Usnic acid, Atranorin, Salazinic acid)	Used for wounds, as an antimicrobial, and in traditional incense	Nepal; Asia, particularly abundant in Nepal and India	IUCN (2004), Pyakurel et al. (2019), Pradhan et al. (2020)
Fagopyrum esculentum Moench [Polygonaceae]	Mitho phapar (Nep)	Leaves	Yellow	Flavonoids (Rutin, Quercetin), Tannins, Naphthodianthrones (Fagopyrins)	Used to strengthen blood vessels and for high blood pressure	1800–4100 m (WCE Nepal); C. Asia, widely cultivated	Malla (1993)

(continued)

Table 1.1 (continued)

Latin names [Family]	Local names	Plant parts	Dye or pigment	Active compounds	Medicinal Uses	Altitudinal range, distribution in Nepal and in the world	References
Fagopyrum megacarpum H. Hara (*Eskemukerjea megacarpum* (H. Hara) H. Hara) [Polygonaceae]	Bhote khair (Nep), Kuchrung, Umma lauri (Nep-Dol), Kayaur, Bhotikhair (She), Chum tsa, Lete (Tib)	Root	Yellow	Flavonoids (Rutin), Tannins	Similar uses to common buckwheat	2400–3000 m; W Nepal	Ghimire et al. (2001)
Ficus benghalensis L. [Moraceae]	Bar (Nep)	Leaves	Reddish, brown	Flavonoids (Leucocyanidin, Quercetin), Tannins, Sterols	Bark and aerial roots used for diabetes, diarrhoea, and dysentery	500–1200 m (WCE Nepal); Nepal, Pakistan, India; widely cultivated	Malla (1993)
Ficus racemosa L (*Ficus glomerata* Roxb.) [Moraceae]	Dumri, Gular (Nep)	Bark	Black	Flavonoids (Leucocyanidin, Quercetin), Tannins, Sterols	Bark and fruit used for diabetes, diarrhoea, and urinary disorders	300 m (WC Nepal); Pakistan, Nepal, India, Ceylon, S. W. China (Yunnan), Indo-China, Malaysia, Australia	Malla (1993)

(continued)

Table 1.1 (continued)

Latin names [Family]	Local names	Plant parts	Dye or pigment	Active compounds	Medicinal Uses	Altitudinal range, distribution in Nepal and in the world	References
Ficus religiosa L. [Moraceae]	Peepal (Nep)	Bark and leaves Dyeing and tanning	Redish pale brown	Flavonoids (Leucocyanidin, Quercetin), Tannins, Sterols	Bark used for diabetes, gout, and as an astringent	150–1500 m (WCE Nepal); India; widely cultivated in S. E. Asia	Malla (1993), Manandhar (2002), Bhandari et al. (2020a)
Flemingia macrophylla (Willd.) Merr. (*Moghania macrophylla* (Willd.) Kuntze) [Fabaceae]	Bhatwasee (Nep)	Pod	Orange for silk	Flavonoids (Flemingin), Pterocarpans, Steroids	Root used for epilepsy, convulsions, and as a sedative	700–1700 m (WCE Nepal); Himalaya, India, Ceylon, S. E. Asia, Malaysia, Australia, China, and cultivated in Africa	Malla (1993), Bhandari et al. (2020a)
Fumaria parviflora Lam. (*Fumaria indica* (Hausskn.) Pugsley) [Papaveraceae]	Dhukure (Nep)	Leaf	Yellow to brown with modrant	Isoquinoline alkaloids (Fumaricine, Protopine, Parfumine), Flavonoids	Used for skin diseases, fever, and as a blood purifier	C Nepal; N. Africa, Europe, C. Asia	Malla (1993)

(continued)

Table 1.1 (continued)

Latin names [Family]	Local names	Plant parts	Dye or pigment	Active compounds	Medicinal Uses	Altitudinal range, distribution in Nepal and in the world	References
Garcinia xanthochymus Hook. f. ex J. Anderson [Clusiaceae]	Chunyal (Nep)	Bark	Yellow	Xanthones (Garciniaxanthone), Benzophenones (Xanthochymol, Isoxanthochymol), Organic acids	Fruit rind used for digestion and as a laxative; also applied to skin eruptions	200–500 m (E Nepal); E. Himalaya (Nepal to Assam), India, Burma, Thailand, S. W. China (Yunnan), Malaya, Andaman and Nicobar Isl	Malla (1993)
Gardenia jasminoides Ellis [Rubiaceae]	Indra kamal (Nep)	Fruit	Yellow, yellow dye to color clothes and food	Iridoid glycosides (Geniposide, Gardenoside), Carotenoids (Crocin), Flavonoids	Fruit used for anxiety, insomnia, and for liver and gallbladder disorders	1000–1500 m (CE Nepal); Native of China and Japan, cultivated in Himalaya and India	Bhandari et al. (2020a)
Garuga pinnata Roxb. [Burseraceae]	Dabdabe (Nwp), Jhengra (Tha)	Leaves	Red	Tannins, Flavonoids, Triterpenoids	Bark and leaves used for stomach ulcers, coughs, and asthma	300–1200 m (WCE Nepal); Subtropical Himalaya (Nepal to Bhutan), India, Burma, Thailand, Indo-China, Malaysia	Malla (1993), Nautiyal and Kaul (2003), Bhandari et al. (2020a)

(continued)

Table 1.1 (continued)

Latin names [Family]	Local names	Plant parts	Dye or pigment	Active compounds	Medicinal Uses	Altitudinal range, distribution in Nepal and in the world	References
Geranium nepalense Sweet [Geraniaceae]	Chunitro ghans (Nep.), Salyan sai (Che), Pueran (Gur)	Root	Red	Tannins, Flavonoids (Geraniin), Essential oils	Used for diarrhoea, kidney stones, and wound healing	500–4000 m (WCE Nepal); Afghanistan, Himalaya, Assam, Tibet, N. Burma, N. Indo-China, W. China	Malla (1993)
Geranium wallichianum D. Don ex Sweet [Geraniaceae]	Rakla mool (Nep)	Root	Yellow	Tannins, Flavonoids, Essential oils	Used for joint pain, wounds, and as an astringent	2100–4200 m (WCE Nepal); Afghanistan, Himalaya (Kashmir to Bhutan)	Rokaya et al. (2010)
Girardinia diversifolia (Link) Friis [Urticaceae]	Allo, Allo sisnu (Nep)	Whole plant	Blue, green	Alkaloids, Flavonoids, Tannins, Histamine, Acetylcholine	Used for joint pain and as a diuretic. Causes a severe sting	1700–3000 m (WCE Nepal); Himalaya (Punjab to Bhutan), India, Ceylon, Burma, east to C. China, Malaysia	Shrestha (1994)
Gmelina arborea Roxb. [Verbenaceae]	Khamari (Nep)	Wood	Yellow or brownish-yellow	Flavonoids (Luteolin, Apigenin), Lignans (Gmelinol, Arboreol), Triterpenoids	Root used for fever, digestive issues, and as a demulcent	200–1100 m (WE Nepal); Himalaya (Nepal to Bhutan), India, Ceylon, Philippines	Malla (1993)

(continued)

Table 1.1 (continued)

Latin names [Family]	Local names	Plant parts	Dye or pigment	Active compounds	Medicinal Uses	Altitudinal range, distribution in Nepal and in the world	References
Gossypium arboreum L. [Malvaceae]	Rukh kapas (Nep)	Flowers	Yellow	Flavonoids (Gossypetin, Quercetin), Tannins, Sesquiterpenoid aldehydes (Gossypol)	Root bark used to induce labour and as an abortifacient (very dangerous); seed oil is demulcent	1500 m (E Nepal); Old World tropics	Malla (1993)
Gossypium herbaceum L. [Malvaceae]	Cotton (Eng), Kapas (Nep)	Flowers	Yellow	Sesquiterpenoid aldehydes (Gossypol), Flavonoids, Tannins	Similar uses to *G. arboreum*; root bark used for menstrual disorders	Cultivated in India; origin in SW Asia	Nautiyal and Kaul (2003), Bhandari et al. (2020a)
Grewia optiva J. R. Drumm. ex Burret (*Grewia oppositifolia* Buch.-Ham. ex D. Don) [Teliaceae]	Syal phusre (Nep)	Leaves	Fine bright shades of dark	Flavonoids, Tannins	Bark and leaves used for fractures and as a demulcent	150–1800 m (WCE Nepal); Himalaya (Kashmir to Nepal)	Malla (1993)
Gymnema tingens (Roxb.) Spreng. [Asclepiadaceae]	Gudmaar (Nep)	Leaves	Green	Triterpene saponins (Gymnemic acids), Pregnane glycosides	Used for diabetes, malaria, and as a diuretic and snakebite antidote	400–1500 m (WE Nepal); Subtropical Himalaya (Kumaun to Sikkim), Assam, W. and S. China	Malla (1993)

(continued)

Table 1.1 (continued)

Latin names [Family]	Local names	Plant parts	Dye or pigment	Active compounds	Medicinal Uses	Altitudinal range, distribution in Nepal and in the world	References
Hedera nepalensis K. Koch [Araliaceae]	Kathe laharo, Dudelaa (Nep)	Fruits	Green	Saponins (Hederasaponins), Flavonoids, Polyacetylenes	Used as an expectorant for coughs and bronchitis; externally for skin infections and burns	2000–3200 m (WCE Nepal); Afghanistan, Himalaya (Kashmir to Bhutan), Assam, Burma, China (var.)	Malla (1993)
Hedychium spicatum Sm. [Zingiberaceae]	Spiked ginger lily (Eng), Pankha phul, Seto saro (Nep)	Root-stock	Red	Volatile oils (1,8-Cineole, Linalool), Diterpenes (Hedychenone), Flavonoids	Rhizome used for asthma, indigestion, and as a deodorant	1200–3200 m (CE Nepal); Bhutan, China, NE India, Myanmar, Sikkim, N Thailand	Malla (1993), Bhandari et al. (2020b)
Heracleum wallichii DC. [Apiaceae]	Chimphing (Nep)	Root	Yellow	Furanocoumarins (Imperatorin, Bergapten), Volatile oils	Used for digestive disorders and as an aphrodisiac. Sap can cause photodermatitis	3600–4100 m (C Nepal); Himalaya (Nepal to Bhutan)	Malla (1993)

(continued)

Table 1.1 (continued)

Latin names [Family]	Local names	Plant parts	Dye or pigment	Active compounds	Medicinal Uses	Altitudinal range, distribution in Nepal and in the world	References
Hibiscus rosa-sinensis L. [Malvaceae]	Ghanti phul, Rakta puspi, Japa kusum (Nep)	Flower petals	Purple	Flavonoids, Tannins, Anthocyanins	Flowers and leaves used for hair growth, as a demulcent, and for bronchitis	900–1400 m (E Nepal); Of uncertain, possibly E. African origin. Widely cultivated throughout the tropics and subtropics. Hybrid and cultivated worldwide	Malla (1993), Manandhar (2002), Bhandari et al. (2020b)
Hibiscus sabdariffa L. [Malvaceae]	Belchandana, Lalambari (Nep)	Fruit	Red	Organic acids (Hibiscus acid, Citric acid), Anthocyanins (Delphinidin, Cyanidin), Flavonoids	Calyces used for hypertension, as a diuretic, and for liver health	1000–1500 m (E Nepal); Native to Central African Republic, Chad, Congo, Gabon, Ghana, Nigeria, Sudan, Zaïre. Cultivated in the tropics	Malla (1993), Bhandari et al. (2020b)

(continued)

Table 1.1 (continued)

Latin names [Family]	Local names	Plant parts	Dye or pigment	Active compounds	Medicinal Uses	Altitudinal range, distribution in Nepal and in the world	References
Holarrhena pubescens (Buch.-Ham.) Wall. ex G. Don (*Holarrhena antidysenterica* Wall. ex DC.) [Apocynaceae]	Ban khirro, Madhisae Khirro (Nep)	Wood ashes	Red, chocolate, and purple	Steroidal alkaloids (Conessine, Conamine, Holarrhimine)	A well-known remedy for amoebic dysentery and diarrhoea; also for intestinal worms and fever	100–1500 m (WCE Nepal); Tropical Himalaya, India, Burma, Indo-China, Malaya	Malla (1993)
Hypericum uralum Buch.-Ham. ex D. Don [Clusiaceae]	Urilo, Yurilo (Nep)	Leaves	Black	Naphthodianthrones (Hypericin, Pseudohypericin), Phloroglucinols (Hyperforin), Flavonoids	Used for wounds, depression (like its relative St. John's Wort), and as an anti-inflammatory	1200–3600 m (WCE Nepal); Himalaya (Kashmir to Bhutan), Burma, Thailand, S. China, Sumatra	Bhandari et al. (2020b)
Impatiens balsamina L	Gulmendi, Tiuree (Nep)	Flowers	Yellow, Red	Naphthoquinones (Lawsone, 2-Methoxy-1,4-naphthoquinone), Flavonoids, Coumarins	Used for warts, snakebites, and burns; also for rheumatism and fractures	1200–1900 m (WCE Nepal); Widely cultivated, native of S. E. Asia	Malla (1993), Bhandari et al. (2020b)
Indigofera atropurpurea Buch.-Ham. ex Hornem. [Fabaceae]	Sagino (Nep)	Leaves	Blue	Flavonoids, Alkaloids, Tannins	Used for epilepsy, nervous disorders, and headaches	700–3200 m (WCE Nepal); Himalaya (Kashmir to Bhutan), India, China	Malla (1993)

(continued)

Table 1.1 (continued)

Latin names [Family]	Local names	Plant parts	Dye or pigment	Active compounds	Medicinal Uses	Altitudinal range, distribution in Nepal and in the world	References
Jasminum humile L. [Oleaceae]	Masino Jaai (Nep)	Roots	Yellow	Flavonoids, Iridoid glycosides, Alkaloids	Leaves used for eye diseases; root for ringworm	1100–3800 m (WCE Nepal); Afghanistan, China, India, Tajikistan; SW Asia	Malla (1993), Ghimire et al. (2001)
Jatropha curcas L. [Euphorbiaceae]	Sajiwan (Nep)	Bark, Juice or fruits	Black	Diterpenes (Jatrophone, Jatrophine), Trypsin inhibitors	Latex applied to wounds and burns; seeds are purgative. All parts, especially seeds, are highly toxic	500–1200 m (WCE Nepal); Native of New World Tropics, cultivated in other tropical areas	Malla (1993)
Juglans regia L. (*Juglans regia* var. *kamaonia* C. DC.) [Juglandaceae]	Haade okhar, Sano okhar (Nep)	Hulls, bark	Brown, black	Naphthoquinones (Juglone), Flavonoids, Phenolic acids	Leaves and bark used for skin diseases, eczema, and as an astringent	500–4000 m (WCE Nepal); Widely distributed in C, E, NW, and SW China, SW Asia to Himalayas, SE Europe	Malla (1993); Ghimire et al. (2001), Lama et al. (2001), Manandhar (2002), Bhandari et al. (2020b)

(continued)

Table 1.1 (continued)

Latin names [Family]	Local names	Plant parts	Dye or pigment	Active compounds	Medicinal Uses	Altitudinal range, distribution in Nepal and in the world	References
Juniperus communis L. [Cupressaceae]	Dhupi, Pamo (Nep)	Berries	Khaki	Volatile oils (α-Pinene, Sabinene, Limonene), Flavonoids, Lignans, Diterpenes	Berries used as a diuretic, for urinary tract infections, and for arthritis	WCE Nepal; Forms dense patches near the tree limit and up to 4200 m. Fruit medicinal; Distribution: Alpine and arctic Europe and Asia, N. Africa, N. America	Malla (1993)
Justicia adhatoda L. (*Adhatoda vasica* Ness) [Acanthaceae]	Malabar Nut (Eng), Asuro (Nep), Vashak (Tam), Asur (Tha.)	Wood, leaves	Yellow	Quinazoline alkaloids (Vasicine, Vasicinone)	For respiratory conditions like bronchitis, asthma, and cough; acts as an expectorant and bronchodilator	500–1600 m (WCE Nepal); Subtropical Himalaya, India, Indo-China, Malaya, sometimes cultivated	Pohle (1990), Malla (1993), Bhandari et al. (2020b)
Lagerstroemia parviflora Roxb. [Lythraceae]	Asarephul, Bot Dhaiyanro (Nep)	Bark, leaves	Black	Tannins, Flavonoids, Triterpenoids	Bark used for diarrhoea and diabetes; leaves for wounds	200–800 m (WCE Nepal); Tropical Himalaya (Kumaun to Sikkim), India, Burma	Malla (1993), Dangol (2002), Manandhar (2002)

(continued)

Table 1.1 (continued)

Latin names [Family]	Local names	Plant parts	Dye or pigment	Active compounds	Medicinal Uses	Altitudinal range, distribution in Nepal and in the world	References
Lannea coromandelica (Houtt.) Merr. (*Odina wodier* Roxb.) [Anacardiaceae]	Dabdabi, Hallunde, Holongre (Nep)	Bark	Brownish Red or Golden & pale brown	Tannins, Flavonoids, Phenolic acids (Gallic acid), Triterpenoids (Oleanolic acid)	Applied to wounds, ulcers, and skin diseases; and also for diarrhoea and dysentery	100–1400 m (WCE Nepal); Himalaya (Swat to Bhutan), Assam, Burma, Indo-China, Ceylon, Andaman Isl., China, Malaysia	Malla (1993), Nautiyal and Kaul (2003)
Larix griffithiana Carriere (*Larix griffithii* Hook. f. & Thomson) [Pinaceae]	Boarga salla, Langtang sallo (Nep)	Cones, needles	Brown	Volatile oils (α-Pinene, β-Phellandrene), Flavonoids, Proanthocyanidins	Resin (turpentine) used externally for rheumatic pain and bruises	1400–3900 m (C Nepal); Himalaya (Nepal, Sikkim, Bhutan), Assam	Malla (1993)
Lawsonia inermis L. (*Lawsonia alba* Lam.) [Lythraceae]	Meharee (Nep)	Leaves	Reddish brown, Orange, Red	Naphthoquinones (Lawsone), Flavonoids, Tannins	Leaves used for skin diseases, wounds, and as a cooling agent; famous as a hair dye	Nepal; C. Asia, India, often cultivated	Malla (1993), Manandhar (2002), Nautiyal and Kaul (2003), Sigdel and Rokaya (2011), Bhandari et al. (2020b)

(continued)

Table 1.1 (continued)

Latin names [Family]	Local names	Plant parts	Dye or pigment	Active compounds	Medicinal Uses	Altitudinal range, distribution in Nepal and in the world	References
Leucaena leucocephala (Lam.) de Wit [Fabaceae]	Seto babul, Ipil-ipil (Nep)	Bark	Brown	Non-protein amino acids (Mimosine), Tannins, Flavonoids	Seeds used as a vermifuge; leaves for diabetes	1500 m (W Nepal); Cultivated and naturalised widely in tropical and subtropical regions	Manandhar (2002)
Luculia gratissima (Wall.) Sweet [Rubiaceae]	Ban kangiyo, Doware phool (Nep)	Wood	Yellow	Iridoid glycosides, Flavonoids, Alkaloids	Used for fever and as an anti-inflammatory	1000–2100 m (WCE Nepal); Himalaya (Nepal to Assam), Tibet, Indo-China, Yunnan	Malla (1993), Luong et al. (2023)
Lycopersicon esculentum Mill. [Solanaceae]	Gol bhenda, Tamatar (Nep)	Whole plant, fruit	Red Brown	Carotenoids (Lycopene, Beta-carotene), Flavonoids, Vitamins (C, E)	Nutritional support; high in antioxidants like lycopene for prostate and heart health	250–1400 m (WCE Nepal); C. and S. America, Mexico, widely cultivated for its fruit (tomatoes) and naturalised elsewhere	Malla (1993)

(continued)

Table 1.1 (continued)

Latin names [Family]	Local names	Plant parts	Dye or pigment	Active compounds	Medicinal Uses	Altitudinal range, distribution in Nepal and in the world	References
Maclura spinosa (Willd.) C. C. Berg (*Plecospermum spinosum* (Willd.) Trécul) [Urticaceae]	Maidal-lara (Hin)	Wood	Yellow	Flavonoids (Morin, Macluraxanthone), Prenylated compounds, Triterpenoids	Used for skin diseases and as an anti-inflammatory	East Himalaya, India, Sri Lanka	Malla (1993)
Maharanga emodi (Wall.) A. DC. [Boraginaceae]	Maharangi (Nep)	Root	Red	Naphthoquinones, Phenolic acids	Used for burns, wounds, and joint pain	2200–4500 m (WCE Nepal); Himalaya (Garhwal to Bhutan), S. Tibet	Malla (1993, 1994)
Mahonia napaulensis DC. (*Berberis nepalensis* (DC.) Spreng.) [Berberidaceae]	Jamane mandro (Nep), Kome (Gur), Sersing, Mendok (Sher), Kyarpa (Tam)	Flowers and Bark	Yellow	Isoquinoline alkaloids (Berberine, Oxyacanthine, Jatrorrhizine)	Used for dysentery, eye ailments, and as an anti-inflammatory	2000–2900 m (WCE Nepal); Himalaya (Nepal to Bhutan), Assam	Malla (1993), Manandharn (2002)

(continued)

Table 1.1 (continued)

Latin names [Family]	Local names	Plant parts	Dye or pigment	Active compounds	Medicinal Uses	Altitudinal range, distribution in Nepal and in the world	References
Mallotus philippensis (Lam.) Mull. Arg. [Euphorbiaceae]	Sindure, Rohini (Nep.), Raini (Dar), Rora (Sat)	Fruit	Golden red; power of petals produce red color for coloring clothes and silk	Phloroglucinol derivatives (Rottlerin, Isorottlerin), Tannins	Glandular hairs on the fruit (Kamala) are used as an anthelmintic to expel worms	150–1800 m (WCE Nepal); Himalaya (Kumaun to Bhutan), India, Ceylon, Indo-China, China, Malaysia, Australia, Polynesia	Malla (1993), Dangol (2002), Bhandari et al. (2020b)
Mangifera indica L. [Anacardiaceae]	Aanp (Nep)	Bark and keaves	Yellow	Xanthonoids (Mangiferin), Phenolic acids (Gallic acid), Tannins, Flavonoids, Carotenoids (Beta-carotene)	Bark for diarrhoea and fever, leaves for diabetes, flowers for diarrhoea, seeds for asthma	300–700 m (WCE Nepal); Tropical Himalaya, India, Ceylon, Burma, Indo-China, Malaysia, widely cultivated and often naturalised in Tropics	Malla (1993), Nautiyal and Kaul (2003)

(continued)

Table 1.1 (continued)

Latin names [Family]	Local names	Plant parts	Dye or pigment	Active compounds	Medicinal Uses	Altitudinal range, distribution in Nepal and in the world	References
Marsdenia tinctoria R. Br. [Asclepiadaceae]	Kali laharo (Nep)	Leaves	Indigo	Flavonoids, Triterpenoids, Phenolic compounds	Used as a febrifuge (reduces fever) and for dyeing; also for skin conditions	500–1200 m (E Nepal); Subtropical Himalaya (Nepal to Bhutan), Assam, east to W. and S. China, Taiwan, Malaysia	Malla (1993)
Melastoma melabathricum L. [Melastomataceae]	Angeri, Chulesi (Nep)	Bark	For blackening, fruits purple dye	Tannins, Flavonoids, Triterpenoids	Leaves used for diarrhoea and wound healing; as a gargle for sore throat	200 m (C Nepal); Tropical Himalaya, India, Ceylon, Burma, Thailand, Indo-China, Malaysia, Australia	Malla (1993), Manandhar (2002)
Mentha sp. [Lamiaceae]	Pudina (Nep)	Leaves	Green	Volatile oils (Menthol, Menthone, 1,8-Cineole), Flavonoids, Phenolic acids	Used for indigestion, nausea, colds, and as a calming agent	Widely cultivated	Malla (1993)

(continued)

Table 1.1 (continued)

Latin names [Family]	Local names	Plant parts	Dye or pigment	Active compounds	Medicinal Uses	Altitudinal range, distribution in Nepal and in the world	References
Michelia champaca L. [Magnoliaceae]	Aule chaanp (Nep)	Flower	Yellow	Volatile oils (Linalool, Indole, Benzyl acetate), Alkaloids, Flavonoids	Bark and flowers used for fever, nausea, and as an anti-inflammatory	600–1300 m (CE Nepal); Nepal, India, Burma, Thailand, Indo-China, S. Yunnan; commonly planted in SE. Asia	Malla (1993), Manandhar (2002), Bhandari et al. (2020b)
Morinda angustifolia Roxb. (*Morinda angustifolia* Roxb. (Banhaldi)) [Rubiaceae]	Banhaldi (Nep)	Root	Yellow	Anthraquinones (Rubiadin, Morindin), Iridoid glycosides	Root used for dysentery and as a tonic	300–800 m (CE Nepal); Himalaya (Nepal to Bhutan), Assam, Burma	Malla (1993)
Moringa oleifera Lam [Moringaceae]	Sheetal Chini (Nep.)	Wood	Blue	Glucosinolates/Isothiocyanates (4-(α-L-Rhamnosyloxy)benzyl isothiocyanate), Flavonoids (Quercetin, Kaempferol), Alkaloids	Leaves and pods are highly nutritious; used for anaemia, inflammation, and diabetes	150–1100 m (CE Nepal); N. W. India, cultivated throughout the tropics	Nautiyal and Kaul (2003), Bhandari et al. (2020b)
Morus macroura Miq. (*Morus alba* var. *laevigala* Bureau) [Moraceae]	Kimbu (Nep)	Fruit	Grey	Flavonoids (Moracin, Kuwanon), Alkaloids, Stilbenes	Fruit used as a tonic and laxative; leaves for fever	1200–1700 m (E Nepal); Himalaya (Kumaun to Bhutan), India, Indo-China, W. and S. China	Malla (1993), Bhandari et al. (2020b)

(continued)

Table 1.1 (continued)

Latin names [Family]	Local names	Plant parts	Dye or pigment	Active compounds	Medicinal Uses	Altitudinal range, distribution in Nepal and in the world	References
Morus serrata Roxb. (*Morus alba* var. *serrata* Bureau) [Moraceae]	Kimbu (Nep)	Fruit	Grey	Flavonoids, Alkaloids, Stilbenes	Fruit used for anaemia and as a tonic; bark for coughs	1600–2400 m (WC Nepal); Himalaya (Kumaun to Nepal)	Bhandari et al. (2020c)
Myrica esculenta Buch.-Ham. ex D. Don [Myricaceae]	Box myrtle, Baybery (Eng), Kafal (Nep), Coboshi (New)	Bark	Yellow	Flavonoids (Myricetin), Tannins, Triterpenoids	Bark used for coughs, asthma, and fever; fruit is astringent	1200–2300 m (WCE Nepal); Himalaya (Kashmir to Bhutan), India, Burma, east to W. and S. China and south to Malaysia	Malla (1993, 1994), Manandhar (2002), Bhandari et al. (2020b)
Nardostachys grandiflora DC. [Valerianaceae]	Bhutle, Jatamasi (Nep), Jermasi (Ger), Pangbu (She)	Inforecence	Red	Sesquiterpenoids (Jatamansone/ Valeranone, Nardostachone), Volatile oils	A renowned sedative and anti-epileptic; used for anxiety, insomnia, and as a tonic	3200–5000 m (WCE Nepal); Himalaya (Garhwal to Bhutan), Tibet, W. China	Malla (1993), Malaisse et al. (n.d.)

(continued)

Table 1.1 (continued)

Latin names [Family]	Local names	Plant parts	Dye or pigment	Active compounds	Medicinal Uses	Altitudinal range, distribution in Nepal and in the world	References
Neolamarckia cadamba (Roxb.) Bosser (*Anthocephalus chinensis* (Lam.) A. Rich. ex Walp., *Anthocephalus cadamba* (Roxb.) Miq.) [Rubiaceae]	Kadam (Nep)	Bark	Brown	Alkaloids (Cadambine), Triterpenoids, Flavonoids	Bark used for fever and as an astringent	290–800 m (CE Nepal); Himalaya (Nepal), India, Ceylon, Burma, Indo-China, S. China, Malaysia	Malla (1993)
Nyctanthes arbor-tristis L. [Oleaceae]	Parijat, Rudhilo (Nep)	Coroll atubes of the flower	Saffron-yellow or orange, yellow	Iridoid glycosides (Arbortristosides, Nyctanthin), Alkaloids, Carotenoids	Leaves used for arthritis, fever, and as a laxative; flowers for hair loss and skin care	200–1200 m (WCE Nepal); Subtropical Himalaya, India, often cultivated for its fragrant flowers	Malla (1993), Manandhar (2002), Nautiyal and Kaul (2003)
Ocimum tenuiflorum L. [Lamiaceae]	Tulasi (Nep)	Leaves	Redish brown	Volatile oils (Eugenol, Methyl eugenol, Caryophyllene), Flavonoids	Used for stress, respiratory infections, fever, and as an adaptogen	400–900 m (CE Nepal); S. W. Asia, Himalaya (Nepal), Bengal, India, Ceylon, Burma, China, Thailand, Malaysia; often cultivated	Bhandari et al. (2020b)

(continued)

Table 1.1 (continued)

Latin names [Family]	Local names	Plant parts	Dye or pigment	Active compounds	Medicinal Uses	Altitudinal range, distribution in Nepal and in the world	References
Onosma bracteatum Wall. [Boraginaceae]	Muktsi (smug-rtsi) (Tib)	Roots	Red	Naphthoquinones (Shikonin derivatives), Phenolic acids	Used for urinary tract disorders, fever, and as a detoxifying agent	3300–5000 m (WC Nepal); Himalaya (Garhwal to Nepal)	Pohle (1990), Ghimire et al. (2001)
Onosma hookeri C. B. Clarke [Boraginaceae]	Maharangi (Nep), Dimok, Muktsi (Tib)	Roots	Red	Naphthoquinones (Shikonin derivatives), Phenolic acids	Used for cough, fever, and skin diseases	4100–4500 m (WC Nepal); Tibet	Ghimire et al. (2001)
Oroxylum indicum (L.) Kurz [Bignoniaceae]	Tatelo (Nep), Pharaha (Che), Sauna (Tha), Laa mendo (Tam)	Stem and roots	Yellow	Flavonoids (Baicalein, Chrysin, Oroxylin A), Glycosides	Bark used for diarrhoea and dysentery; root for fever and arthritis	400–1400 m (WCE Nepal); Tropical Himalaya, India to Indo-China, Malaysia, W. and S. China	Sanai et al. (2023)
Oryza sativa L. [Poaceae]	Chamal, Dhaan (Nep)	Husks	Pale yellowish brown	Vitamins (B vitamins, Tocotrienols), Phenolic acids (Ferulic acid), γ-Oryzanol	Nutritional support; used in Ayurvedic preparations as a base and for balancing	200–1200 m (C Nepal); Widely cultivated	Malla (1993)

(continued)

Table 1.1 (continued)

Latin names [Family]	Local names	Plant parts	Dye or pigment	Active compounds	Medicinal Uses	Altitudinal range, distribution in Nepal and in the world	References
Oxalis corniculata L. [Oxalidaceae]	Chari amilo (Nep)	Flowers	Yellow dye	Flavonoids (Vitexin, Isovitexin), Organic acids (Oxalic acid, Tartaric acid), Tannins	Used for indigestion, fever, and dysentery; applied to skin for inflammation	300–2900 m (WCE Nepal); Almost cosmopolitan	Singh (2017)
Oxalis latifolia Humb. [Oxalidaceae]	Chari amilo (Nep)	Root	Golden yellow to yellow-green dye	Flavonoids, Organic acids (Oxalic acid), Tannins	Similar to *Oxalis corniculata*; used for fever and as a diuretic	1200–1600 m (CE Nepal); C. and S. America; naturalised in S. Europe, India, Malaysia	Manandhar (2002)
Paspalum scrobiculatum L. [Poaceae]	Kodee, Janai ghans (Nep)	Seeds	Yellow	Alkaloids, Flavonoids, Tannins	A nutritious grain; used for diabetes and as a digestive	200–2200 m (CE Nepal); S. E. Asia to Polynesia and Australia	Malla (1993)
Pedicularis longiflora var. *tubiformis* Rudolph [Scrophulariaceae]	Jumre Jhar (Nep)		Redish	Iridoid glycosides, Phenylpropanoid glycosides, Flavonoids	Used for weakness, fever, and as a tonic	2500–4100 m (WCE Nepal); Himalaya (Kashmir to Bhutan), S. E. Tibet, W. China	Pandey (2006)

(continued)

Table 1.1 (continued)

Latin names [Family]	Local names	Plant parts	Dye or pigment	Active compounds	Medicinal Uses	Altitudinal range, distribution in Nepal and in the world	References
Perilla frutescens (L.) Britton (*Perilla ocimoides* L.) [Lamiaceae]	Bhangriya, Silam (Nep)	Leaves	Red and blue in alkaline condition (Fujiwara et al. 2018)	Volatile oils (Perillaldehyde, Limonene), Flavonoids, Phenolic acids	Used for asthma, nausea, and food poisoning	600–2400 m (WCE Nepal); Himalaya (Kashmir to Bhutan), India, China, Burma, Malaysia, Japan (introduced); frequently cultivated	Fujiwara et al. (2018)
Phyllanthus emblica L (*Emblica officinalis* Gaertn.) [Euphorbiaceae]	Amala (Nep), Tigi (Gur), Ghwarmeth (Mag.), Korosi (Rai), Meral (Sat), Aonla (Tha.)	Fruits and bark	Black grey and brown	Tannins (Emblicanin A & B, Gallic acid), Flavonoids, Alkaloids, Vitamin C	A key component of "*Triphala*"; used as a rejuvenating tonic, for immunity, digestion, and hair growth	150–1400 m (WCE Nepal); India, Himalaya (Kumaun to Bhutan), Assanl, N. Burma, S. China, Indo-China, Malaysia	Malla (1993), Dangol (2002), Nautiyal and Kaul (2003), Bhandari et al. (2020b)
Phytolacca acinosa Roxb. [Phytolaccaceae]	Hokling (She)	Fruits	Red (Cheng et al. 2017; Zhao et al 2014)	Triterpenoid saponins, Alkaloids	Root used for rheumatism and as an emetic. All parts are toxic	2200–3200 m (WC Nepal); Himalaya (Kashmir to NEFA), Assam, Laos, W. China	Malla (1993), Magray et al. (2025)

(continued)

Table 1.1 (continued)

Latin names [Family]	Local names	Plant parts	Dye or pigment	Active compounds	Medicinal Uses	Altitudinal range, distribution in Nepal and in the world	References
Pinus roxburghii Sarg. [Pinaceae]	Long-leaved pine Chir (Eng), Khote sallo (Nep)	Bark, Cone	Brown to brownish yellow (Avinc et al. 2013); bark yellow (Nautiyal and Kaul 2003)	Volatile oils (α-Pinene, β-Pinene, Longifolene), Flavonoids, Tannins	Resin used for wounds, boils, and for respiratory ailments	1100–2100 m (WCE Nepal); Afghanistan, Himalaya (Kashmir to Bhutan)	Malla (1993), Nautiyal and Kaul (2003)
Pinus wallichiana A. B. Jacks. [Pinaceae]	Gobre sallo (Nep)	Bark, Cone	Brown to brownish yellow (Avinc et al. 2013); yellow to deep yellow (Nautiyal and Kaul 2003)	Volatile oils (α-Pinene, δ-Carene), Flavonoids, Tannins	Resin used for wounds and joint pain; needles for coughs	1800–3300 m (WCE Nepal); Afghanistan, Himalaya (Kashmir to Nepal)	Malla (1993), Nautiyal and Kaul (2003)
Piper longum L. [Piperaceae]	Pipla (Nep)	Wood and roots	Pale Brown	Alkaloids (Piperine, Piperlongumine), Volatile oils (β-Caryophyllene)	Used for asthma, cough, indigestion, and as a rejuvenating tonic	200–800 m (WCE Nepal); Himalaya (Nepal to Bhutan), India, Ceylon, Malaysia	Malla (1993)

(continued)

Table 1.1 (continued)

Latin names [Family]	Local names	Plant parts	Dye or pigment	Active compounds	Medicinal Uses	Altitudinal range, distribution in Nepal and in the world	References
Pistacia chinensis subsp. *integerrima* (J. L. Stewart) Rech. f. [Anacardiaceae]	Kaakadsinghi (Nep)	Leaf, flower	Yellow	Tannins, Flavonoids, Volatile oils (Pinene, Limonene), Resins	Used for its anti-inflammatory properties, especially for coughs, asthma, and liver disorders	2100 m (W Nepal); Afghanistan, Himalaya (Chitral, Punjab, Nepal)	Bhandari et al. (2020c)
Plantago major L. [Plantaginaceae]	Luhuriya (Nep)	Root	Red	Mucilage (Polysaccharides), Iridoid glycosides (Aucubin), Tannins, Flavonoids	Leaves used for wounds, coughs, and as a diuretic	C Nepal; Europe and W. Asia, widely introduced elsewhere	Malla (1993)
Pogostemon benghalensis (Burm. f.) Kuntze [Lamiaceae]	Rudhilo (Nep), Basdam, Nampani (Che), Utajara (Dan), Rutili (Tam), Kali-bant (Tha)	Leaves, shoot	Blackish blue (Shigwan et al. 2013)	Volatile oils (Patchouli alcohol, Sesquiterpenes), Flavonoids	Used for asthma, coughs, and as an insect repellent	150–1300 m (WCE Nepal); Himalaya (Kumaun to Nepal), India	Dangol (2002, 2008)

(continued)

Table 1.1 (continued)

Latin names [Family]	Local names	Plant parts	Dye or pigment	Active compounds	Medicinal Uses	Altitudinal range, distribution in Nepal and in the world	References
Populus deltoides Marshall [Salicaceae]	Lahare pipal (Nep)	Leaves, catkins	Leaves-Grey to yellowish; catkins-grey to yellow (Harmon and Siddika 2022)	Flavonoids/Phenolic glycosides (Populin, Salicin, Salireposide), Tannins	Buds used for colds, coughs, and as an anti-inflammatory	Cultivated	Malla (1993)
Potentilla nepalensis Hook. [Rosaceae]	Bajradanti (Nep)	Root	Red	Tannins, Flavonoids, Triterpenoids	Used for diarrhoea, fever, and as an astringent	2300–2600 m (W Nepal); Himalaya (Kashmir to Nepal)	Malla (1993)
Prinsepia utilis Royle [Rosaceae]	Dhatelo (Nep)	Fruits	Deep purple	Cyanogenic glycosides, Flavonoids, Tannins	Seed oil used for rheumatism and skin diseases	1500–2900 m (WCE Nepal); Himalaya (Hazara to Bhutan), S. Tibet, Assam, W. China	Manandhar (2002)
Prunus cornuta (Wall. ex Royle) Steud. (*Prunus padus* Brandis) [Rosaceae]	Arupate (Nep)	Bark	Leaves-green and fruit-dark grey to green (Grae 1974)	Cyanogenic glycosides, Flavonoids, Tannins	Bark used as a febrifuge and astringent	2100–3500 m (WCE Nepal); Afghanistan, Himalaya (Kashmir to Bhutan), Assam, S. Tibet, W. China	Malla (1993)

(continued)

Table 1.1 (continued)

Latin names [Family]	Local names	Plant parts	Dye or pigment	Active compounds	Medicinal Uses	Altitudinal range, distribution in Nepal and in the world	References
Prunus persica (L.) Batsch	Aaru (Nep)	Fruit, leaf	Fruit and leaves-dark grey to green; Fruits used as mordant	Cyanogenic glycosides, Flavonoids, Triterpenoids	Leaves and kernels used for coughs and constipation. Kernels contain cyanide	W Nepal; China, widely cultivated in Asia and Europe	Malla (1993), Önal et al. (2021)
Punica granatum L. [Punicaceae]	Anar (Nep)	Fruit rind, Flower, bark	Red	Tannins (Punicalagins, Ellagic acid), Flavonoids (Anthocyanins), Alkaloids (Pelletierine)	Fruit rind used for diarrhoea and dysentery; seeds as a tonic	700–2700 m (WC Nepal); C. and W. Asia, W. Himalaya; widely cultivated in E. Himalaya, S. Europe, Asia	Malla (1993), Manandhar (2002), Nautiyal and Kaul (2003), Bhandari et al. (2020b)
Quercus leucotrichophora A. Camus (*Quercus incana* Roxb.) [Fabaceae]	Banjh (Nep)	Bark	Black brown	Tannins, Flavonoids, Triterpenoids	Bark used as an astringent for diarrhoea, wounds, and haemorrhoids	1500–2400 m (WC Nepal); W. Pakistan, Himalaya (Kumaun to Nepal), Ceylon, ?Burma	Malla (1993)

(continued)

Table 1.1 (continued)

Latin names [Family]	Local names	Plant parts	Dye or pigment	Active compounds	Medicinal Uses	Altitudinal range, distribution in Nepal and in the world	References
Ranunculus sceleratus L. [Ranuculaceae]	Nakkore (Nep)	Flowers	Yellow	Toxic glycosides (Protoanemonin), Alkaloids, Flavonoids	Used for warts and skin diseases. Highly toxic and a strong skin irritant	800–1700 m (WCE Nepal); Europe, C. Asia, Himalaya, N. India, Siberia, Mongolia, China, Japan, N. America	Malla (1993)
Rheum acuminatum Hook. f. & Thomson ex Hook. [Polygonaceae]	Padamchaal (Nep)	Rhizome, Roots	Yellow	Anthraquinones (chrysophanol, emodin, aloe-emodin, physcion and rhein) and two stilbenes (piceatannol and resveratrol)	Root used as a purgative and for digestive disorders	3300–4200 m (CE Nepal); Himalaya (Nepal to Bhutan), Assam, N. Burma	Rokaya et al. (2012)
Rheum australe D. Don (*Rheum emodi* Wall. ex Meisn.) [Polygonaceae]	Padamchaal (Nep)	Rhizome, Roots	Yellow	Anthraquinones (Emodin, Rhein, Chrysophanol), Stilbenes (Picetannol), Tannins	Root used as a purgative and for digestive disorders	3200–4200 m (WC Nepal); Himalaya (Lahul to Nepal, ?Bhutan), S. Tibet	Malla (1993, 1994), Ghimire et al. (2001), Bhandari et al. (2020b)
Rheum moorcroftianum Royle [Polygonaceae]	Padamchal (Nep)	Roots	Red	Anthraquinones (Emodin, Rhein), Stilbenes, Tannins	Root used for wounds and as a purgative	3600–4400 m (WC Nepal); Himalaya (Kumaurl, Nepal)	Malla (1993)

(continued)

Table 1.1 (continued)

Latin names [Family]	Local names	Plant parts	Dye or pigment	Active compounds	Medicinal Uses	Altitudinal range, distribution in Nepal and in the world	References
Rhus punjabensis J. L. Stewart [Anacardiaceae]	Titari (Nep)	Bark	Brown	Tannins (Gallotannins, Ellagitannins), Biflavonoids, Phenolic acids (Gallic acid)	Used for diarrhoea, dysentery, and as a gargle for sore throats	2100–2300 m (WC Nepal); Himalaya (Kashmir to Bhutan)	Malla (1993)
Ricinus communis L. [Euphorbiaceae]	Ander (Nep), Andi (Dan), Yanyan (Gur), Iradam (Sat), Dhandarobi (Tam.)	Castor oil	Red	Ricin, Alkaloids (Ricinine), Triglycerides (Ricinoleic acid)	Castor oil from seeds is a powerful laxative. The seeds themselves are deadly due to the ricin toxin	150–2400 m (WCE); Believed to be native of N. E. Tropical Africa; widely cultivated and occasionally naturalised throughout the tropics	Malla (1993)

(continued)

Table 1.1 (continued)

Latin names [Family]	Local names	Plant parts	Dye or pigment	Active compounds	Medicinal Uses	Altitudinal range, distribution in Nepal and in the world	References
Rubia manjith Roxb. ex Fleming (*Rubia cordifolia* L.) [Rubiaceae]	Majitho (Nep)	Root	Red	Anthraquinones (Purpurin, Munjistin, Rubiadin), Iridoid glycosides	A premier blood purifier; used for skin diseases, jaundice, and as a colouring agent	1200–2100 m (WCE Nepal); Himalaya (Simla to Bhutan), Khasia	Malla (1993, 1994), Ghimire et al. (2001), Manandhar (2002), Kunwar et al. (2009), Bhandari et al. (2020b)
Rubia sikkimensis Kurz [Rubiaceae]	Sikkim madder (Eng.)	Root	Red	Anthraquinones, Iridoid glycosides	Similar uses to *Rubia manjith*; used as a blood purifier	600–1000 m (E Nepal); E. Himalaya (Nepal to NEFA), N. Assam, Manipur	Malla (1993)
Rumex acetosa L. [Polygonaceae]	Amile ghans, Hale (Nep)	Whole plant, flower	Different colours depending on mordant	Anthraquinones (Emodin, Chrysophanol), Oxalic acid, Flavonoids	Leaves used as a diuretic and for skin inflammation	2100–4100 m (WC Nepal); Europe, W. Asia, Himalaya (Kashmir to Nepal), Tibet, Siberia, China, Japan, N. America	Malla (1993)

(continued)

Table 1.1 (continued)

Latin names [Family]	Local names	Plant parts	Dye or pigment	Active compounds	Medicinal Uses	Altitudinal range, distribution in Nepal and in the world	References
Rumex nepalensis Spreng. [Polygonaceae]	Halhale saag (Nep)	Roots	Yellow	Anthraquinones (Nepodin, Emodin), Tannins, Flavonoids	Root used for diarrhoea, dysentery, and skin diseases	1200–4200 m (WCE Nepal); S. W. Europe, W. Asia, Himalaya, India, east to W. & C. China, ?Java	Pohle (1990), Ghimire et al. (2001)
Schima wallichii (DC.) Korth. [Theaceae]	Chilaune (Nep)	Bark	Brownish black	Tannins, Flavonoids, Triterpenoid saponins	Bark used for diarrhoea and as an anthelmintic	900–2100 m (CE Nepal); Himalaya (Nepal to Bhutan), Assam. S. Tibet, east to W. China	Malla (1993)
Schleichera oleosa (Lour.) Oken (*Schleichera trijuga* Willd.) [Sapindaceae]	Kusum (Nep)	Flowers	Pinkish-brown (Sarkar et al. 2022)	Tannins, Flavonoids, Triterpenoids	Seed oil used for skin diseases and rheumatism	200–300 m (WCE Nepal); Tropical Himalaya (Punjab to Nepal), India, Ceylon, Burma, Thailand, Indo-China, Malaysia	Malla (1993)

(continued)

Table 1.1 (continued)

Latin names [Family]	Local names	Plant parts	Dye or pigment	Active compounds	Medicinal Uses	Altitudinal range, distribution in Nepal and in the world	References
Semecarpus anacardium L. f. [Anacardiaceae]	Bhalayo, Kag bhalayo (Nep)	Fruit	Black	Biflavonoids, Phenolic lipids (Anacardic acid), Alkaloids	Used in Ayurveda for rejuvenation and treating tumours, arthritis, and skin diseases (often after specific purification processes)	150–1200 m (WCE Nepal); Himalaya (Sirmore to Sikkim), India, Burma, Malaysia, N. Australia	Malla (1993), Dangol (2002)
Sesamum orientale L. (*Sesamum indicum* L.) [Pedaliaceae]	Teel, Til (Nep)	Oil	Pale orange	Lignans (Sesamin, Sesamolin), Phytosterols, Tocopherols	Seeds and oil used for skin and hair; considered a nutritive tonic	600–2400 m (WCE Nepal); Widely cultivated in tropical Africa, W. Pakistan, India, Ceylon, E. Asia	Malla (1993)

(continued)

Table 1.1 (continued)

Latin names [Family]	Local names	Plant parts	Dye or pigment	Active compounds	Medicinal Uses	Altitudinal range, distribution in Nepal and in the world	References
Shorea robusta Gaertn. [Diptarocarpaceae]	Saal (Nep), Sakhuwa (Dan), Sasingba (Lim), Phoksing (Mag), Solam (Rai), Koraila (Tha)	Bark	Red and Black	Triterpenoids (Shorea acid), Tannins, Flavonoids, Sterols	Resin (Sal Dammar) used for diarrhoea, dysentery, and skin diseases	150–1500 m (WCE Nepal); Subtropical Himalaya (Garhwal to Assam), India	Malla (1993, 1994), Dangol (2002), Nautiyal and Kaul (2003)
Solanum surattense Burm. f. (*Solanum xanthocarpum* Schrad. and Wendl.) [Solanaceae]	Kantakari (Nep)	Leaves	Light red	Steroidal alkaloids (Solasodine, Solamargine), Flavonoids, Sterols	Used for cough, asthma, and chest congestion	300–900 m (WCE Nepal); Himalaya, N. India, S. E. Asia, Malaysia, Australia, Polynesia	Bhandari et al. (2020b)
Solidago virga-aurea L. [Asteraceae]	Pinja phool, Sonali (Nep)	Flowers	Yellow	Flavonoids (Quercetin, Rutin), Saponins, Phenolic glycosides, Diterpenes	Used for urinary tract infections, kidney stones, and as an anti-inflammatory for the throat and mouth	2300–3400 m (WCE Nepal); Himalaya (Kashmir to Nepal), Khasia, China, Japan, Bering Sea Is., Europe	Malla (1993)

(continued)

Table 1.1 (continued)

Latin names [Family]	Local names	Plant parts	Dye or pigment	Active compounds	Medicinal Uses	Altitudinal range, distribution in Nepal and in the world	References
Spinacia oleracea L. [Amaranthaceae]	Palungo (Nep)	Leaves	Green; Yellow presence of alum, dark green to light green	Carotenoids (Lutein, Beta-carotene), Flavonoids, Vitamins (K, C, Folate), Glycoglycerolipids	Nutritional supplement for anaemia and vitamin deficiency; also used for eye health	Native to Afghanistan, Iran, Kazakhstan, Kirgizstan, Pakistan, Tadzhikistan, Turkmenistan, Uzbekistan; widely cultivated	Malla (1993)
Symplocos paniculata (Thunb.) Miq. (*Symplocos crataegoides* Buch.-Ham. ex D. Don) [Symplocaceae]	Lodh (Nep)	Bark	Yellow	Triterpenoids, Flavonoids, Alkaloids	Bark used for eye diseases, diarrhoea, and as a cardiac tonic	1000–2500 m (WCE Nepal); Himalaya (Kashmir to NEFA), Burma, Indo-China, China, Korea, Japan	Malla (1993), Manandhar (2002)
Symplocos theifolia D. Don (*Symplocos phyllocalyx* C. B. Clarke) [Symplocaceae]	Bhauli, Ghole, Lal Chandan (Nep)	Leaves	Pale yellow	Triterpenoids, Flavonoids, Alkaloids	Similar uses to *S. paniculata*; used for gum diseases and fever	1500–3000 m (WCE Nepal); Himalaya, Burma, Thailand, Indo-China, S. China, Malaysia	Malla (1993)

(continued)

Table 1.1 (continued)

Latin names [Family]	Local names	Plant parts	Dye or pigment	Active compounds	Medicinal Uses	Altitudinal range, distribution in Nepal and in the world	References
Syzygium cumini (L.) Skeels (*Eugenia jambolana* Lam.) [Myrtaceae]	Jamun (Nep)	Bark, fruit	Brown	Tannins (Ellagic acid, Gallic acid), Flavonoids (Anthocyanins, Quercetin)	Seeds and fruit used for diabetes and diarrhoea; bark for sore throat	300–1200 m (WCE Nepal); Subtropical Himalaya, India, Ceylon, Malaysia, Australia	Malla (1993), Dangol (2002), Bhandari et al. (2020b)
Tabernaemontana divaricata (L.) R. Br. ex Roem. & Schult. [Apocynaceae]	Chandanee (Nep)	Pulp	Red	Indole alkaloids (Coronaridine, Voacangine, Apparicine)	Used for eye ailments, headaches, and as an anti-inflammatory; the root is used for toothache	250–1200 m (CE Nepal); Native of tropical Asia; cultivated throughout the tropics	Malla (1993)
Tagetes erecta L. [Asteraceae]	Sayapatree (Nep)	Flowers	Yellow	Carotenoids (Lutein), Thiophene derivatives, Flavonoids, Essential oils	Used for colds, coughs, and as a diuretic; also for skin problems and as an insect repellent	800–2000 m (C Nepal); A garden plant of Mexican origin	Malla (1993), Bhandari et al. (2020b)

(continued)

Table 1.1 (continued)

Latin names [Family]	Local names	Plant parts	Dye or pigment	Active compounds	Medicinal Uses	Altitudinal range, distribution in Nepal and in the world	References
Tagetes patula L. [Asteraceae]	Barhamase sayaptree (Nep)	Roots	Yellow	Carotenoids, Thiophene derivatives, Flavonoids, Essential oils	Similar uses to *Tagetes erecta*; also for stomach aches and as an antiseptic	900–2000 m (WCE Nepal); A garden plant of Mexican origin	Malla (1993)
Tamarindus indica L. [Fabaceae]	Amili, Emali (Nep)	Leaves	Red	Organic acids (Tartaric acid, Acetic acid), Flavonoids	Fruit pulp used as a digestive, laxative, and for fever	200–400 m (CE Nepal); Pantropics in cultivation, possibly native in tropical Africa	Malla (1993)
Tamarix dioica Roxb. ex Roth [Tamaricaceae]	Jhauwa, Lai (Nep)	Bark, flowers	Gray blue, orange-brown, greens depends on mordant	Tannins, Flavonoids (Tamarixin), Phenolic acids	Used for liver and spleen disorders, and as an astringent	100–400 m (WE Nepal); C. Asia, Afghanistan, Tropical Himalaya, N. India, Burma	Malla (1993)
Taraxacum officinale Wigg. [Asteraceae]	Tuki phool (Nep)	Flowers	Yellow	Sesquiterpene lactones (Taraxinic acid), Triterpenes (Taraxasterol), Phenolic acids (Cichoric acid), Polysaccharides (Inulin)	Used as a diuretic ("pee-the-bed"), for liver detoxification, and as a digestive tonic	Nepal. Native to Europe and Asia, and was originally imported to America as a food crop. It is now naturalized throughout the world	Malla (1993)

(continued)

Table 1.1 (continued)

Latin names [Family]	Local names	Plant parts	Dye or pigment	Active compounds	Medicinal Uses	Altitudinal range, distribution in Nepal and in the world	References
Taxus contorta Griff. [Taxaceae]	Dhangre Salla, Launth Sallo (Nep)	Bark	Red	Diterpenoids (Paclitaxel/Taxol, 10-Deacetylbaccatin III), Alkaloids	Source of the anti-cancer drug paclitaxel (Taxol). All parts except the aril are highly toxic	2300–3400 m (WC Nepal); Afghanistan, Himalaya (Kashmir to Bhutan), Assam, N. Burma, Indo-China, W. China, Malaysia	Poudel et al. (2013)
Taxus wallichiana Zucc. [Taxaceae]	Dhangre Salla, Launth Sallo (Nep)	Bark	Red	Diterpenoids (Paclitaxel/Taxol, 10-Deacetylbaccatin III), Alkaloids	Same as above; a primary source of paclitaxel. Highly toxic	2300–3400 m (CE Nepal); Afghanistan, Himalaya (Kashmir to Bhutan), Assam, N. Burma, Indo-China, W. China, Malaysia	Malla (1993)
Tectona grandis L. f. [Verbenaceae]	Teak, saguan (Nep)	Leaf	Yellow	Quinones (Lapachol, Tectoquinone), Volatile oils, Tannins	Leaves and flowers used for digestive problems and as a diuretic	Below 900 m (WCE Nepal); native to India, China, Indonesia, Malaysia, and Myanmar	Nautiyal and Kaul (2003) Kunwar et al. (2009)

(continued)

Table 1.1 (continued)

Latin names [Family]	Local names	Plant parts	Dye or pigment	Active compounds	Medicinal Uses	Altitudinal range, distribution in Nepal and in the world	References
Terminalia alata Heyne ex Roth (*Pentaptera tomentosa* Roxb.) [Combretaceae]	Asna, Saaj (Nep), Darsing (Che), Sas (Tam), Asna (Tha)	Bark	Red	Tannins, Flavonoids, Triterpenoids (Arjunic acid, Arjunetin)	Bark used for diarrhoea, wounds, and as a cardiac tonic	200–1400 m (WCE Nepal); Himalaya, India, Ceylon, Burma, Thailand, Indo-China	Malla (1993), Dangol (2002), Bhandari et al. (2020b)
Terminalia bellirica (Gaertn.) Roxb. [Combretaceae]	Barro (Nep), Tupsi (Che), Haratun (Gur), Godling (Mag), Lopong (Sat), Baheri (Tha)	Bark, fruit	Bark gives yellow dye (Malla et al. 1993), fruit blue (Bhandari et al. 2020c, a, b), yellow (Nautiyal and Kaul 2003)	Tannins (Gallic acid, Ellagic acid), Lignans, Triterpenoids	One of the three fruits in "*Triphala*"; used for constipation, respiratory infections, and as a tonic	300–1100 m (CE Nepal); India, Ceylon, Burma, Thailand, Indo-China, Malaysia	Malla (1993), Dangol (2002) Nautiyal and Kaul (2003), Bhandari et al. (2020b)
Terminalia chebula Retz. [Combretaceae]	Harro (Nep), Leesi (Che), Thopsing (Mag), Arubharelu (She)	Fruits	Yellow and black; bark pinkish red (Bhandari et al. 2020c, a, b)	Tannins (Chebulic acid, Chebulinic acid, Gallic acid), Flavonoids, Triterpenoids	The "King of Medicines" in Ayurveda; used as a rejuvenating tonic, for digestion, and for constipation	150–1100 m (CE Nepal); Kumaun, India, Ceylon, Burma (but not wild in Malaysia, fide Exell in Fl. Males. 4 (1954))	Malla (1993, 1994), Dangol (2002), Kunwar et al. (2009), Bhandari et al. (2020b)

(continued)

Table 1.1 (continued)

Latin names [Family]	Local names	Plant parts	Dye or pigment	Active compounds	Medicinal Uses	Altitudinal range, distribution in Nepal and in the world	References
Thespesia lampas (Cav.) Dalzell & Gibson [Malvaceae]	Ban kapas (Nep)	Flowers	Yellow	Flavonoids (Gossypetin, Quercetin), Tannins, Sesquiterpenoid aldehydes (Gossypol)	Used for skin diseases, diarrhoea, and as an anti-inflammatory	200–1100 m (WC Nepal); Africa, S. and S. E. Asia. Widely cultivated as ornamental plant	Dangol (2002)
Toddalia asiatica (L.) Lam. [Asteraceae]	Main kanda (Nep)	Flowers	Yellow	Quinoline alkaloids (Skimmianine, γ-Fagarine), Coumarins, Flavonoids	Used for malaria, fever, and rheumatism; the root is used for indigestion and cough	450–1700 m (CE Nepal); Africa, Himalaya (Nepal to Bhutan), India, Ceylon, Burma, east to China, Malaysia	Malla (1993)
Toona ciliata M. Roem. (*Cedrela toona* Roxb. ex Rottler & Willd.) [Meliaceae]	Tunee, Tuna (Nep)	Flower, seed, leaf	Flower, seed give Red, Yellow; leaf gives black colour (Nautiyal and Kaul 2003)	Limonoids (Toonacilin, Toonafolin), Tannins, Volatile oils	Bark used for diarrhoea, dysentery, and fever	200–1700 m (WCE Nepal); Afghanistan, Himalaya, India, Ceylon, east to China, Malaysia, Australia	Malla (1993), Dangol (2002), Nautiyal and Kaul (2003), Bhandari et al. (2020b)

(continued)

Table 1.1 (continued)

Latin names [Family]	Local names	Plant parts	Dye or pigment	Active compounds	Medicinal Uses	Altitudinal range, distribution in Nepal and in the world	References
Trapa quadrispinosa Roxb. (*Trapa natans* forma *quadrispinosa* (Roxb.) Makino; *Trapa bispinosa* Roxb.) [Trapaceae]	Singada (Nep)	Shoot	Saffron-yellow or orange, yellow	Flavonoids, Tannins, Polysaccharides	Nut is a cooling; nutritive food used during fasting; astringent	600–800 m (C Nepal); Himalaya, India	Malla (1993)
Urtica dioica L. [Urticaceae]	Sisnu (Nep)	Stalk leaves	Yellow	Flavonoids (Quercetin, Kaempferol), Amines (Histamine, Serotonin), Lignans	Used as a diuretic for urinary issues, for arthritis, and allergies. Fresh plant stings	3000–4500 m (WC Nepal); Europe, N. Africa, W. Siberia, C. Asia, Himalaya, Tibet, W. China, naturalised widely in other temperate regions	Malla (1993)
Usnea thomsonii Stirt. [Parmeliaceae]	Jyau (Nep)	Whole plant	Light brown	Lichen acids (Usnic acid), Depsidones	Used as an antimicrobial for wounds and respiratory infections	Nepal; Australia, the Americas, West Africa and India	IUCN (2004), Pyakurel et al. (2019)

(continued)

Table 1.1 (continued)

Latin names [Family]	Local names	Plant parts	Dye or pigment	Active compounds	Medicinal Uses	Altitudinal range, distribution in Nepal and in the world	References
Viburnum mullaha Buch.-Ham. ex D. Don [Caprifoliaceae]	Molo (Nep)	Fruits	Light brown	Triterpenoids, Flavonoids, Tannins	Fruit used as a digestive; root and bark for fever and asthma	2500–4000 m (WCE Nepal); Bhutan, India (Kashmir to Assam), Indo-China	Manandhar (2002)
Vitex negundo L. [Verbenaceae]	Nirgandhi, Simali (Nep)	Leaves	Light yellow	Flavonoids (Casticin, Vitexin, Isovitexin), Iridoid glycosides (Agnuside), Volatile oils	Used for arthritis, asthma, and as an anti-inflammatory	100–3200 m (WCE Nepal); Pakistan, India, W. Asia and N. Africa; introduced and widely cultivated as an ornamental elsewhere	Malla (1993)
Woodfordia fruticosa (L.) Kurz [Apocynaceae]	Ras dhainyaro, dhainyaro (Nep); Darsing (Che); Chyuhuwa (Dan); Dhaiyar (Dar); Dhurseli (Gur)	Juice	Yellow	Tannins (Ellagitannins, Woodfordin C), Flavonoids (Quercetin), Anthocyanins	Flowers are used for diarrhoea, dysentery, haemorrhoids, and as a uterine tonic	200–1800 m (WCE Nepal); Africa, W. Asia, Subtropical Himalaya, India, Ceylon, Burma, east to China	Malla (1993), Dangol (2002), Nautiyal and Kaul (2003)

(continued)

Table 1.1 (continued)

Latin names [Family]	Local names	Plant parts	Dye or pigment	Active compounds	Medicinal Uses	Altitudinal range, distribution in Nepal and in the world	References
Wrightia arborea (Dennst.) Mabb. (*Wrightia tomentosa* Roem. & Schult.) [Apocynaceae]	Raani Khirro (Nep)	Leaf	Brownish yellow	Triterpenoids (Ursolic acid, β-amyrin), Flavonoids, Cardiac glycosides, Alkaloids	Used for fever, diarrhoea, and skin diseases; the latex is applied to sores	450–950 m (CE Nepal); E. Himalaya, India, Ceylon, Burma, Thailand, Indo-China, S. China	Malla (1993)
Xanthium strumarium L. [Asteraceae]	Bhende kuro (Nep)	Leaves	Yellow	Thiophene derivatives, Glycosides (Carboxyatractyloside), Sesquiterpene lactones (Xanthanolides)	Used for sinusitis, headaches, and rheumatism. Seeds are highly toxic and can be fatal	100–2500 m (WCE Nepal); Eurasia	Malla (1993)
Zingiber officinale Rosc. [Zingiberaceae]	Aduwa (Nep)	Rhizome	Brown	Phenolic compounds (Gingerols, Shogaols), Volatile oils (Zingiberene, Sesquiphellandrene)	Used for nausea, indigestion, colds, and as an anti-inflammatory	Native to India to S. Central China; cultivated elsewhere	Bhandari et al. (2020c)
Zizyphus mauritiana Lam. (*Zizyphus jujuba* (L.) Gaertn.) [Rhamnaceae]	Bayar, Thulo bayar (Nep)	Bark, Fruit, flower	Reddish pink	Alkaloids/Saponins (Jujubosides, Sanjoinine), Flavonoids, Triterpenoids (Betulin)	Fruit used as a blood purifier and for sore throat; seeds for anxiety and insomnia	200–1200 m (WCE Nepal); Tropical Asia, Australia, widely cultivated	Malla (1993), Nautiyal and Kaul (2003)

Table 1.2 List of plant species that are imported from other countries and used in Nepal for obtaining dye or pigment. The plant species are arranged alphabetically with any synonyms in parentheses (), followed by the family in square brackets []. The available vernacular names are given for each plant and are arranged in alphabetical order: English (Eng.), Hindi (Hindi), and Nepali (Nep). The plant parts used to obtain dye or pigment are followed by active compounds found in each plant species, their medicinal uses, distribution ranges, and references

Latin names [Family]	Local names	Plant parts	Dye or pigment	Active compounds	Medicinal Uses	Distribution ranges	References
Aleurites moluccana Willd. [Euphorbiaceae]	Jangli Akhrot (Nep)	Root	Brown	Fatty Acids: Linoleic acid, oleic acid; Phorbol Esters; Flavonoids: Vitexin, isovitexin; Tannins	The oil (kukui nut oil) is used for topical treatment of burns, wounds, and skin conditions; seeds used as a laxative; bark for stomach tumours	Tropical and Subtropical Asia to W. Pacific	Nautiyal and Kaul (2003)
Andropogon virginicus L. [Poaceae]	Broomsedge bluestem, yellowsedge bluestem (Eng)	Stalk and leaves	Yellowish	Essential Oils: Geraniol, citronellol; Alkaloids; Flavonoids; Tannins	Used as a diuretic and for digestive problems; traditional use for fever and inflammation; has insect-repellent properties	Native to North America and southern New England. Introduced in west America, Europe and E. Asia	Malla (1993), Nautiyal and Kaul (2003)
Bassia eriophora (Schrad.) Asch. [Amaranthaceae]	Mahua (Nep)	Bark as adjunct	Black	Saponins; Flavonoids; Phenolic Acids; Triterpenoids	Used in traditional medicine for its antipyretic (fever-reducing) and anti-inflammatory properties; also as a diuretic	The native range of this species is E. Medit. to Mongolia and Arabian Peninsula. It is an annual and grows primarily in the subtropical biome	Malla (1993), Nautiyal and Kaul (2003)

(continued)

Table 1.2 (continued)

Latin names [Family]	Local names	Plant parts	Dye or pigment	Active compounds	Medicinal Uses	Distribution ranges	References
Biancaea sappan (L.) Tod. (*Caesalpinia sappan* L.) [Fabaceae]	Sappan wood (Eng)	Wood	Red	Flavonoids: Brazilein, brasilein (primary pigments); Phenolic Compounds: Protosappanin; Tannins	Used to improve blood circulation and relieve pain; treats diarrhoea and dysentery; antimicrobial for skin and wounds; gives red colour	Indian Subcontinent to Indo-China	Nautiyal and Kaul (2003)
Chrysanthemum cinerariifolium (Trevir.) Vis. (*Tanacetum cinerariifolium* (Trevir.) Sch.Bip.) [Asteraceae]	Pyrethrum daisy (Eng)	Blossoms	Yellow	Pyrethrins: Pyrethrin I and II, cinerins (primary active compounds); Sesquiterpenes	Source of the insecticide pyrethrum; not typically used internally in medicine due to neurotoxicity, but effective against lice	The plant is native to the Balkans and is important commercially as a source of pyrethrin	Malla (1993)

(continued)

Table 1.2 (continued)

Latin names [Family]	Local names	Plant parts	Dye or pigment	Active compounds	Medicinal Uses	Distribution ranges	References
Chukrasia tabularis Adr. Juss. [Meliaceae]	Lal Devdari (Hindi)	Flowers	Red, Yellow	Limonoids: Chukrasins; Tannins; Flavonoids: Aromadendrin	Used for diarrhoea and dysentery; treatment of intestinal worms; antipyretic (fever-reducing) and anti-inflammatory properties	Bangladesh, Bhutan, China, India, Indonesia, Laos, Malaysia, Nepal, Sri Lanka, Thailand, Vietnam	Malla (1993)
Cordia myxa L. [Boraaginaceae]	Lasora (Hindi)	Leaves	Yellow	Tannins; Flavonoids: Quercetin; Triterpenoids: Cordialin A & B	The fruit is demulcent and used for coughs, chest complaints, and sore throat; treats urinary and lung infections; laxative	Pakistan, India Sri-Lanka. Cultivated	Malla (1993)

(continued)

Table 1.2 (continued)

Latin names [Family]	Local names	Plant parts	Dye or pigment	Active compounds	Medicinal Uses	Distribution ranges	References
Crocus sativus L. [Iridaceae]	Saffron (Eng)	Flower	Orange, Yellow	Carotenoids: Crocin, crocetin (for color); Monoterpene Aldehydes: Safranal (for aroma); Flavonoids: Kaempferol	Used for depression, anxiety, and mood disorders; antioxidant; potential aphrodisiac; relieves menstrual pain and asthma	Cultivated	Nautiyal and Kaul (2003), Bhandari et al. (2020b)
Diospyros peregrina (Gaertn.) Gürke [Ebenaceae]	Indian persimmon (Eng)	Fruit	Brown	Tannins: Betulin, oleanolic acid; Naphthoquinones: Diospyrin; Flavonoids	Used for diarrhoea, dysentery, and skin diseases; fruit is astringent; seeds and bark used for fever and parasitic infections	W. and Central Malesia	Nautiyal and Kaul (2003)
Eucalyptus globulus Labill. [Myrtaceae]	Masala (Eng)	Bark	Brown to black (Maulik et al. 2021)	Essential Oils: 1,8-Cineole—primary compound; Tannins; Flavonoids	Used as a decongestant for colds, coughs, and sinusitis; inhaled for asthma and bronchitis; topical antiseptic for wounds	Native to Australia, Introduced in different countries around the world	Malla (1993)

(continued)

Table 1.2 (continued)

Latin names [Family]	Local names	Plant parts	Dye or pigment	Active compounds	Medicinal Uses	Distribution ranges	References
Garcinia pedunculata Roxb. [Clusiaceae]	Malabar Tamarind (Eng)	Fruit	Blue	Organic Acids: Hydroxycitric acid (HCA)—primary active compound; Xanthones: Garcinol, isogarcinol; Flavonoids: Luteolin, apigenin	Used as a digestive aid and to treat diarrhoea and dysentery; believed to have anti-obesity properties; applied to sores and skin infections	South and Southeast Asia, including Bangladesh, Laos, Myanmar, Tibet, and Thailand	Nautiyal and Kaul (2003)
Gomphrena celosioides Mart. [Amaranthaceae]	Gomphrena weed (Eng)	Whole plant	Reddish-purple	Flavonoids: Gomphrenol; Betacyanins; Phenolic Acid	Used for urinary tract infections, kidney problems, and as a diuretic; also for coughs and colds; applied to wounds	Pantropical	Malla (1993)
Hymenodictyon orixense (Roxb.) Mabberley (*Hymenodictyon excelsum* (Roxb.) Wall) [Rubiaceae]	Bridal Couch Tree (Eng)	Leaves	Brown	Alkaloids: Quinoline alkaloids; Iridoids; Tannins	Used for fever (especially malaria), and as a tonic; treats skin diseases and inflammation; bark is astringent	Cambodia, China, India, Indonesia (Java), Kashmir, Laos, Malaysia, Myanmar, Philippines, Thailand, Vietnam	Malla (1993)

(continued)

Table 1.2 (continued)

Latin names [Family]	Local names	Plant parts	Dye or pigment	Active compounds	Medicinal Uses	Distribution ranges	References
Indigofera sphaerocarpa A.Gray (*Indigofera tinctoria* Chapm.) [Fabaceae]	Sonoran indigo (Eng)	Leaf	Blue	Alkaloids: Indigotine; Flavonoids; Tannins	Traditionally used for inflammation, epilepsy, and nervous disorders; also for treating wounds and skin ailments	Native to Arizona, Mexico Northeast, Mexico Northwest, New Mexico	Bhandari et al. (2020b)
Memecylon umbellatum Burm. f. [Melastomataceae]	Iron wood tree (Eng)	Flowers	Yellow	Flavonoids;Tannins; Triterpenoids	Used for eye diseases and conjunctivitis; treats diarrhoea and dysentery; anti-inflammatory and antimicrobial for skin	S. India to W. Malesia	Nautiyal and Kaul (2003)
Morinda citrifolia L. [Rubiaceae]	Indian Mulberry (Eng)	Root	Red	Iridoids: Asperuloside, deacetylasperuloside; Fatty Acids: Caproic acid, caprylic acid; Lignans; Alkaloids: Xeronine (proposed)	Used as a general tonic and analgesic; boosts immunity; treats arthritis, diabetes, and skin infections; known as Noni	Tropical and Subtropical Asia to N. Australia	Nautiyal and Kaul (2003)

(continued)

Table 1.2 (continued)

Latin names [Family]	Local names	Plant parts	Dye or pigment	Active compounds	Medicinal Uses	Distribution ranges	References
Morinda pubescens Sm. (*Morinda tinctoria* Roxb.) [Rubiaceae]	Indian Mulberry (Eng)	Bark	Red	Anthraquinones: Nordamnacanthal, rubiadin; Iridoids; Tannins	Used for skin diseases, wounds, and ulcers; root bark is astringent and used for dysentery; provides a dye	Sri Lanka, India to Indo-China, Jawa	Nautiyal and Kaul (2003)
Onosma frutescens Lam. (*Onosma echioides* Sm.) [Boraginaceae]	Maharangi (Nep)	Root	Red	Naphthoquinones: Shikonin, acetylshikonin; Phenolic Acids: Rosmarinic acid	Used for its anti-inflammatory and wound-healing properties; treats burns and skin infections; has antimicrobial activity	East Aegean Is., Greece, Lebanon-Syria, Palestine, Turkey	Malla (1993)
Polygonum tortuosum D. Don (*Aconogonon tortuosum* (D. Don) H. Hara) [Polygonaceae]	Knotweed (Eng)	Plant	Yellow	Flavonoids: Quercetin, rutin; Tannins; Stilbenes	Used for digestive disorders and as an astringent; treats skin diseases and inflammation; has diuretic properties	Afghanistan, China, N India, Pakistan; SW Asia	Malla (1993)

(continued)

Table 1.2 (continued)

Latin names [Family]	Local names	Plant parts	Dye or pigment	Active compounds	Medicinal Uses	Distribution ranges	References
Pterocarpus santalinus L.f. [Fabaceae]	Raktachandan (Nep)	Fruit rind	Yellow, Red	Isoflavonoids: Santalin (red pigment), pterocarpol, homopterocarpin; Tannins	Used for inflammation, fever, and headaches; treats digestive ailments and skin diseases; astringent and cooling agent	Native to India, Bangladesh, Sri Lanka	Malla (1993), Bhandari et al. (2020b)
Rhamnus davurica Pall. [Rhamnaceae]	Chato (Hin)	Bark, fruit	Bark, fruit-Yellow; leaves, bark and branches-green	Anthraquinones: Emodin, chrysophanol; Naphthalene Glycosides; Flavonoids	Used as a purgative and laxative for constipation; also for treating skin diseases and promoting bile flow	Native to S. Siberia to Russian Far East and Korea; Introduced in different parts of the world	Malla (1993)
Rhizophora mucronata Lam [Rhizophoraceae]	Mangrove (Eng)	Bark	Chocolate	Tannins: Condensed tannins (proanthocyanidins); Flavonoids; Triterpenoids	Used as an astringent for diarrhoea and haemorrhages; treats angina and fever; applied to wounds and ulcers	NE. Tropical and S. Africa to W. Pacific, N. Northern Territories and N. Queensland to NE. New South Wales	Nautiyal and Kaul (2003)

(continued)

Table 1.2 (continued)

Latin names [Family]	Local names	Plant parts	Dye or pigment	Active compounds	Medicinal Uses	Distribution ranges	References
Sarcochlamys pulcherrima Gaudich. [Urticaceae]	Dogal tree (Eng)	Leaf	Brown	Flavonoids; Alkaloids; Tannins; Phenolic Compounds	Used for stomach aches and digestive issues; topical application for wounds and skin infections; anti-inflammatory	E. Himalaya to China (NW. Yunnan) and Indo-China	Nautiyal and Kaul (2003)
Senegalia rugata (Lam.) Britton & Rose (*Acacia sinuata* (Lour) Merril) [Fabaceae]	Ritha (Nep)	Bark	Brown	Saponins: Acaciaside; Alkaloids: Saponins; Flavonoids; Tannins	Primarily used as a natural detergent (*Shikakai*) for hair to promote shine and strength; also used as a laxative and for treating skin diseases	China (W. and SW. Yunnan, S. Guangdong) to Tropical Asia	Nautiyal and Kaul (2003)

(continued)

Table 1.2 (continued)

Latin names [Family]	Local names	Plant parts	Dye or pigment	Active compounds	Medicinal Uses	Distribution ranges	References
Senegalia senegal (L.) Britton [Leguminosae]	Gum acacia, gum arabic tree, Sudan gum and Sudan gum arabic (Eng)	Leaves and bark	Black	Polysaccharides: Arabinogalactan (Source of Gum Arabic); Tannins; Flavonoids	Gum is a demulcent used for coughs and sore throat; soothes intestinal inflammation; used in food and pharmaceuticals	0–2000 m (? Nepal); Angola, Benin, Botswana, Burkina, Cameroon, Central African Republic, Chad, Eritrea, Ethiopia, Gambia, Ghana, India, Ivory Coast, Kenya, KwaZulu-Natal, Malawi, Mali, Mauritania, Mozambique, Namibia, Niger, Nigeria, Northern Provinces, Oman, Pakistan, Rwanda, Saudi Arabia, Senegal, Somalia, Sudan, Swaziland, Tanzania, Uganda, Yemen, Zambia, Zaïre, Zimbabwe	Malla (1993)

(continued)

Table 1.2 (continued)

Latin names [Family]	Local names	Plant parts	Dye or pigment	Active compounds	Medicinal Uses	Distribution ranges	References
Senna auriculata (L.) Roxb. (*Cassia auriculata* L.) [Fabaceae]	Tanner's Cassia (Eng)	Flowers	Yellow	Flavonoids: Kaempferol, quercetin; Tannins: Gallic acid; Polysaccharides; Glycosides	Used for diabetes management ("Avaram tea"); treats skin diseases and liver disorders; used as an astringent for conjunctivitis	Indian Subcontinent to Myanmar	Nautiyal and Kaul (2003)
Sphagneticola calendulacea (L.) Pruski (*Wedelia calendulacea* Less.) [Asteraceae]	Bhansa (Nep)	Leaf juice	Blue black	Flavonoids: Wedelolactone, demethylwedelolactone; Saponins; Triterpenoids	Used for liver disorders (hepatoprotective), particularly jaundice; anti-inflammatory and used for wound healing	Nepal; Bangladesh, Borneo, Cambodia, China Southeast, India, Japan, Indonesia, Korea, Laccadive Is., Laos, Malaya, Manchuria, Myanmar, Philippines, Sri Lanka, Taiwan, Thailand, Vietnam	Malla (1993)

(continued)

Table 1.2 (continued)

Latin names [Family]	Local names	Plant parts	Dye or pigment	Active compounds	Medicinal Uses	Distribution ranges	References
Symplocos acuminata (Blume) Miq. (*Symplocos spicata* Roxb.) [Symplocaceae]	Laurel Sapphire Berry (Eng)	Flowers	Pale yellow; mordant	Alkaloids: Loturine; Triterpenoids; Tannins; Flavonoids	Used for diarrhoea, dysentery, and as an astringent; treats eye diseases and fever; bark used for dental problems	Bangladesh, Borneo, Cambodia, China South-Central, China Southeast, East Himalaya, India (Andaman Is., Assam), Indonesia, Laos, Malaya, Myanmar, Sri Lanka, Thailand, Vietnam	Malla (1993)
Terminalia anogeissiana Gere & Boatwr. (*Anogeissus latifolia* Wall.) [Combretaceae]	Axlewood tree (Eng)	Leaf	Black		Used for diarrhoea and dysentery (strong astringent); treatment of wounds and burns;	Indian Subcontinent	
Terminalia arjuna (Roxb. ex DC.) Wight & Arm [Combretaceae]	Arjun (Nep)	Fruits	Light brown to warm copper oranges and deep browns, it is also used as mordant	Triterpenoids: Arjunolic acid, arjunic acid; Flavonoids: Arjunone; Tannins: Pyrocatechols	Cardioprotective; used for angina, hypertension, and heart failure; promotes wound healing; mild diuretic	India and Ceylon; cultivated in the plains of Pakistan as a shade tree	Malla (1994)

(continued)

Table 1.2 (continued)

Latin names [Family]	Local names	Plant parts	Dye or pigment	Active compounds	Medicinal Uses	Distribution ranges	References
Thespesia populnea (L.) Sol. ex Corr. [Rutaceae]	Parsipu (Hin)	Root and leaves	Yellow	Flavonoids: Gossypol, thespesone; Tannins; Triterpenoids	Used for skin diseases, scabies, and psoriasis; treats diabetes and inflammation; antiviral and antimicrobial properties	Cambodia, China, India, Japan (Okinawa, Ryukyu Islands), Philippines, Sri Lanka, Thailand, Vietnam; Africa; widely distributed in the tropics	Malla (1993)
Vachellia leucophloea (Roxb.) Maslin, Seigler & Ebinger [Fabaceae]	White bark acacia (Eng)	Bark	Red	Tannins: Gallic acid, ellagic acid; Flavonoids: Quercetin, leucophleol; Alkaloids: Tryptamine, N-methyltryptamine; Triterpenoids: β-amyrin, oleanolic acid	Used for diabetes, diarrhoea, and dysentery; treats inflammation and fever; applied to wounds and skin ulcers; bark used as an astringent	Indian Subcontinent, Indo-China, Jawa, Lesser Sunda Islands	Nautiyal and Kaul (2003)

Table 1.3 List of plant species that are imported from other countries and used for obtaining mordant during colouring and used in Nepal. The plant species are arranged alphabetically with any synonyms in parentheses (), followed by the family in square brackets []. The available vernacular names are given for each plant and are arranged in alphabetical order: Chepang (Che), Danuwar (Dan), Darai (Dar), English (Eng), Gurung (Gur), Hindi (Hin), Limbu (Lim), Magar (Mag), Moonsoor (Moo), Nepali (Nep.), Newari (Newa.), Rai (Rai), Sherpa (She), and Tamang (Tam.). The plant parts used to obtain dye or pigment are followed by active compounds found in each plant species, their medicinal uses, and altitudinal ranges in Nepal (E as East, C as Centre, and W as West Nepal) and distribution ranges in Nepal and the world. In the end, references are given

Latin names [Family]	Local names	Plant parts	Active compounds	Medicinal uses	Altitudinal range, distribution in Nepal and the world	References
Achyranthes aspera L. [Amaranthaceae]	Apamarga, Datiun (Nep), Churut jahar (Che), Ulta Chirchri (Dan), Naksirkas (Dar), Datiwan (Mag)	Ash-a source of alkali	Alkaloids: Achyranthine, betaine; Saponins: Oleanolic acid; Flavonoids: Quercetin, kaempferol; Phenolic Acids: Gallic acid, caffeic acid; Other: Ecdysterone (a phytoecdysteroid)	Used for renal dropsy and kidney stones; treatment for cough and respiratory problems; applied to insect bites and skin diseases; used in facilitating childbirth and as an emmenagogue	800–2300 m (C Nepal); Bhutan, China, Cambodia, India, Indonesia, Laos, Malaysia, Myanmar, Nepal, Philippines, Sikkim, Sri Lanka, Thailand, Vietnam; Africa, SW Asia, Europe	Malla (1993)
Begonia picta Sm. [Begoniaceae]	Magar kanche (Nep), Namkimro (Gur)	Whole plant	Flavonoids: Quercetin, rutin; Tannins: Gallotannins; Phenolic Acids: Gallic acid, protocatechuic acid; Sterols: β-sitosterol; Organic Acids: Oxalic acid, malic acid, citric acid	Used for respiratory ailments like bronchitis and asthma; juice applied to wounds and burns to soothe irritation; treats urinary tract problems and fever	600–2800 m (WCE Nepal); Himalaya (Punjab to Bhutan), Assam	Manandhar (2002)

(continued)

Table 1.3 (continued)

Latin names [Family]	Local names	Plant parts	Active compounds	Medicinal uses	Altitudinal range, distribution in Nepal and the world	References
Citrus limon (L.) Burm. f. (*Citrus medica* var. *limonum* Hook. f.) [Rutaceae]	Kagati, Nibua (Nep)	Lemon peel	Volatile Compounds/ Monoterpenes: Limonene (major component), γ-terpinene, β-pinene; Polyphenols: Eriocitrin, hesperidin, diosmin; Organic Acids: Citric acid, ascorbic acid (Vitamin C)	Used to treat scurvy (Vitamin C deficiency); aids digestion and relieves nausea; sore throat relief (gargle); skin cleansing and antioxidant tonic	1600 m (C Nepal); India, China. Widely cultivated throughout the subtropics	Malla (1993)
Citrus medica L. [Rutaceae]	Bimiro (Nep)	Leaves	Volatile Compounds: Limonene, citral; Coumarins and Furanocoumarins: Bergapten, limettin; Flavonoids: Hesperidin, diosmin; Polyphenols: Caffeic acid, ferulic acid	Used for digestive issues, nausea, and constipation; relieves asthma and other respiratory conditions; considered a general tonic for the heart and brain	700–1200 m (WCE Nepal); Himalaya (Garhwal to Sikkim), India, Burma, Indo-China, cultivated	Malla (1993)

(continued)

Table 1.3 (continued)

Latin names [Family]	Local names	Plant parts	Active compounds	Medicinal uses	Altitudinal range, distribution in Nepal and the world	References
Cynodon dactylon (L.) Pers. [Poaceae]	Dubo (Nep), Dub (Che), Chame chamba (Gur), Cheri (Rai), Panja (She), Narkabam (Tam)	Aerial parts	Phytosterols: β-sitosterol, campesterol; Flavonoids: Luteolin, apigenin, tricin; Alkaloids: Carpine	Used to control heavy menstrual bleeding and urinary tract infections; treatment for hysteria and epilepsy; topical application for wounds and skin itching; diuretic	100–3000 m (WCE Nepal); Widely distributed in all warm countries	Bhandari et al. (2020b)
Cyperus rotundus L. [Cyperaceae]	Mothe kaseru, Nagar mothe (Nep), Motha Moo), Maalaahaa (New)	Whole part	Essential Oils/Sesquiterpenes: Cyperene, α-cyperone, patchoulenone, sugeonol; Flavonoids: Catechin, quercetin; Triterpenoids: Cyperenol; Phenolic Acids: Ferulic acid, p-coumaric acid	Used for treating digestive disorders like diarrhoea and indigestion; regulates menstrual cycles and relieves menstrual cramps; used for fever, wounds, and inflammatory conditions	300–2400 m (WCE Nepal); Cosmopolitan	Malla (1993)

(continued)

Table 1.3 (continued)

Latin names [Family]	Local names	Plant parts	Active compounds	Medicinal uses	Altitudinal range, distribution in Nepal and the world	References
Engelhardia spicata Lesch. ex Blume [Juglandaceae]	Yakpohama (Lim), Mauwa (Nep), Bokto, Okshipu (Rai)	Bark	Diarylheptanoids: Engelhardione, engelhardic acid; Flavonoids: Myricetin, quercetin; Triterpenoids: Oleanolic acid, betulin; Tannins: Ellagitannins	Used for fevers, headaches, and inflammatory conditions; treatment for diarrhoea and dysentery; applied to skin for rashes and fungal infections	400–2500 m (WCE Nepal); Bhutan, China, India, Indonesia, Laos, W Malaysia, Myanmar, Pakistan, Philippines, Sikkim, Thailand, Vietnam	Chaudhary et al. (2002)
Hippophae salicifolia D. Don [Elaeagnaceae]	Dale chuk, Chugo (Nep), Kyungsing (Gur)	Decoction	Vitamins and Carotenoids: High Vitamin C, Vitamin E, Carotenoids (β-carotene, lycopene); Flavonoids: Isorhamnetin, quercetin, kaempferol; Lipids: Palmitoleic acid, palmitic acid; Phenolic Acids: Gallic acid, caffeic acid; Sterols: β-sitosterol	Used for treating gastric ulcers, skin burns, and wounds; alleviates inflammation; boosts immunity due to high antioxidant content; supports cardiovascular health	2200–3500 m (WC Nepal); Himalaya (Punjab to Bhutan), S. Tibet	Pyakurel and Baniya (2011)

(continued)

Table 1.3 (continued)

Latin names [Family]	Local names	Plant parts	Active compounds	Medicinal uses	Altitudinal range, distribution in Nepal and the world	References
Hippophae tibetana Schlecht. [Elaeagnaceae]	Chuk (Nep)	Fruit juice	Vitamins and Carotenoids: Vitamin C, Carotenoids; Flavonoids: Isorhamnetin, quercetin; Lipids: A unique fatty acid profile including palmitoleic acid; Phenolic Acids: Simple phenolic acids; Tocopherols: Alpha-tocopherol	Traditionally used for blood circulation disorders, cough, and asthma; applied topically for skin ailments and radiation burns; improves digestion and general vitality	3800–4500 m (WCE Nepal); Himalaya (Punjab to Bhutan), S. Tibet, N.W. China	Pyakurel and Baniya (2011)
Impatiens racemosa DC. [Balsaminaceae]	Anchirna (Nep)	Seed paste	Naphthoquinones: Lawsone, methylene-3,3'-bilawsone; Flavonoids: Kaempferol, quercetin; Phenolic Acids: Caffeic acid, vanillic acid	Primarily used for skin diseases, ringworm, and scabies; applied to warts and fungal infections; used in the treatment of inflammation and rheumatism	1300–3900 m (WCE Nepal); Himalaya (Kashmir to Bhutan), Assam, S. Tibet	Malla (1993), Pyakurel and Baniya (2011)

(continued)

Table 1.3 (continued)

Latin names [Family]	Local names	Plant parts	Active compounds	Medicinal uses	Altitudinal range, distribution in Nepal and the world	References
Impatiens tricornis Lindl. [Balsaminaceae]	Tiuri jhar (Nep)	Seed paste	Naphthoquinones: Lawsone and its derivatives; Flavonoids: Various glycosides of kaempferol and quercetin; Phenolic Compounds: Simple phenolic acids	Similar to I. racemosa, used for dermatitis, eczema, and other skin ailments; juice applied to wounds and burns; also used for its anti-inflammatory properties	1000–3600 m (WCE Nepal); Himalaya (Kashmir to Bhutan)	Manandhar (2002), Pyakurel and Baniya (2011)
Mesua ferrea L. [Clusiaceae]	Nag keshar, Rukh keshar (Nep)	Flower buds	Coumarins: Mesuol, mammeigin, mesuone; Xanthones: Mesuaxanthone A and B; Triterpenoids and Sterols: α-amyrin, β-sitosterol; Volatile Oils: Containing numerous sesquiterpenes	Used for bleeding piles, dysentery, and coughs; treatment for skin diseases and excessive sweating; seed oil applied for rheumatism and *vata* disorders in Ayurveda	400–900 m (CE Nepal); India to Vietnam and Malay Peninsula, Ceylon, Andaman Isl	Malla (1993)

(continued)

Table 1.3 (continued)

Latin names [Family]	Local names	Plant parts	Active compounds	Medicinal uses	Altitudinal range, distribution in Nepal and the world	References
Musa paradisiaca L. [Musaceae]	Kera (Nep)	Ashes of leaves	Carotenoids: α-carotene, β-carotene (especially in ripe fruit); Flavonoids: Catechin, epicatechin, myricetin; Phenolic Acids: Ferulic acid, sinapic acid; Sterols: β-sitosterol; Other: Serotonin, dopamine (in peel and pulp)	The fruit is a digestive aid and used for peptic ulcers; flower is used for diabetes and dysentery; sap is applied to wounds and insect bites; anti-ulcerogenic	Widely cultivated	Malla (1993)
Passiflora edulis Sims [Passifloraceae]	Jhumke laharo (Nep)	Fruit, leaf	Flavonoids: Vitexin, isovitexin, chrysin; Alkaloids: Harman, harmol, harmine (trace amounts in some species); Cyanogenic Glycosides: Gynocardin; Phenolic Acids: p-Coumaric acid, protocatechuic acid	Primarily used as a sedative and anxiolytic to treat anxiety, nervousness, and insomnia; digestive tonic; topical application for skin inflammation and burns	1300–1700 m (E Nepal); Brazil, often cultivated	Malla (1993)

(continued)

Table 1.3 (continued)

Latin names [Family]	Local names	Plant parts	Active compounds	Medicinal uses	Altitudinal range, distribution in Nepal and the world	References
Tamilnadia uliginosa (Retz.) Tirveng. and Sastre (*Catunaregam uliginosa* (Retz.) Sivar.) [Rubiaceae]	Brilliant gardenia (Eng), Pirar (Nep)	Fruits	Iridoid Glycosides: Dideroside, gardenoside; Tannins: Gallotannins, ellagitannins; Flavonoids: Rutin, quercetin; Triterpenoids: Oleanolic acid, ursolic acid	Used for liver disorders and jaundice; treatment for fever and infectious diseases; fruit is used for dysentery and diarrhoea; bark used in toothache	~300 m (WCE Nepal), India, Sri Lanka, Indo-China; Cultivated	Manandhar (2002)
Verbascum thapsus L. [Scrophulariaceae]	Bandarpuchhre (Nep)	Ashes	Saponins: Verbascosaponin; Iridoid Glycosides: Aucubin, catalpol; Flavonoids: Hesperidin, verbascoside; Mucilage: alpha-D-Galacturonic acid, β -Glucan; Phenylethanoids: Verbascoside	Primarily used for respiratory health (dry cough, bronchitis, asthma, sore throat); demulcent that soothes mucous membranes; topical application for earaches, warts, and haemorrhoids	1800–4000 m (WCE Nepal); Himalaya (Kashmir to Bhutan), Tibet, W. and C. China	Malla (1993)

Fig. 1.1 Chemical structures of alkaloid compounds (1–36)

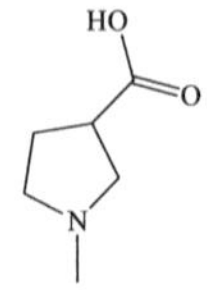

Achyranthine (1)

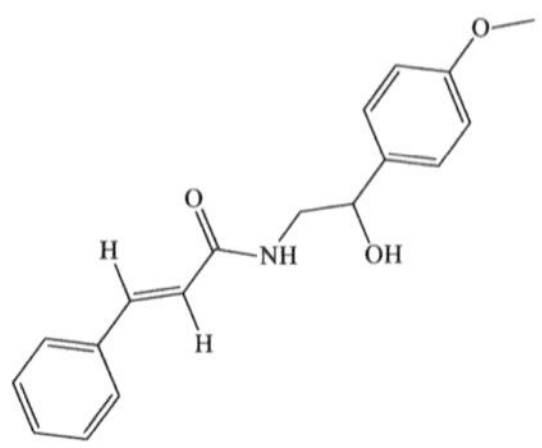

Aegeline (2)

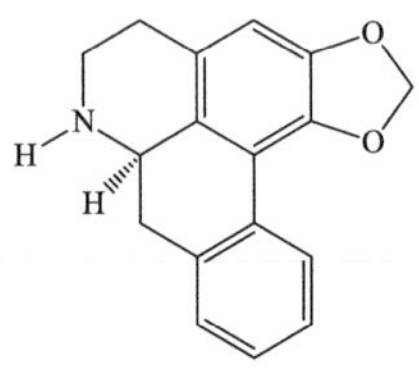

Anonaine (3)

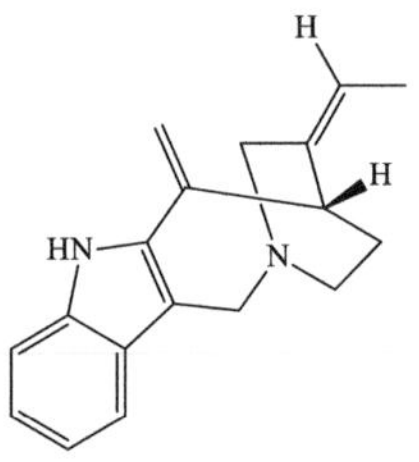

Apparicine (4)

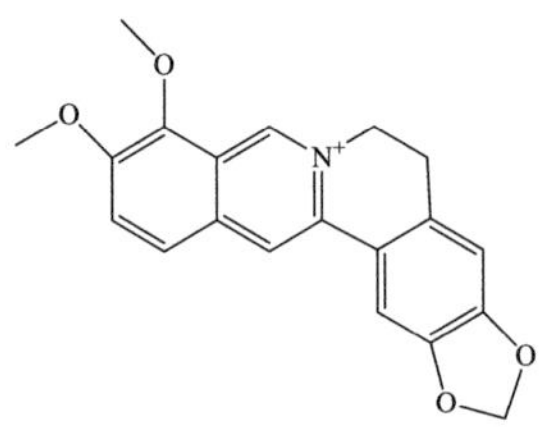

Berberine (5)

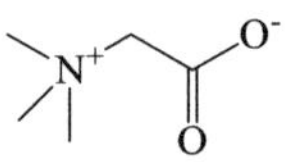

Betaine (6)

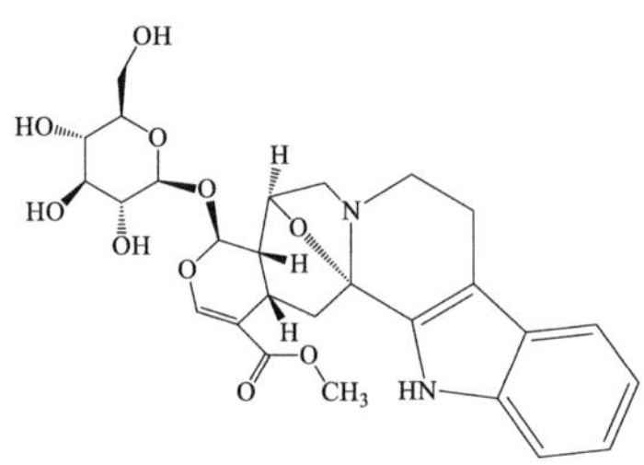

Cadambine (7)

which is the same as the compound found in the indigo plant (*Indigofera tinctoria* L., Fabaceae), and gives a blue colour without using any mordant.

1.3.2.1 List of Dye-Yielding Animal Species and Compounds

The main animals that are used for obtaining colours are.

Fig. 1.1 (continued)

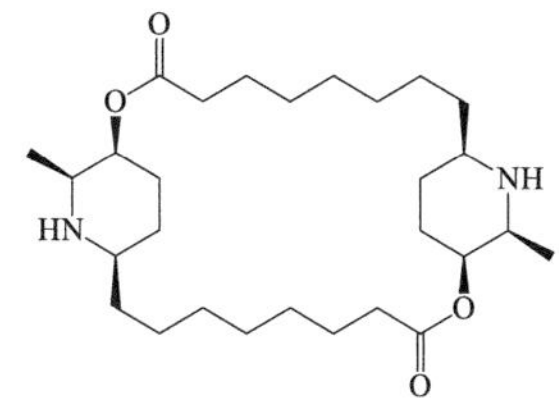

Carpine (8)

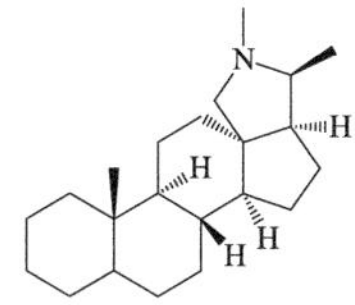

Conamine (9)

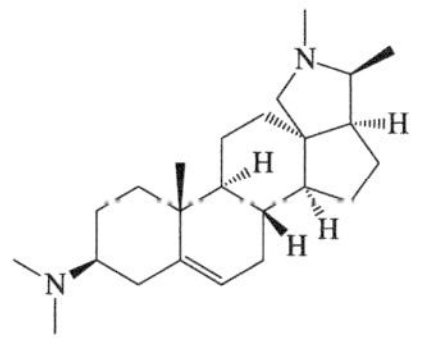

Conessine (10)

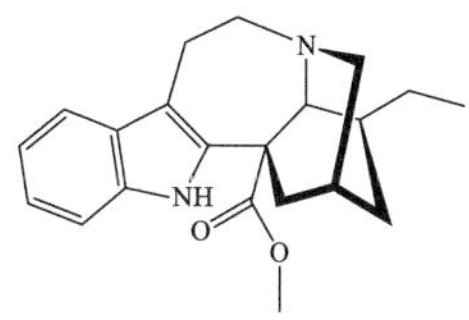

Coronaridine (11)

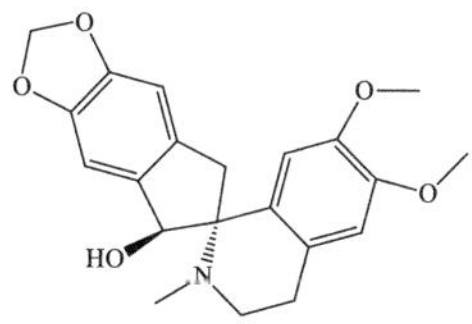

Fumaricine (12)

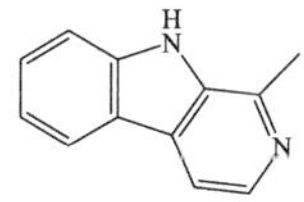

Harman (13)

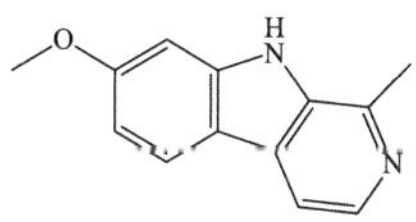

Harmine (14)

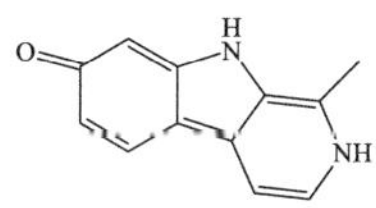

Harmol (15)

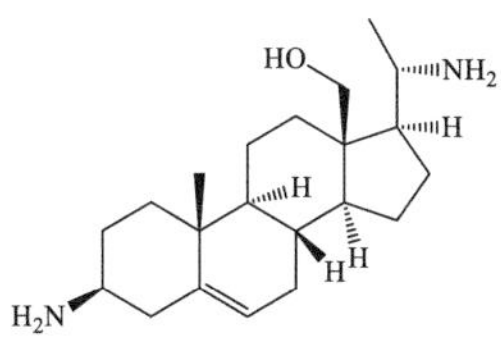

Holarrhimine (16)

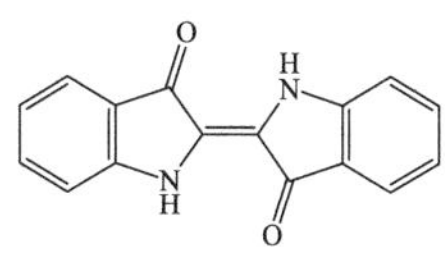

Indigotine (17)

Fig. 1.1 (continued)

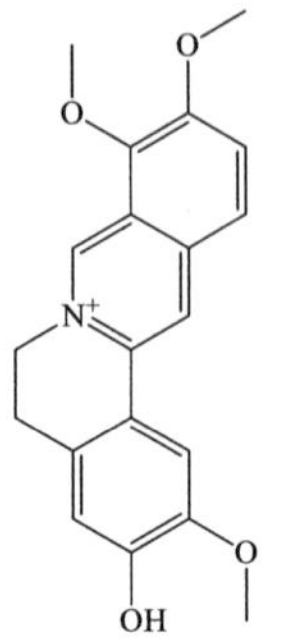

Jatrorrhizine (18)

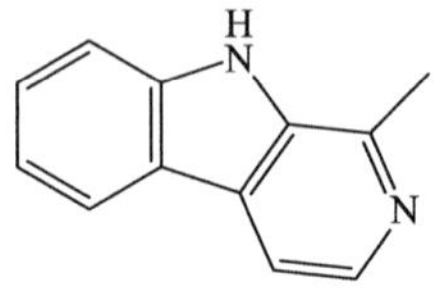

Loturine (19)

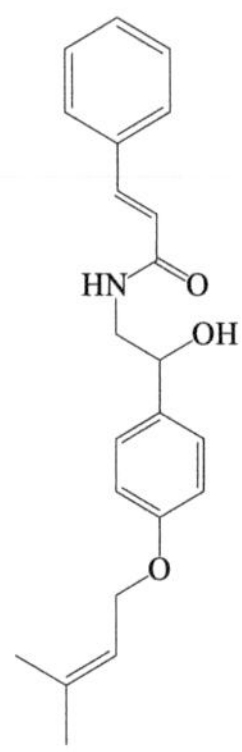

Marmeline (20)

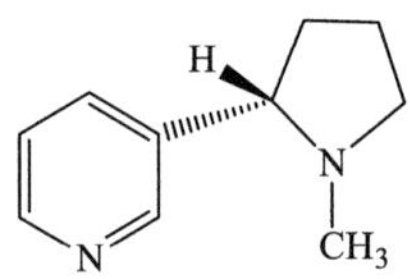

Nicotine (21)

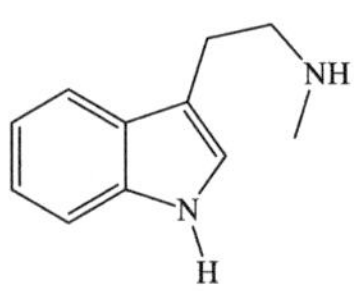

N-methyltryptamine (22)

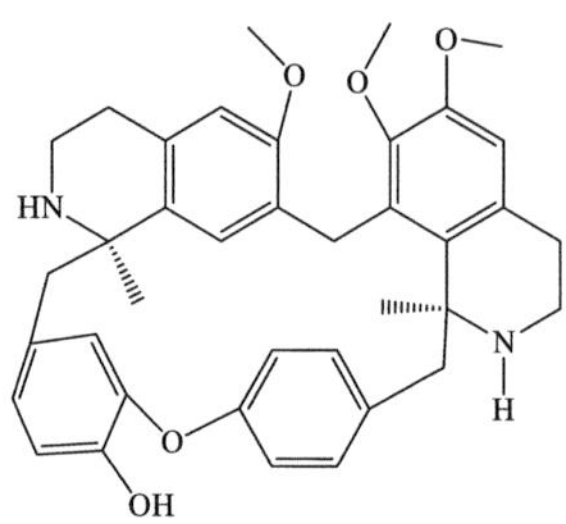

Oxyacanthine (23)

Fig. 1.1 (continued)

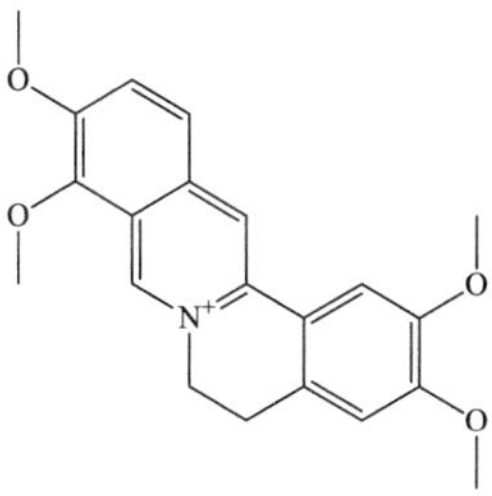

Palmatine (24)

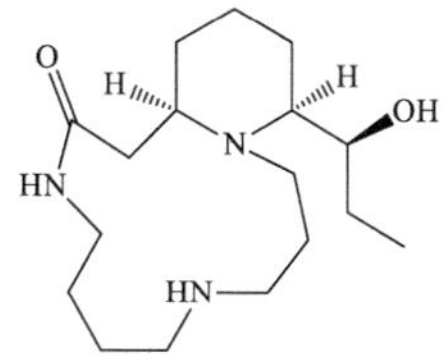

Palustrine (25)

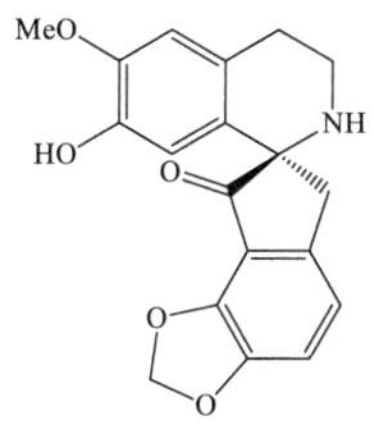

Parfumine (26)

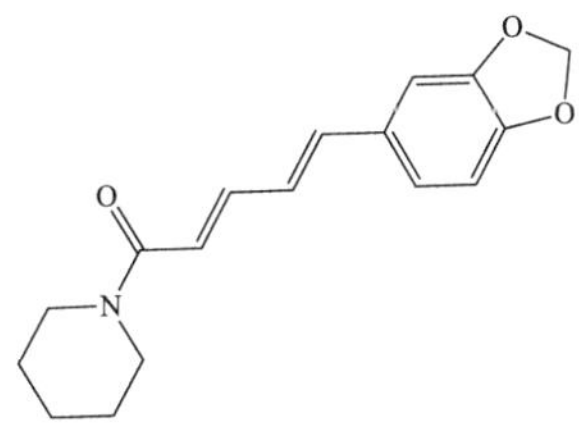

Piperine (27)

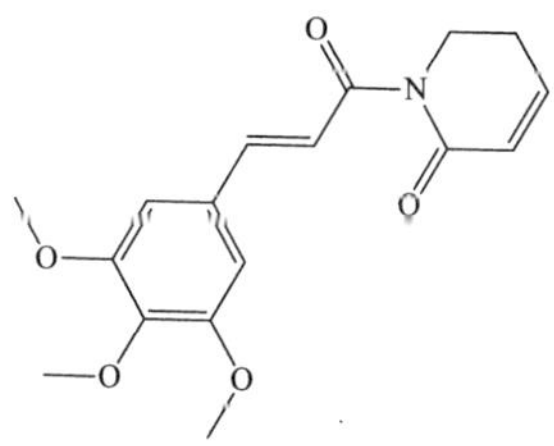

Piperlongumine (28)

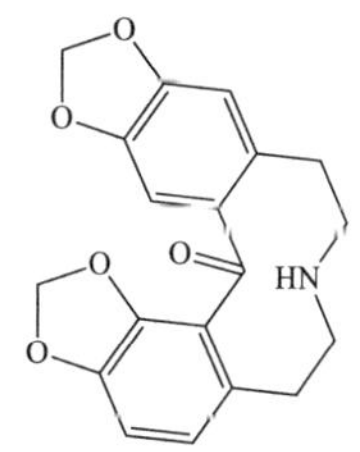

Protopine (29)

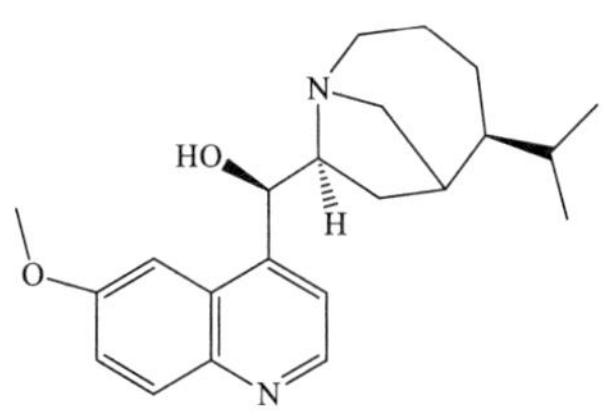

Quinoline alkaloids (30)

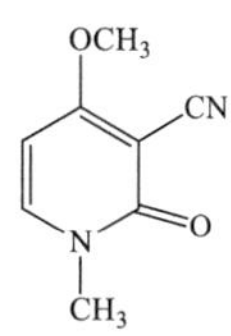

Ricinine (31)

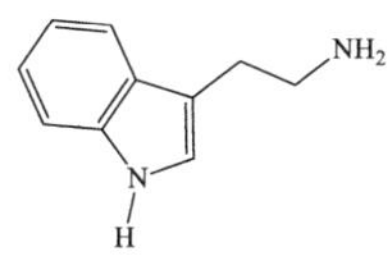

Tryptamine (32)

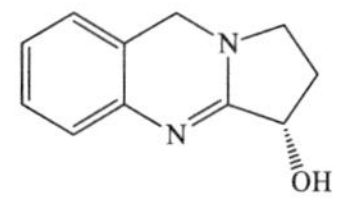

Vasicine (33)

Fig. 1.1 (continued)

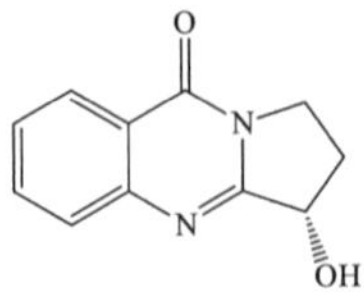

Vasicinone (34)

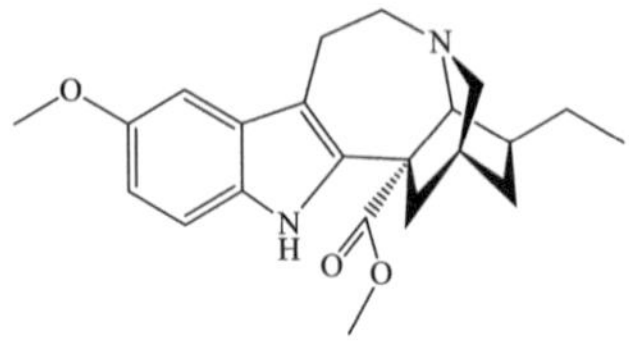

Voacangine (35)

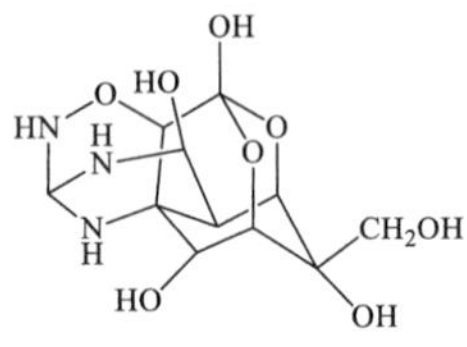

Xeronine (36)

- Cochineal Insect (*Dactylopius coccus*): This insect is the source of the red colour, and dried female cochineal insects are used for colour extraction. These insects feed on red cactus fruits and accumulate red pigment. Until 1600 A.D., the production of cochineal dye was a vast industry, and approximately 500,000 pounds of cochineal dye were imported to Spain annually from the American continent (Degano et al. 2009).
- Kermes oak trees provide habitat for *Kermes vermilio* insects, and they are the main source of the kermes dye, a red dye. It was widely used prior to the use of cochineal dye in Europe, and it was considered a symbol of wealth (Baqri and Omkar 2023).
- Lac Insect (*Kerria lacca*): Similar to cochineal insects, the lac insects (*Kerria lacca*) produce a resinous secretion on plants and are processed to yield a lac dye that gives red and violet colours (Sharma et al. 2020). This dye is used for dyeing different items used in our daily life (Sharma et al. 2020; Baqri and Omkar 2023).
- Murex snails (*Bolinus brandaris* and other snails) give purple and indigo blue dyes. This colour was a legendary Tyrian purple, or indigo blue, that was discovered near Tyre on the Mediterranean coast. In the ancient world, this dye was the ultimate symbol of status and power in society. Thus, it was exclusively used by Roman emperors. However, a large number of snails are needed to release a very small amount of dye. For example, one gram of pure dye can be obtained by crushing 12,000 shells, and it was costly due to the laborious nature of the work required to produce it (Kenoyer 1984).
- Another mollusc, known as the *Hexaplex trunculus*, is also used as the main source of dye, and the colour of such dye ranges from blue to purple (Smith 2018). This blue dye was highly preferred for its use in the sacred *Tekhelet*, Jewish religious items like the fringes (tzitzit) on garments and the high priest's clothing, in ancient Judaism (Sagiv 2015).
- Cow urine is another important source of dye known as *Indian yellow* (Hassan 2024). Similarly, ink from octopus and cuttlefish provides a sepia brown dye,

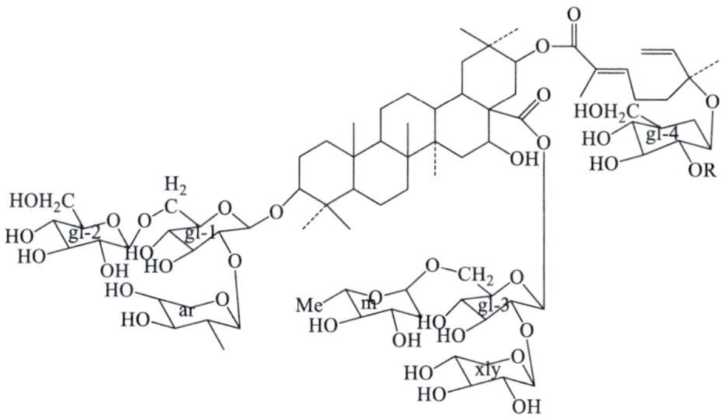

Acaciaside (37): Acaciaside A R= H; Acaciaside B R= Xyl (p)

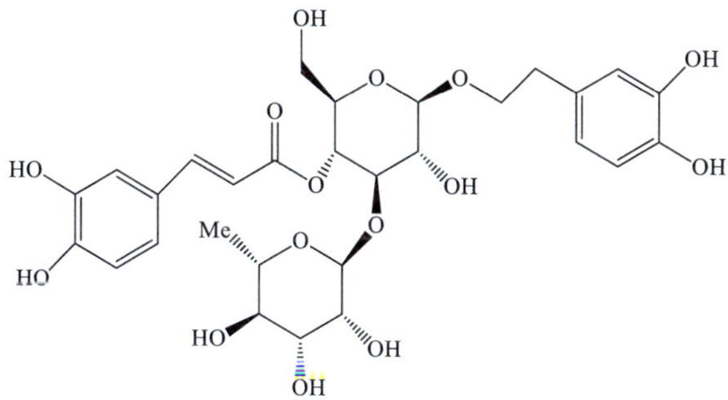

Acteoside (38)

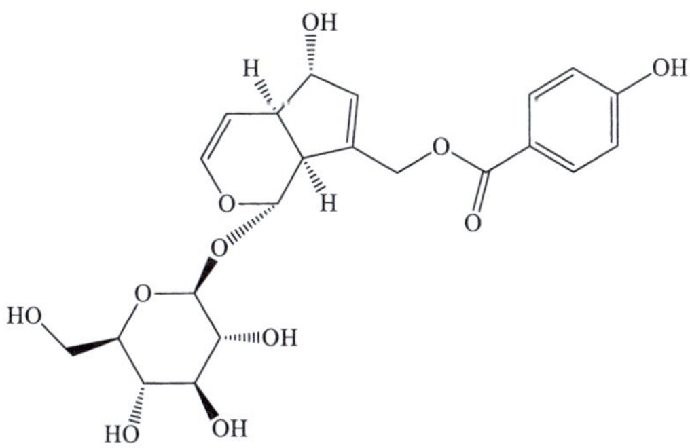

Agnuside (39)

Fig. 1.2 Chemical structures of glycosides (37–58)

Fig. 1.2 (continued)

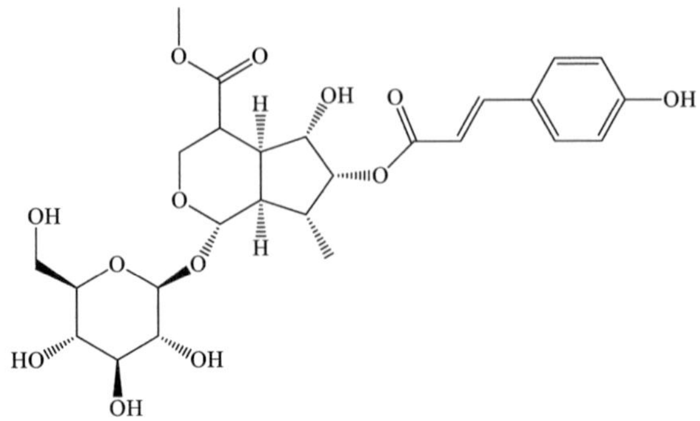

Arbortristosides (40)

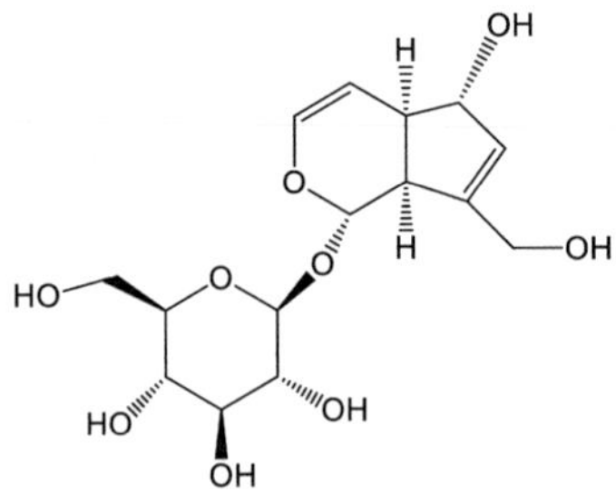

Aucubin (41)

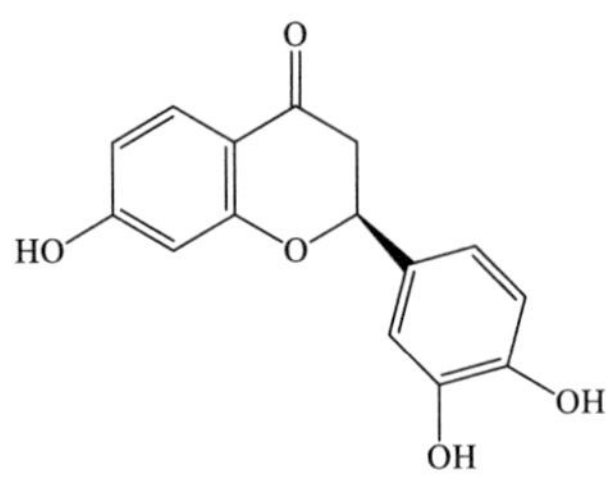

Butin (42)

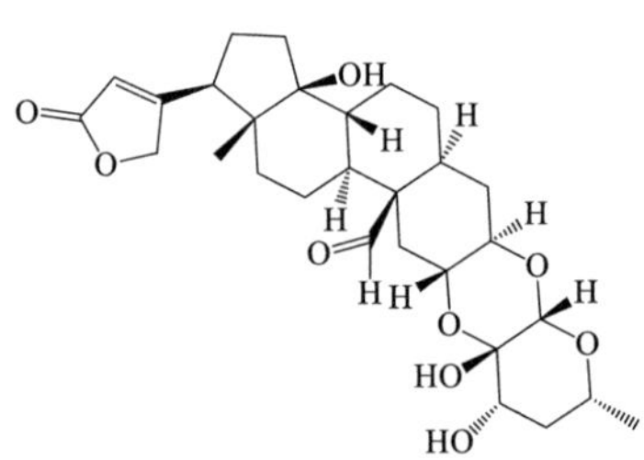

Calotropin (43)

Fig. 1.2 (continued)

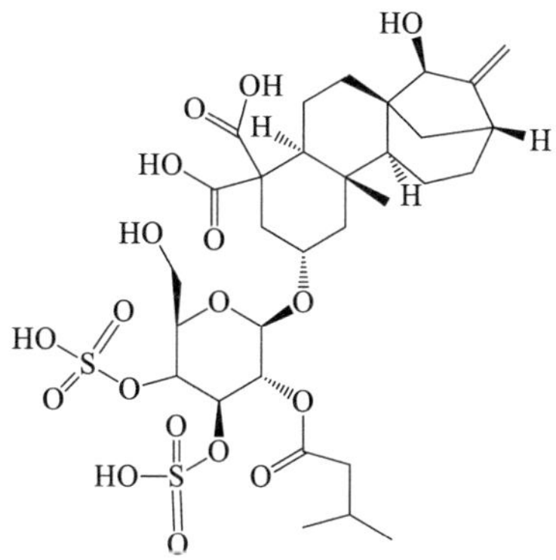

Carboxyatractyloside (44)

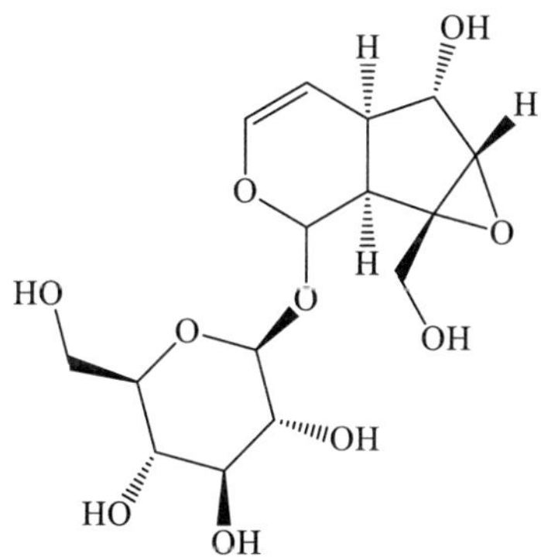

Catalpol (45)

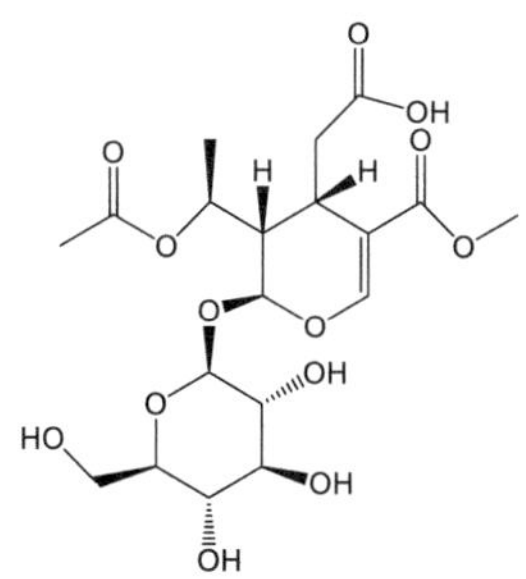

Dideroside (46)

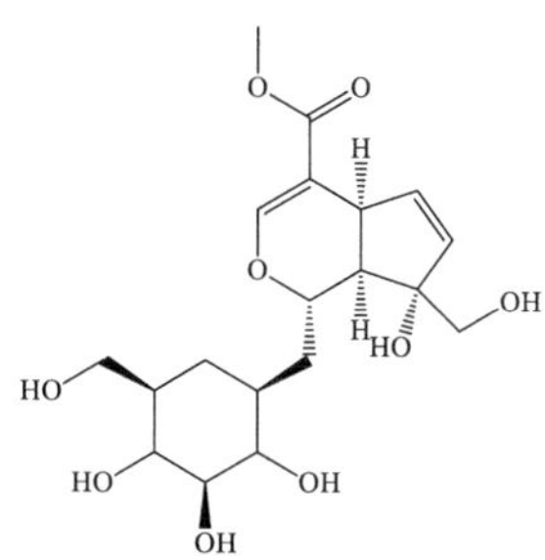

Gardenoside (47)

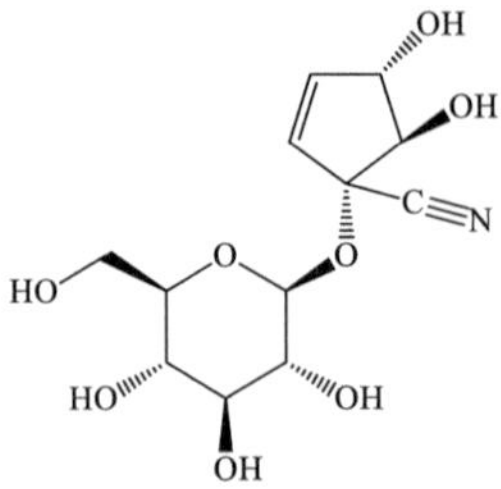

Gynocardin (48)

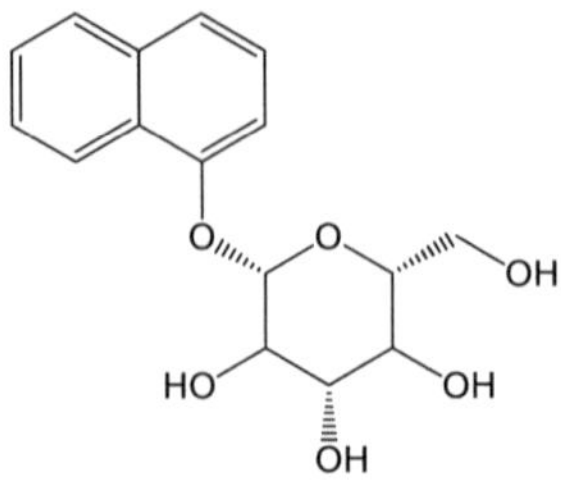

Naphthalene glycosides or Naphthylglucoside (49)

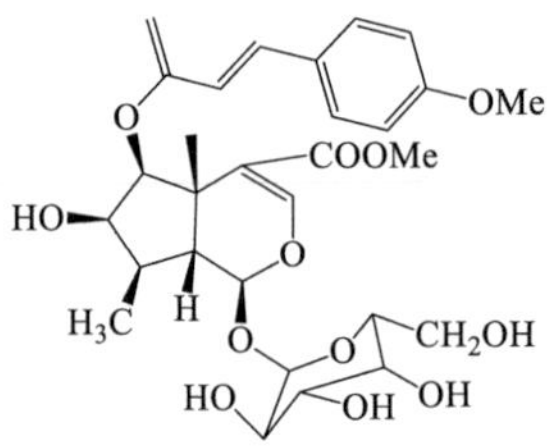

Nyctanthin (50)

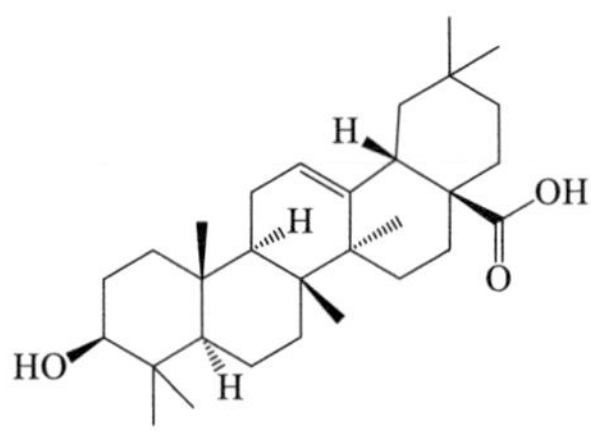

Oleanolic acid (51)

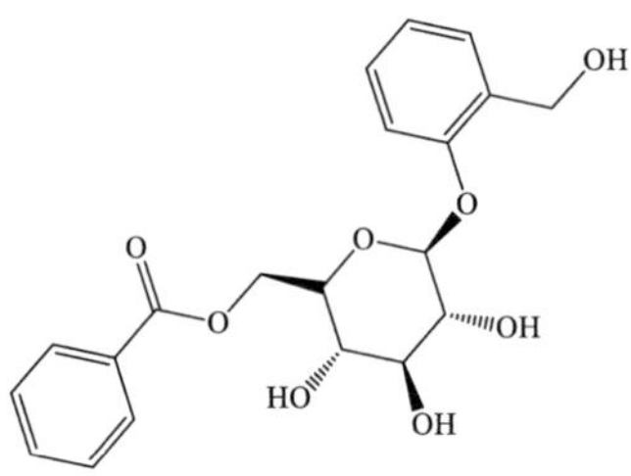

Populin (52)

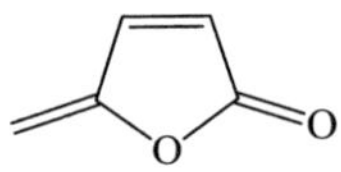

Protoanemonin (53)

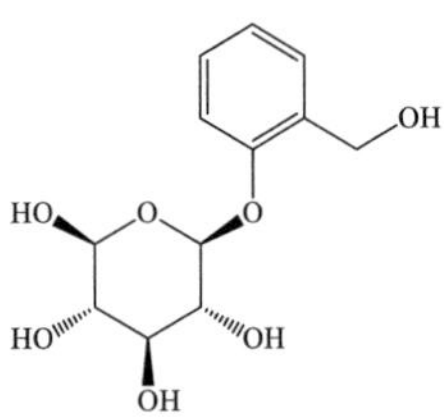

Salicin (54)

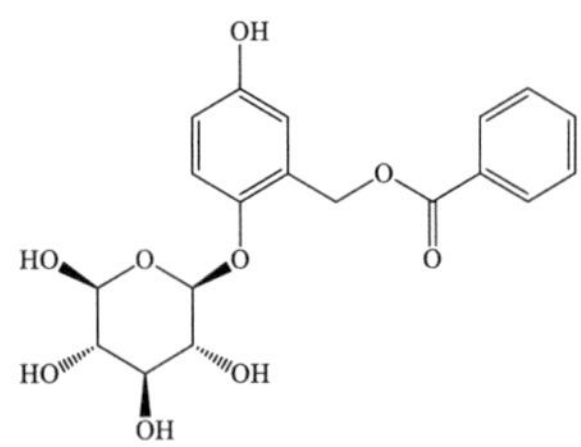

Salireposide (55)

Fig. 1.2 (continued)

Fig. 1.2 (continued)

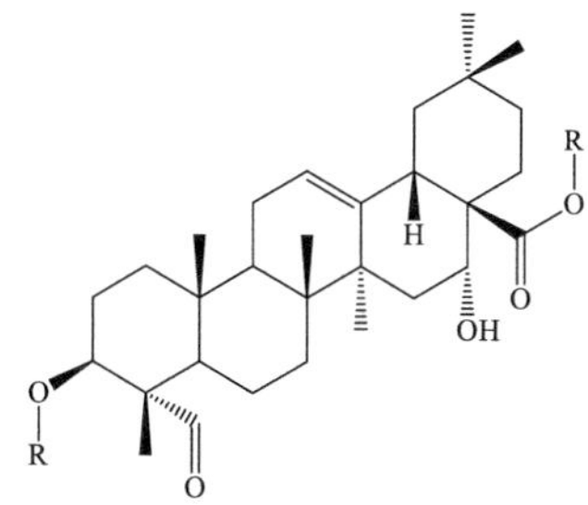

Saponins (56)

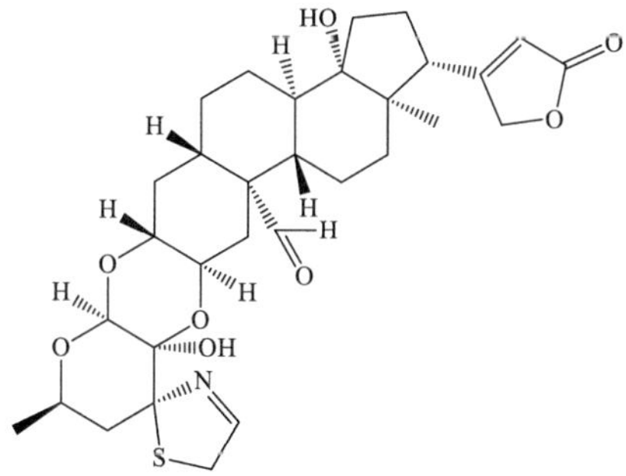

Uscharin (57)

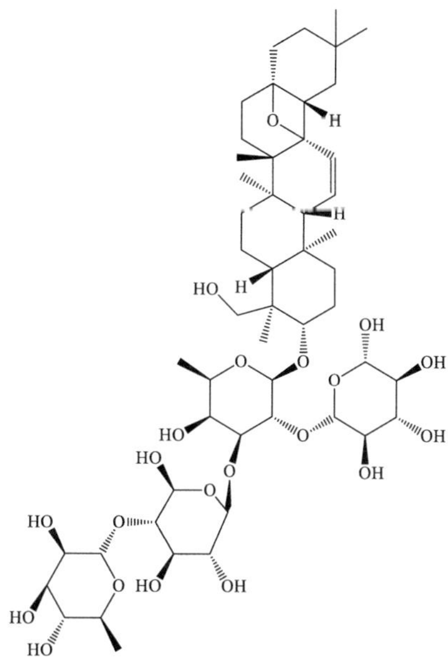

Verbascosaponin (58)

Fig. 1.3 Chemical structures of lipids and fatty acids (59–64)

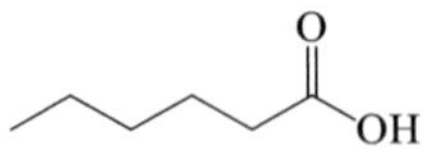

Caproic acid (59)

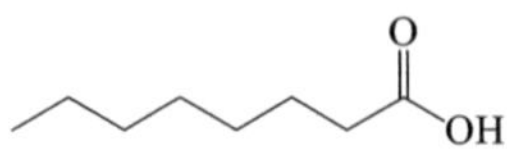

Caprylic acid (60)

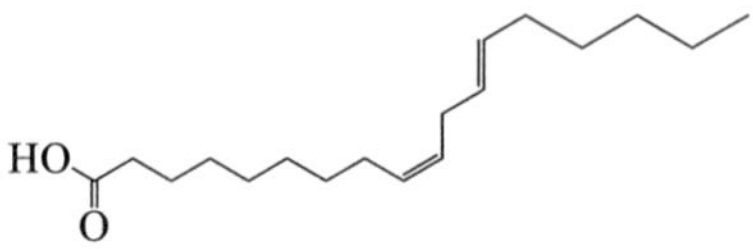

Linoleic acid (61)

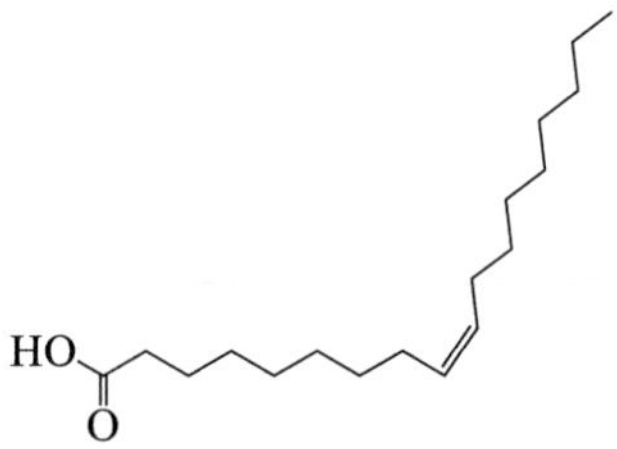

Oleic acid (62)

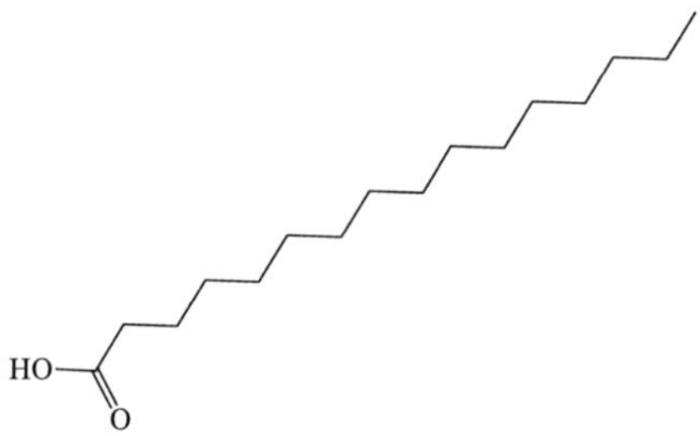

Palmitic acid (63)

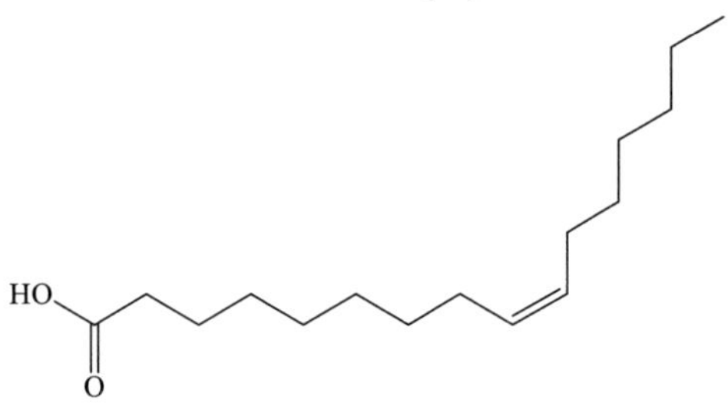

Palmitoleic acid (64)

named after the genus of cuttlefish, and was used for centuries as ink and watercolour (Derby 2014).

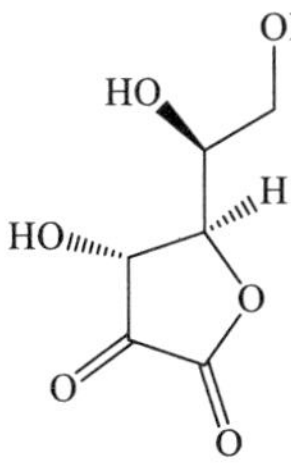

Ascorbic acid or vitamin C (65)

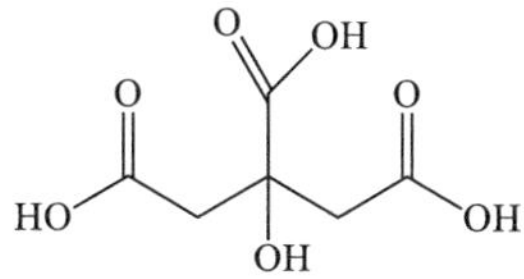

Citric acid (66)

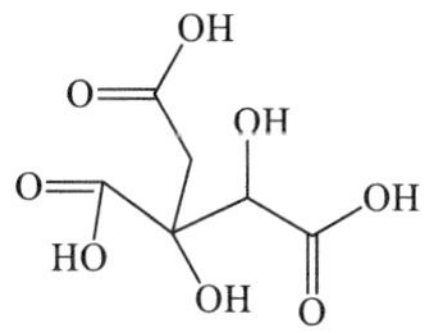

Hydroxycitric acid, HCA (67)

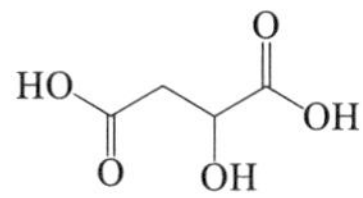

Malic acid (68)

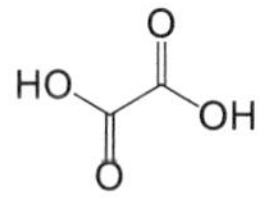

Oxalic acid (69)

Fig. 1.4 Chemical structures of organic acids (65–69)

1.3.3 Minerals

Minerals are substances found in the Earth's crust, and they have a specific chemical composition and are crystalline in structure. These minerals are important because they are an important source of different dyes and pigments. In addition, these minerals are important nutrients that are required by our bodies for proper functioning.

1.3.3.1 List of Dye-Yielding Minerals and Compounds

Mostly inorganic pigments are obtained from naturally occurring clay, earth oxides, and several minerals. However, the dyes or pigments obtained from minerals are often artificially manufactured to increase production and fulfil their high demands. Unlike the complex organic molecules found in natural sources, mineral dyes often consist of simple metallic compounds. Examples of minerals from which dyes or pigments are obtained are chromium green (a compound of chromium and oxygen), chromium red (a compound of chromium and lead), chromium yellow (from chromic acid and lead), and prussian blue (a compound of iron and cyanide) (Sanda and Liliana 2021).

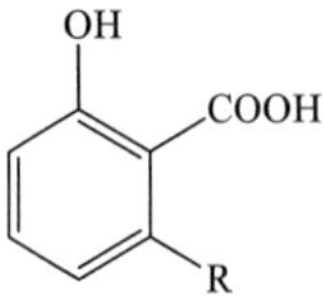

$R = (CH_2)_nCH_3$

Anacardic acid (70)

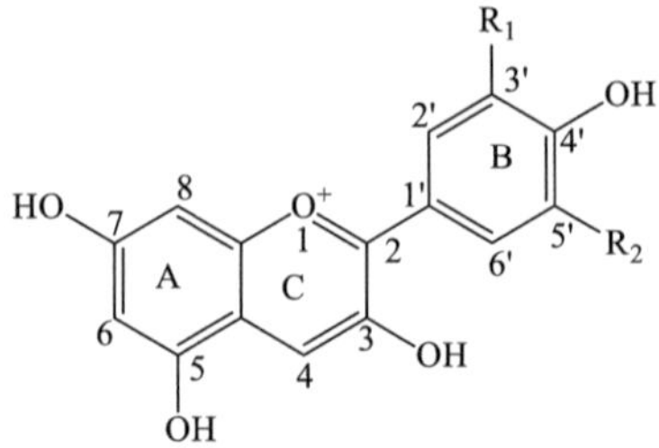

Anthocyanins (71)

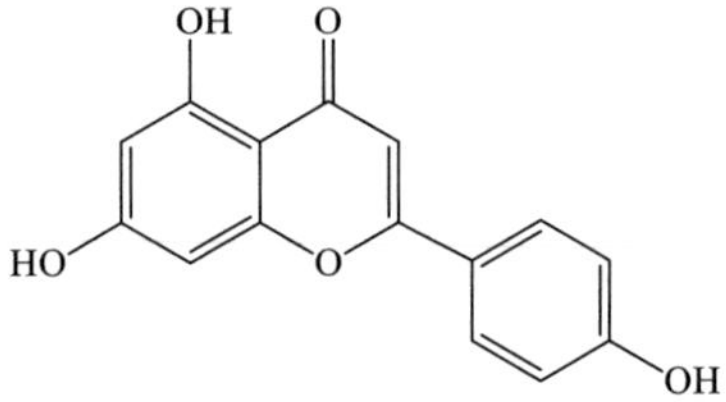

Apigenin (72)

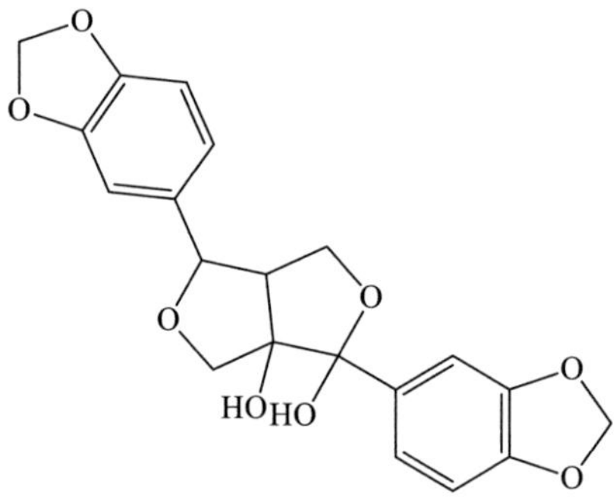

Arboreol (73)

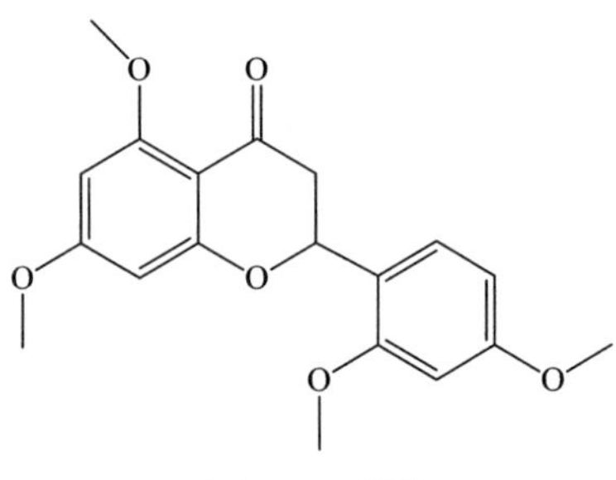

Arjunone (74)

Fig. 1.5 Chemical structures of phenolics (70–160)

Fig. 1.5 (continued)

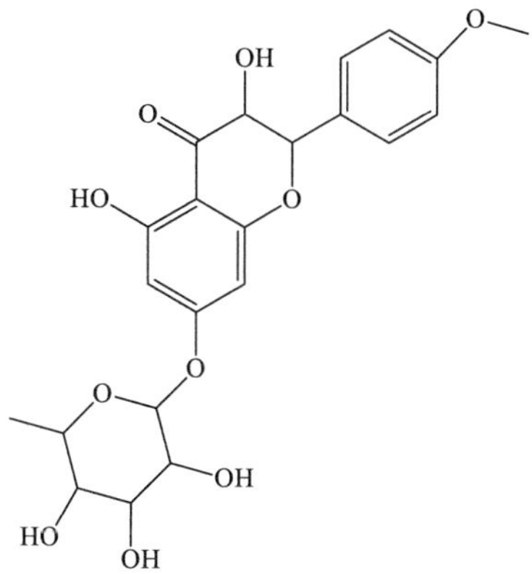

Aromadendrin (75)

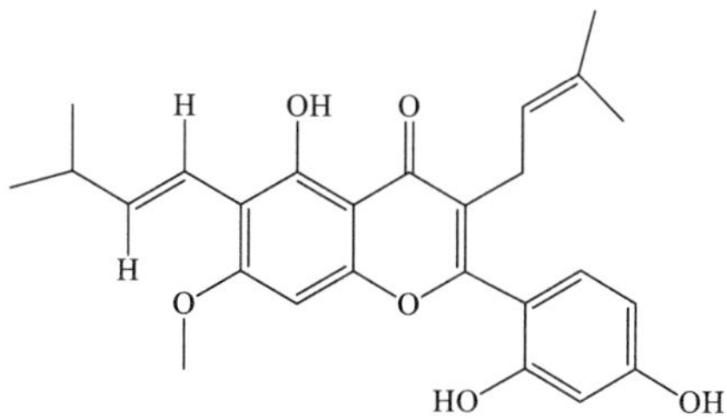

Artocarpin (76)

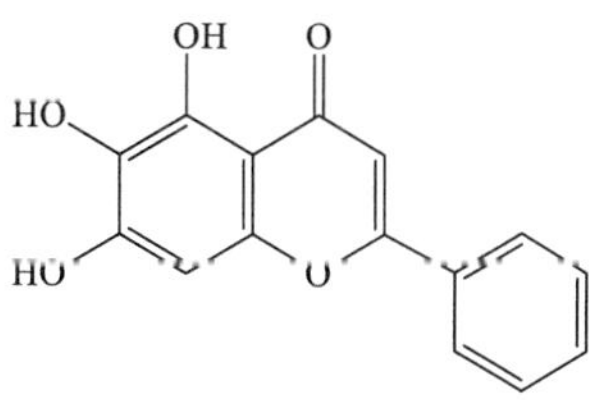

Baicalein (77)

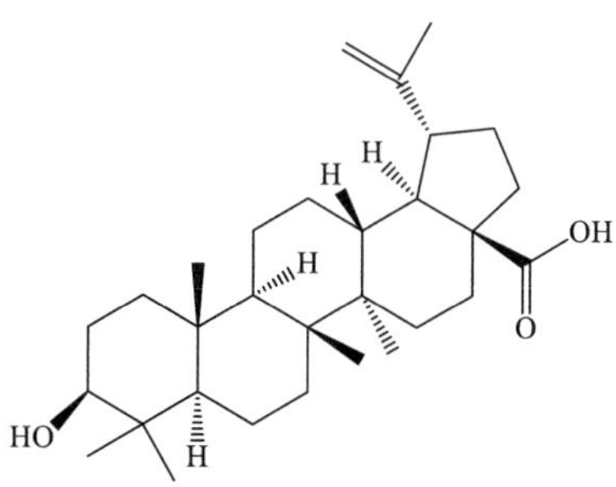

Betulinic acid (78)

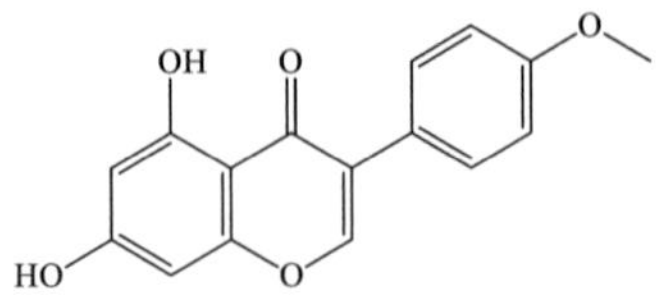

Biochanin a (79)

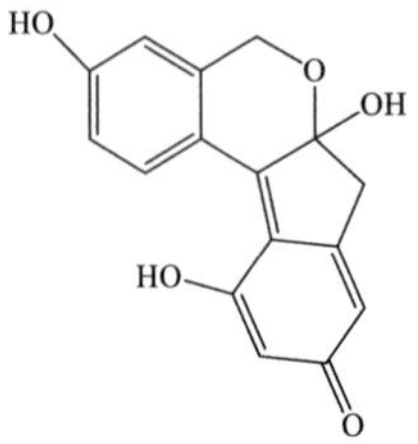

Brasilein (80)

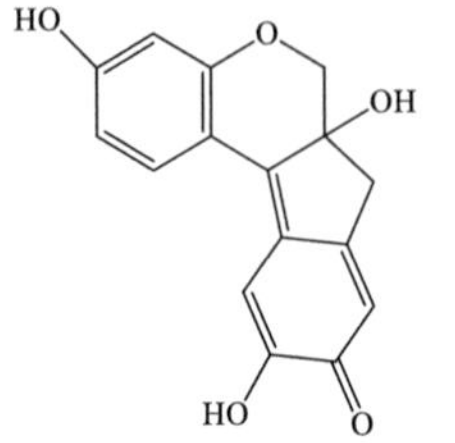

Brazilein (81)

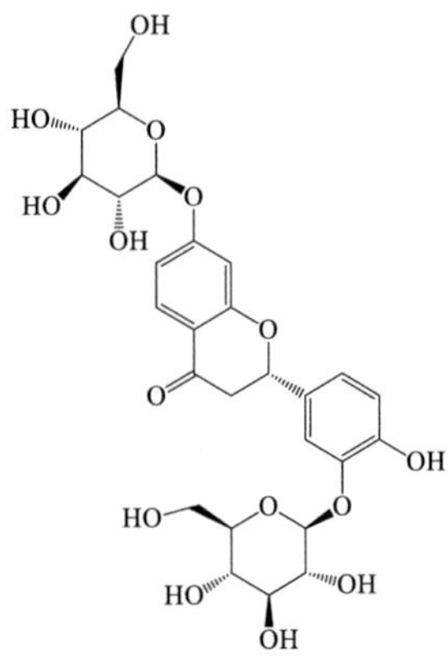

Butrin (82)

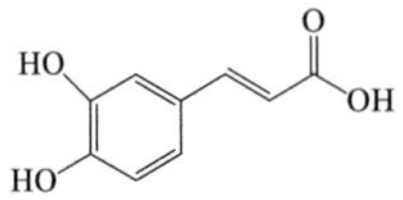

Caffeic acid (83)

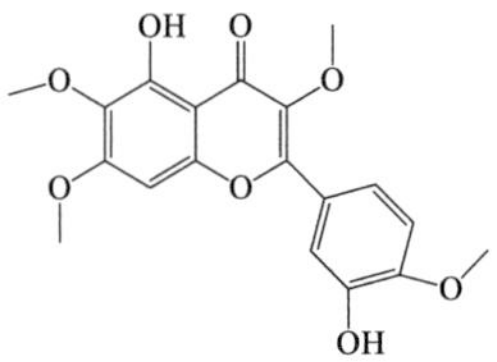

Casticin (84)

Fig. 1.5 (continued)

Fig. 1.5 (continued)

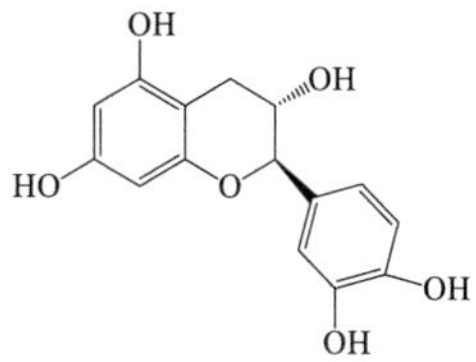

Catechin (85)

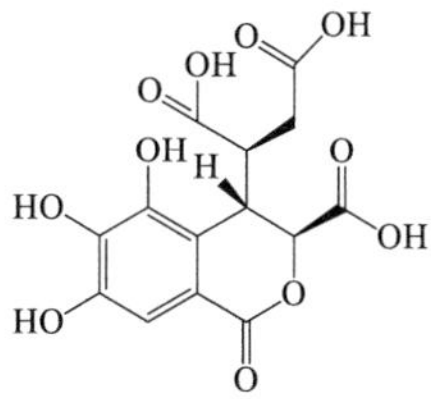

Chebulic acid (86)

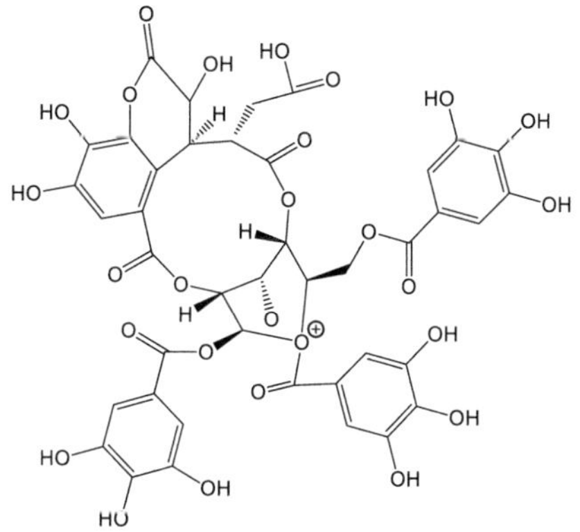

Chebulinic acid (87)

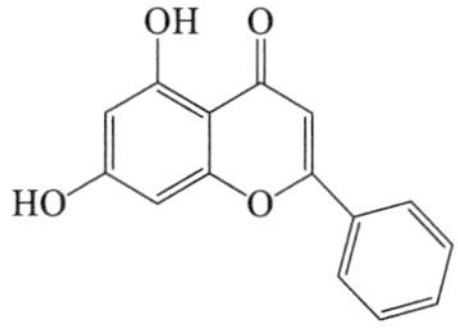

Chrysin (88)

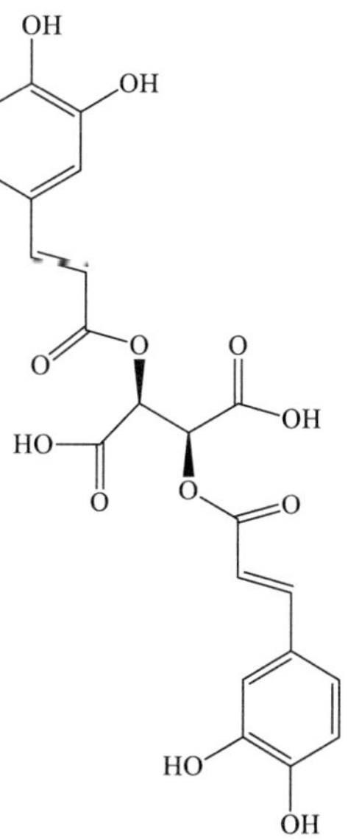

Cichoric acid (89)

Fig. 1.5 (continued)

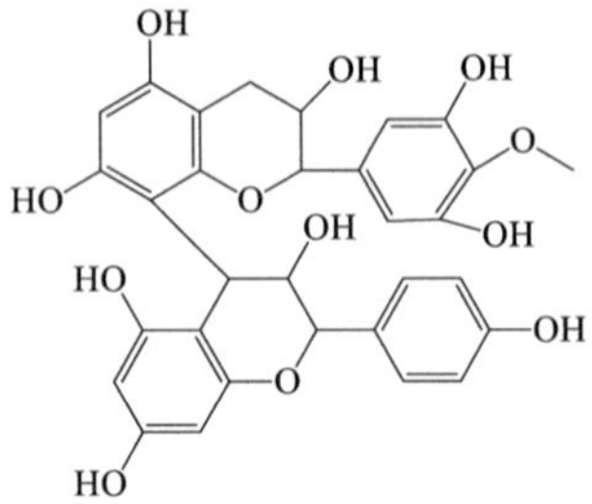

Proanthocyanidins (Condensed tannins) (90)

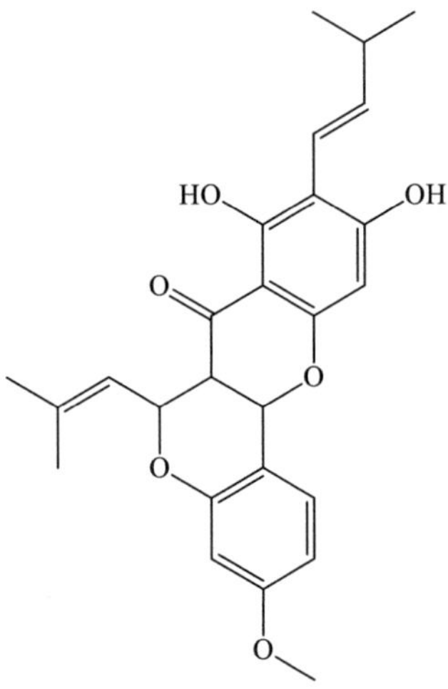

Cycloartocarpin (91)

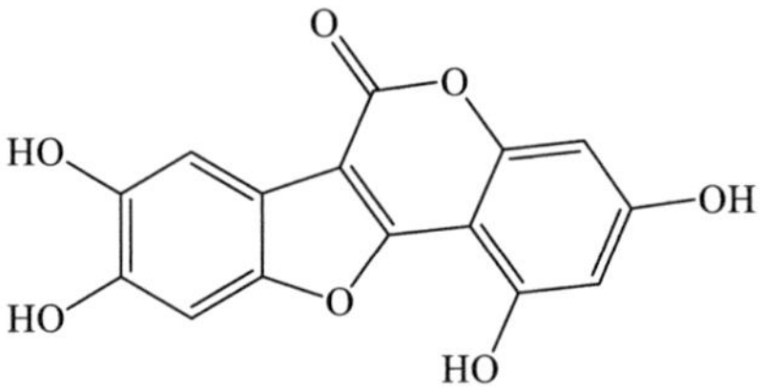

Demethylwedelolactone (92)

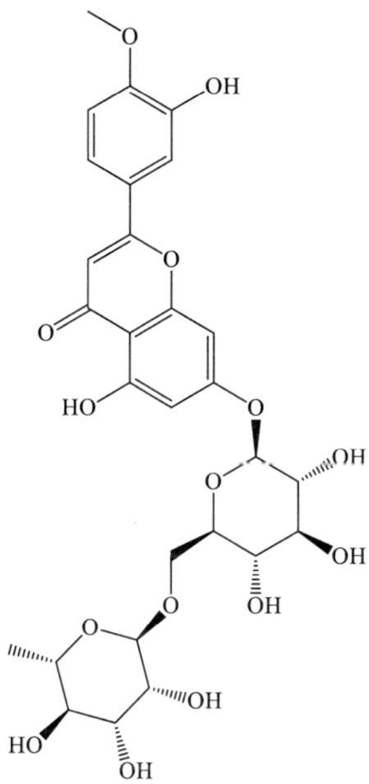

Diosmin (93)

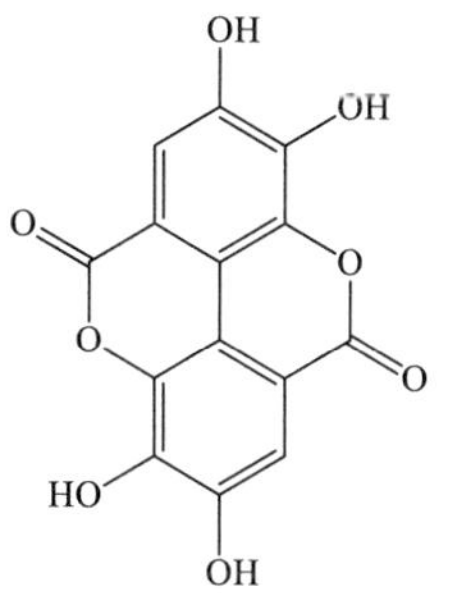

Ellagic acid (94)

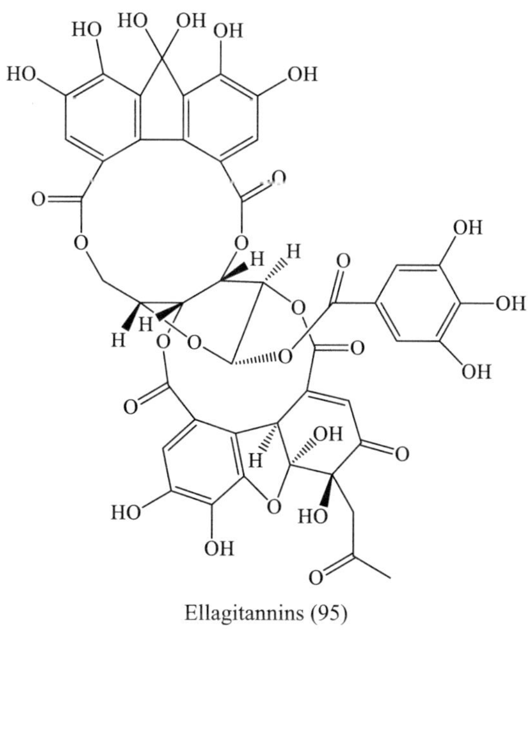

Ellagitannins (95)

Fig. 1.5 (continued)

Fig. 1.5 (continued)

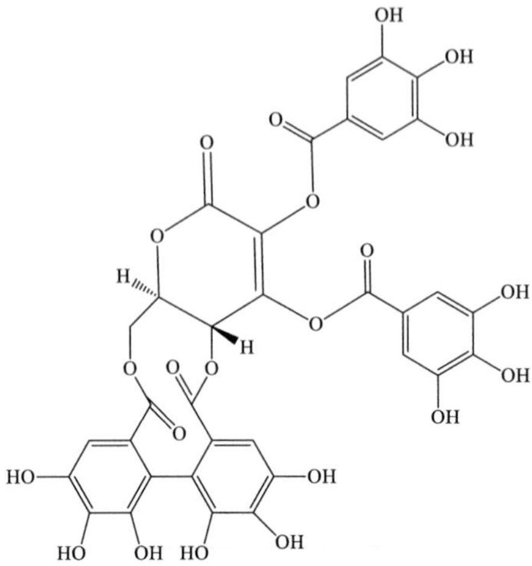

Emblicanin a (96)

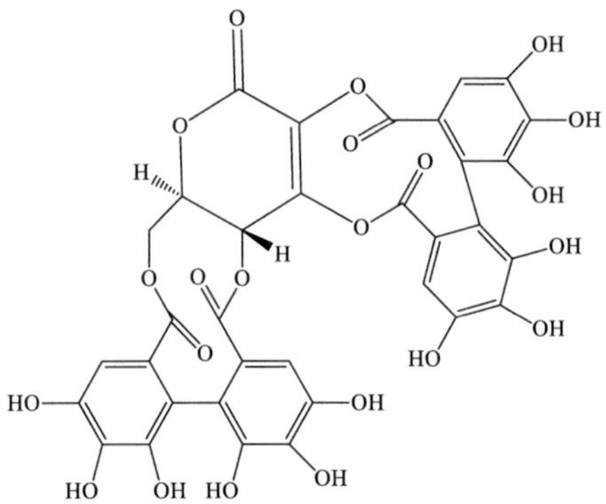

Emblicanin b (97)

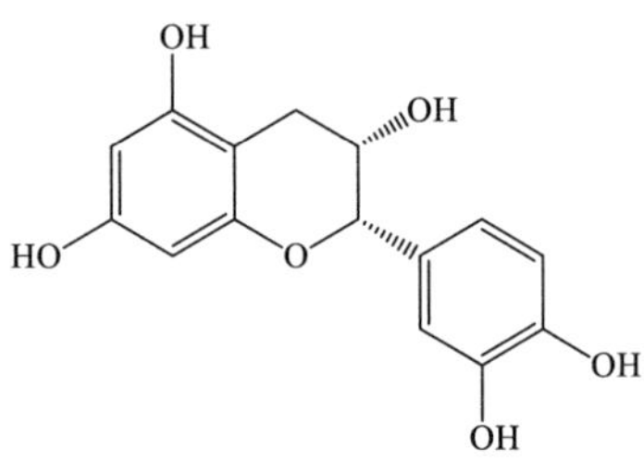

Epicatechin (98)

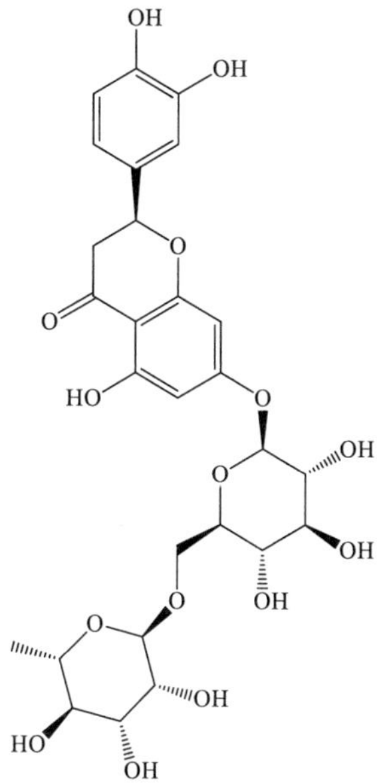

Eriocitrin (99)

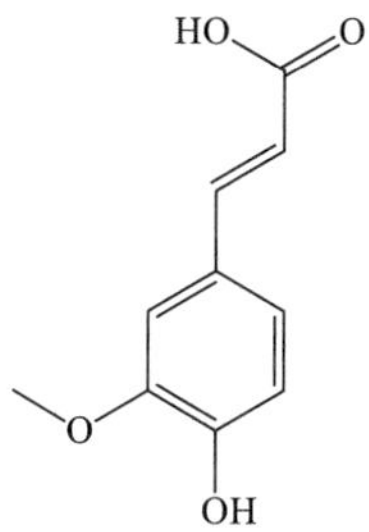

Ferulic acid (100)

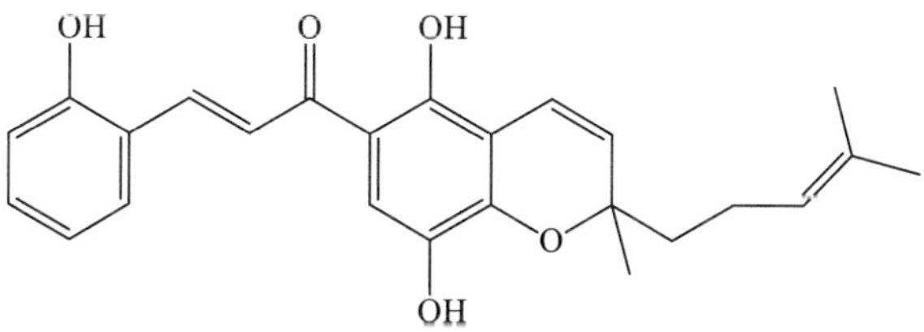

Flemingin (101)

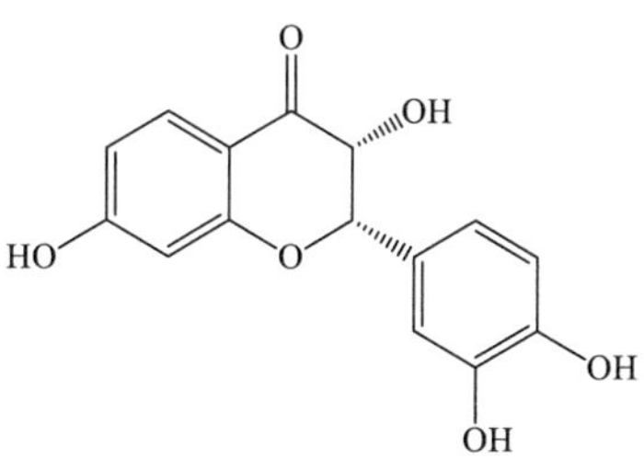

Fustin (102)

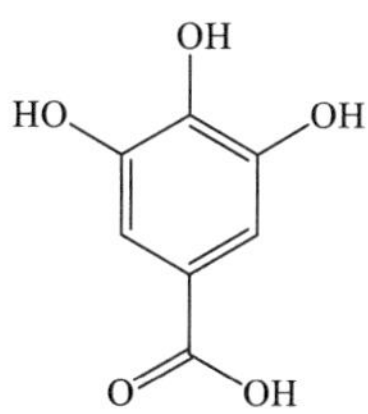

Gallic acid (103)

Fig. 1.5 (continued)

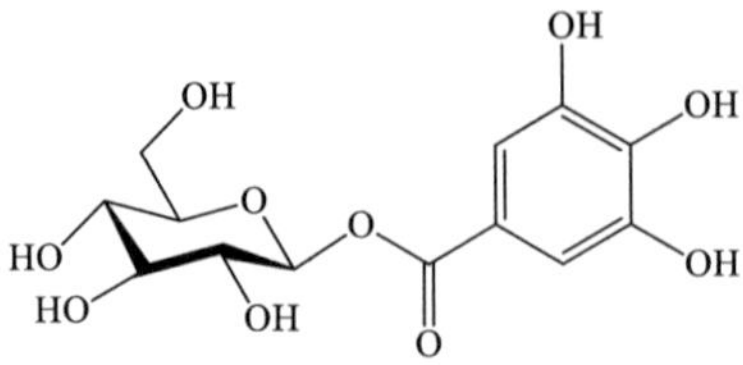

Gallotannins (104)

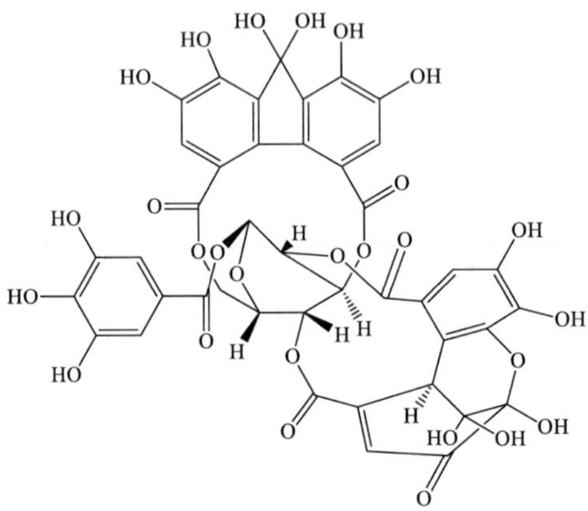

Geraniin (105)

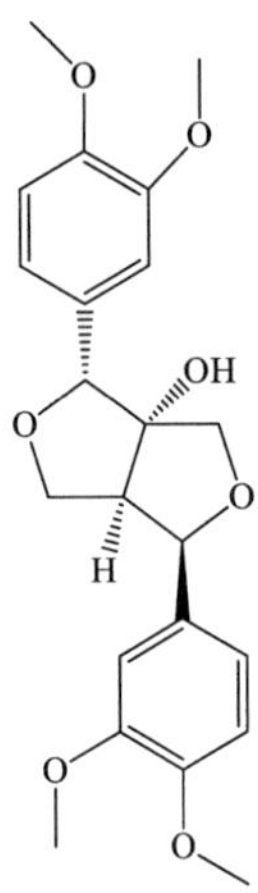

Gmelinol (106)

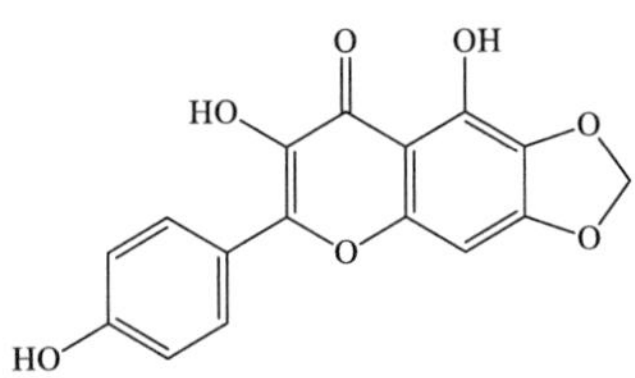

Gomphrenol (107)

Fig. 1.5 (continued)

Fig. 1.5 (continued)

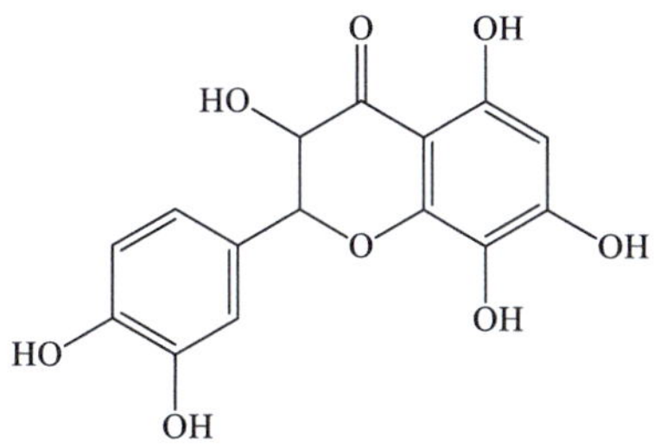

Gossypetin (108)

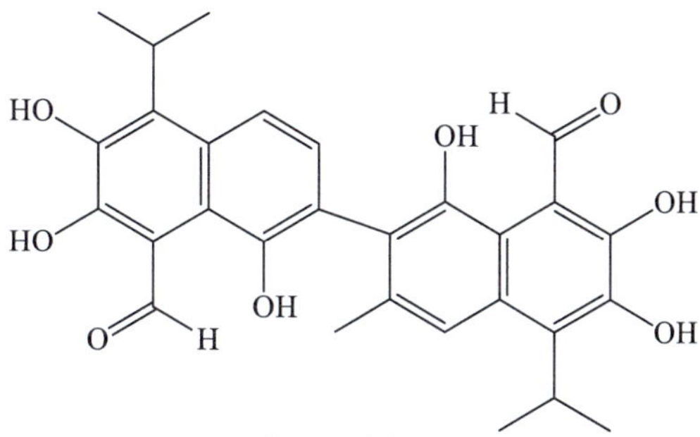

Gossypol (109)

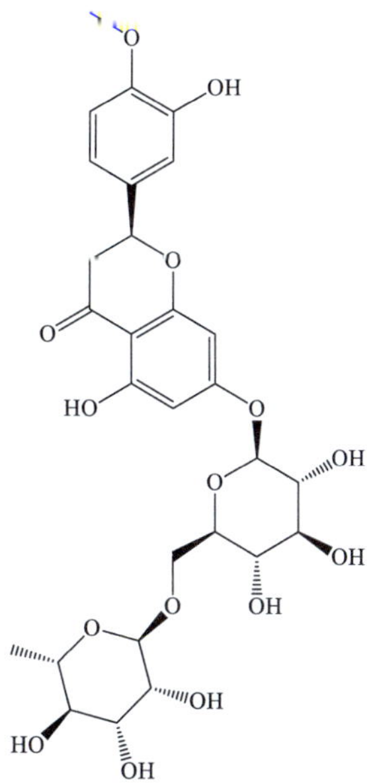

Hesperidin (110)

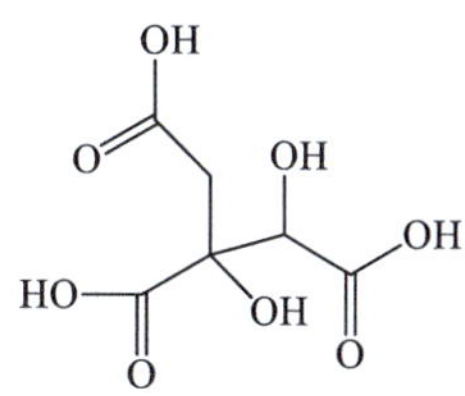

Hibiscus acid (111)

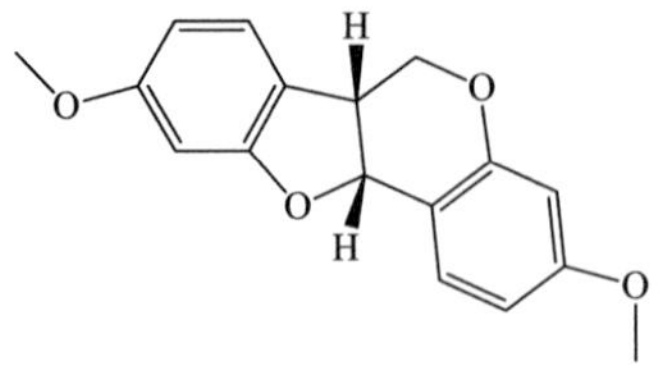

Homopterocarpin (112)

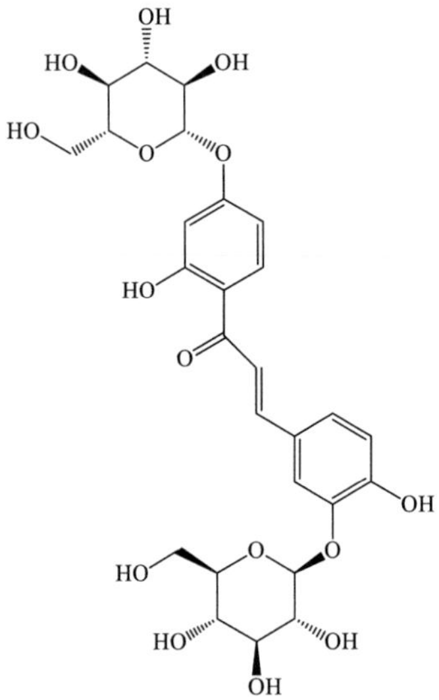

Isobutrin (113)

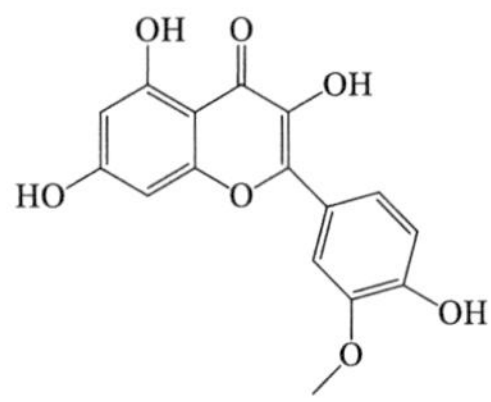

Isorhamnetin (114)

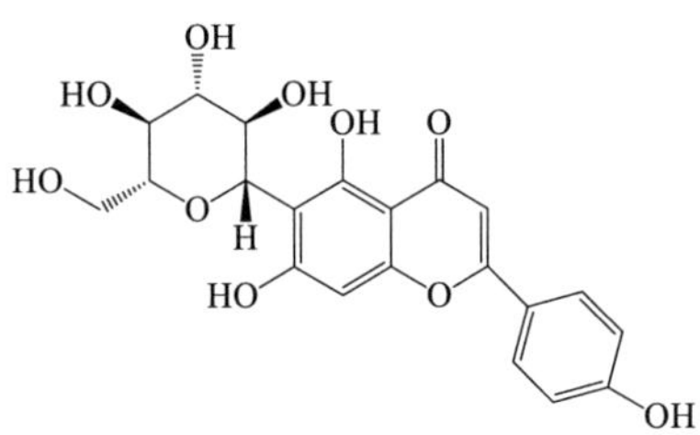

Isovitexin (115)

Fig. 1.5 (continued)

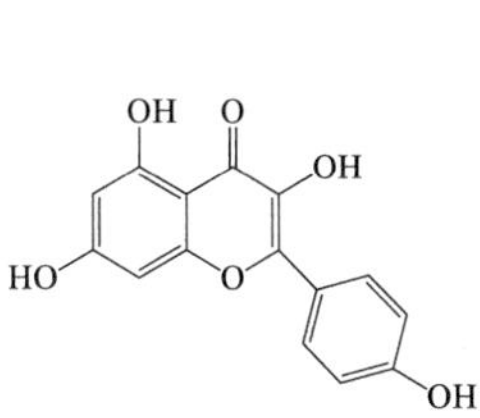

Kaempferol (116)

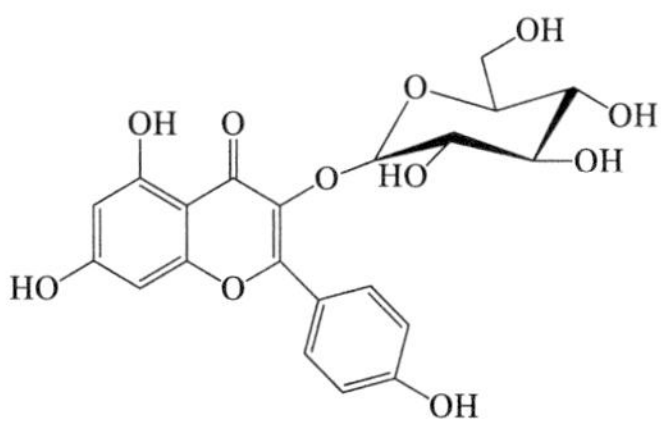

Kaempferol glycosides (117)

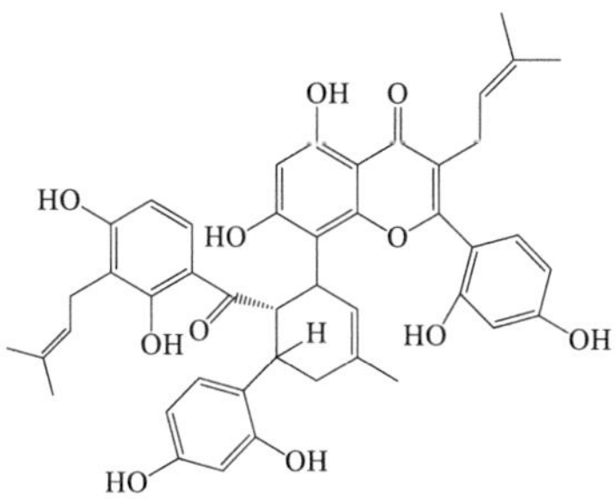

Kuwanon (118)

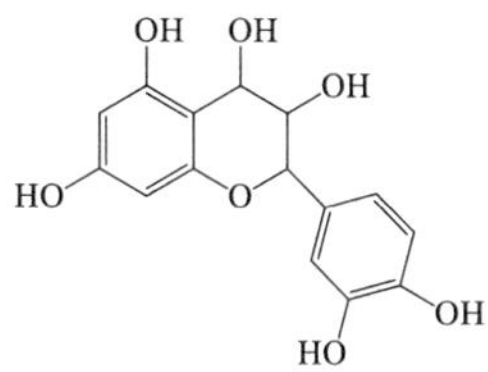

Leucocyanidin (119)

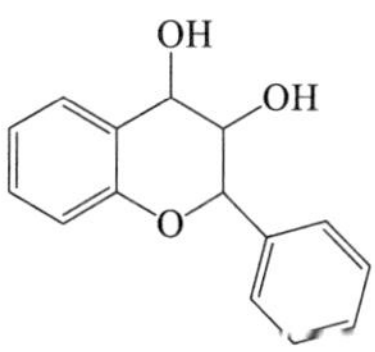

Leucophleol (2,4- flavandiol) (120)

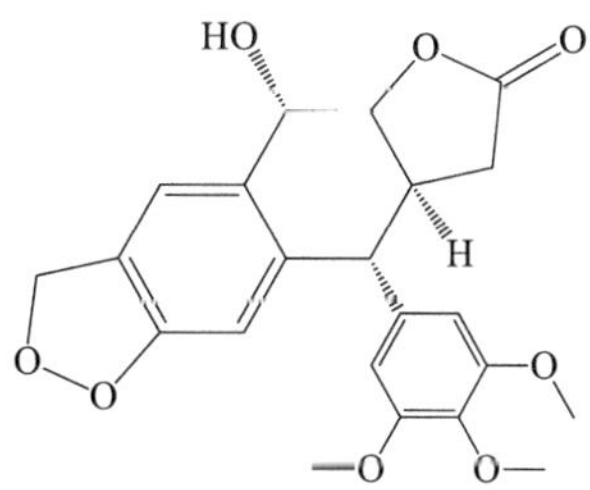

Lignans (121)

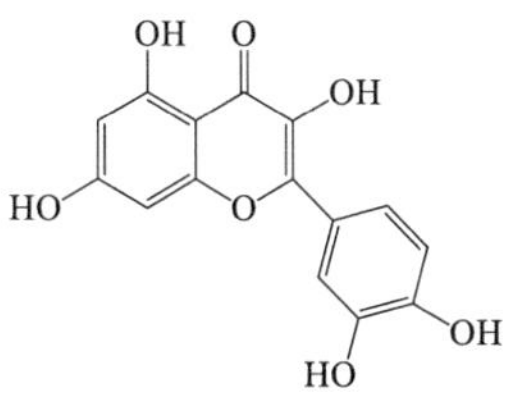

Luteolin (122)

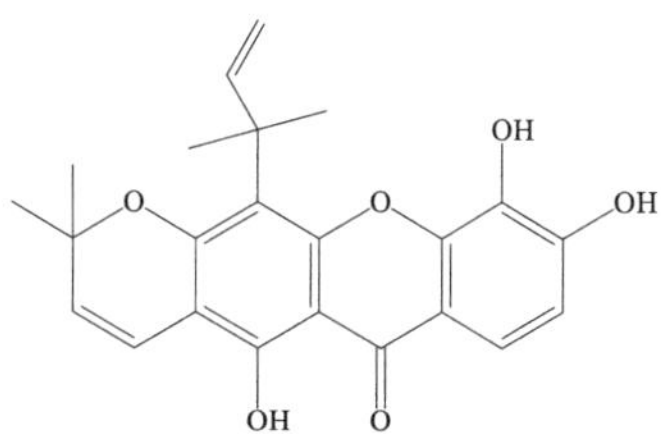

Macluraxanthone (123)

Fig. 1.5 (continued)

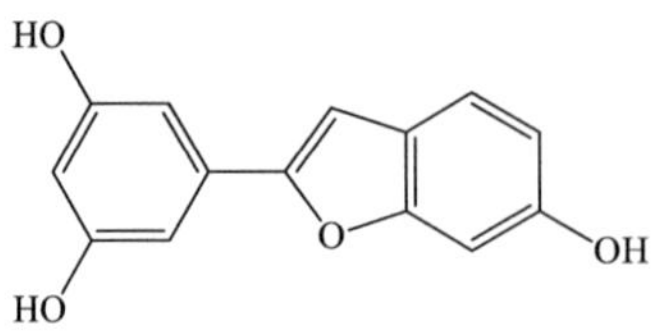

Moracin M (124)

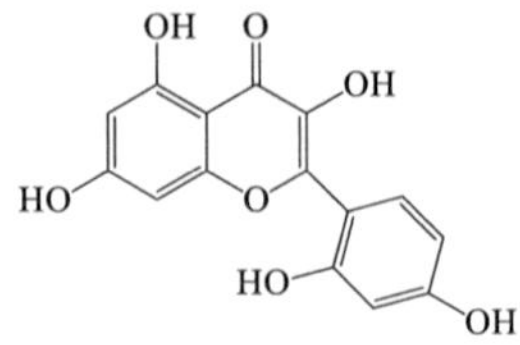

Morin (125)

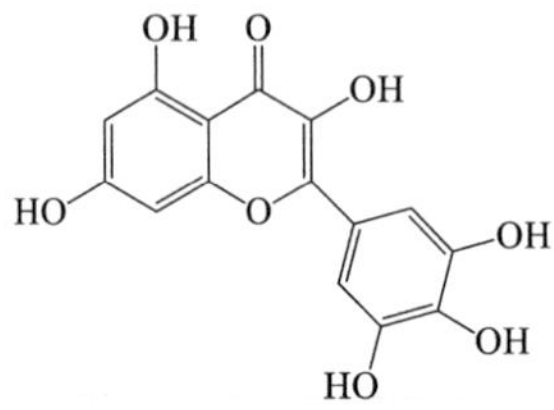

Myricetin (126)

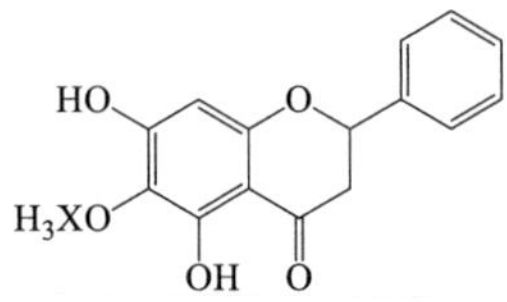

Oroxylin a (127)

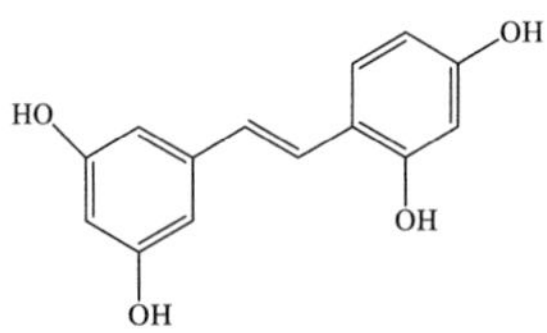

Oxyresveratrol (128)

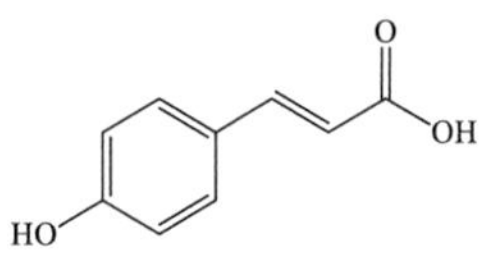

P-coumaric acid (129)

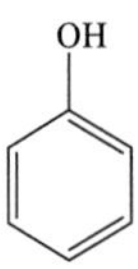

Phenolic compounds (130)

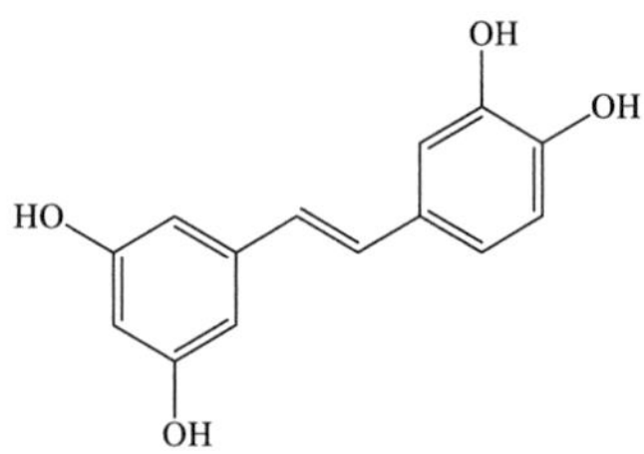

Piceatannol (131)

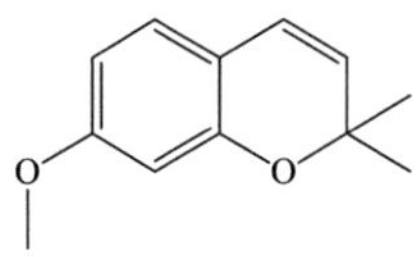

Precocene I (132)

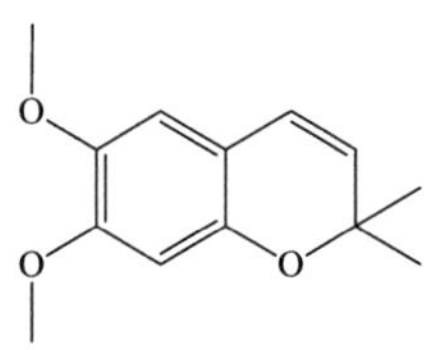

Precocene II (133)

Fig. 1.5 (continued)

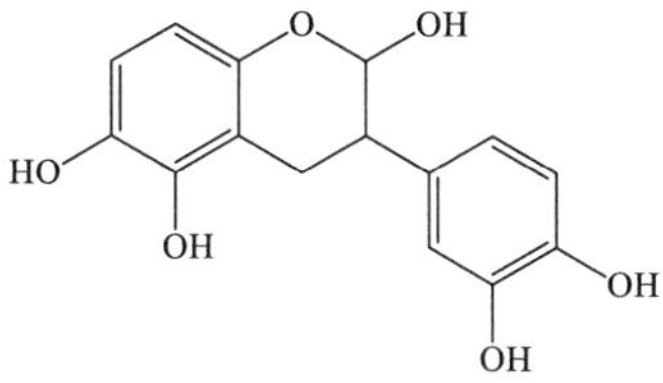

Proanthocyanidins (134)

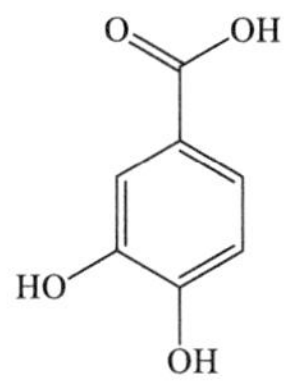

Protocatechuic acid (135)

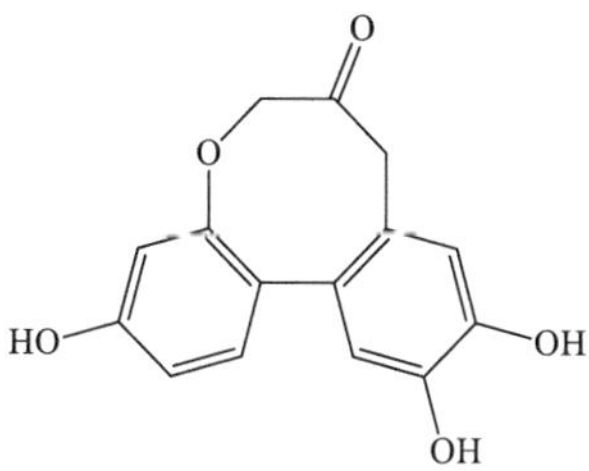

Protosappanin (136)

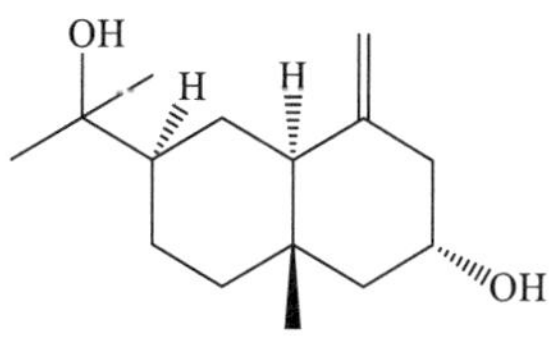

Pterocarpol (137)

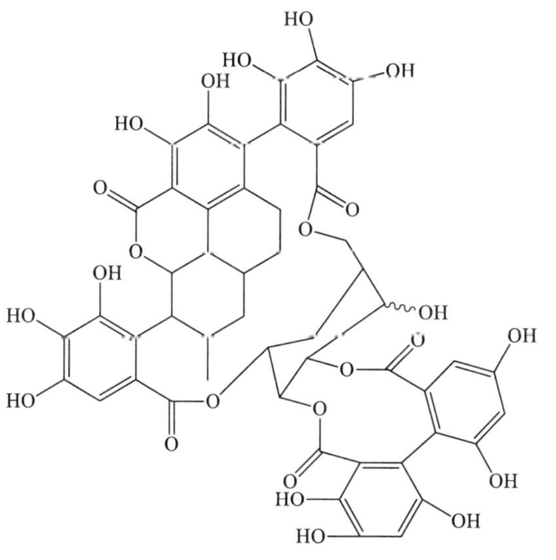

Punicalagins (138)

Fig. 1.5 (continued)

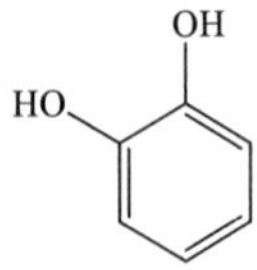

Pyrocatechols (139)

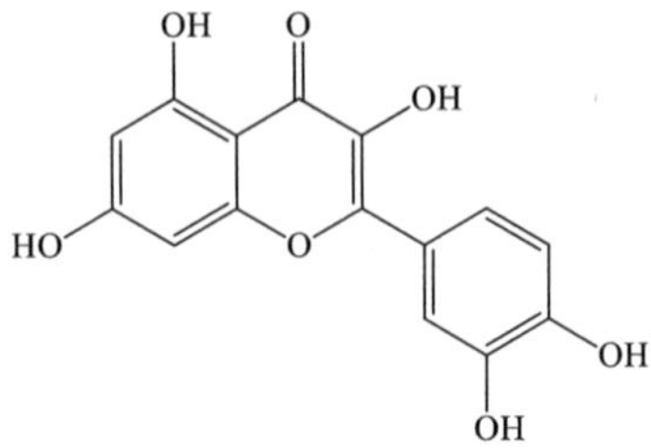

Quercetin (140)

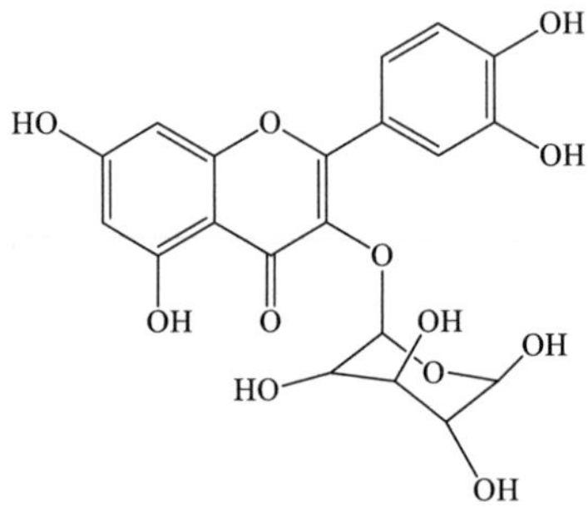

Quercetin glycosides (141)

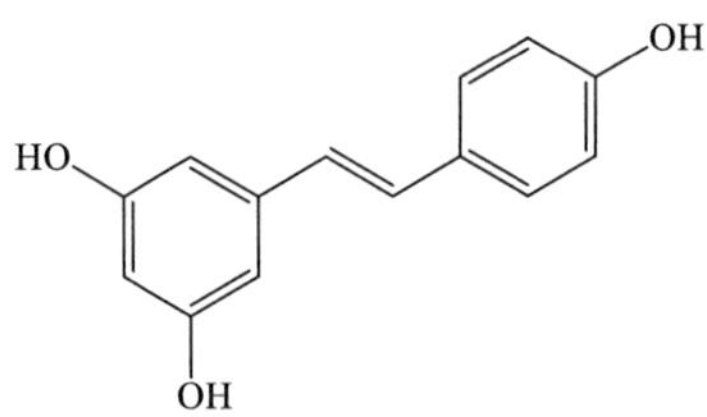

Resveratrol (142)

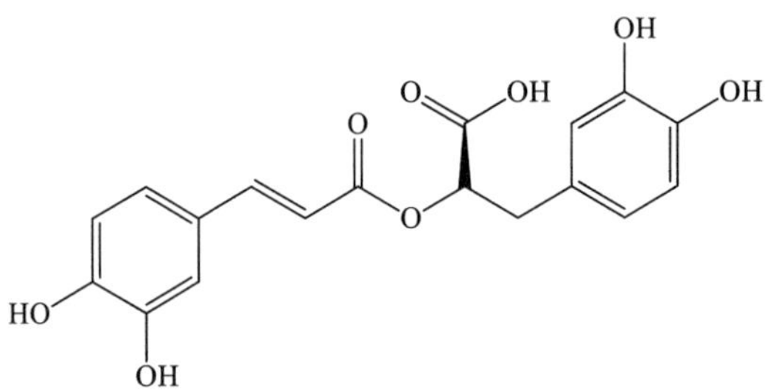

Rosmarinic acid (143)

Fig. 1.5 (continued)

Fig. 1.5 (continued)

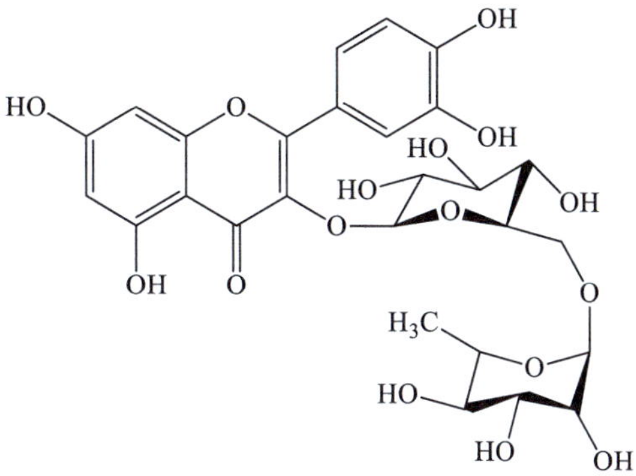

Rutin (144)

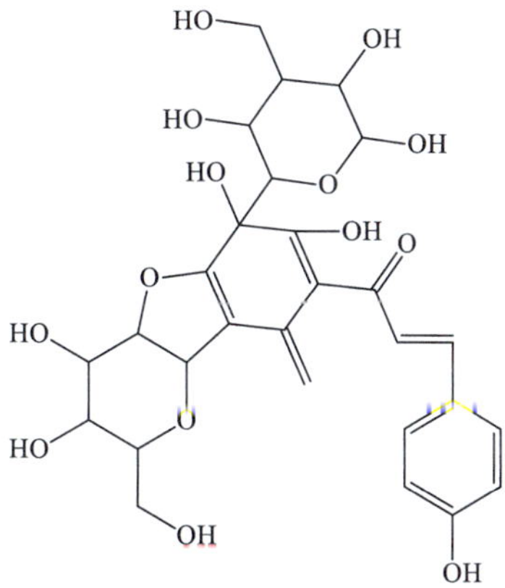

Safflor yellow (145)

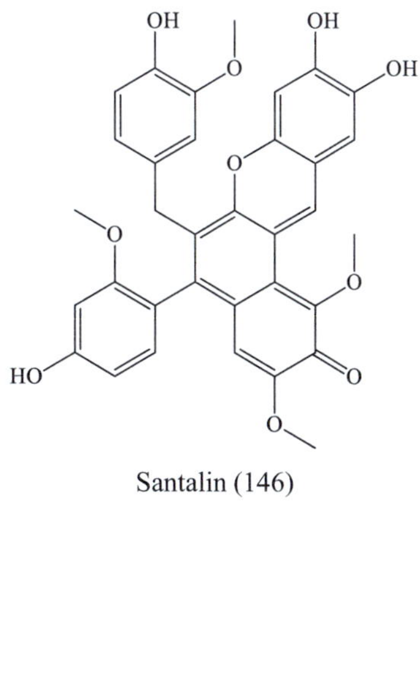

Santalin (146)

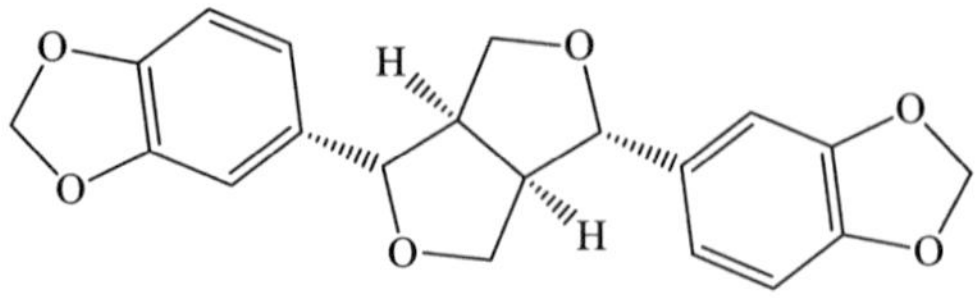

Sesamin (147)

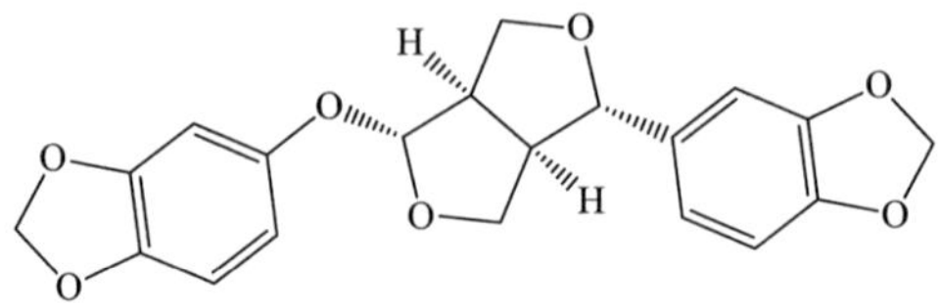

Sesamolin (148)

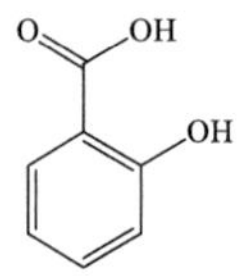

Simple phenolic acids (149)

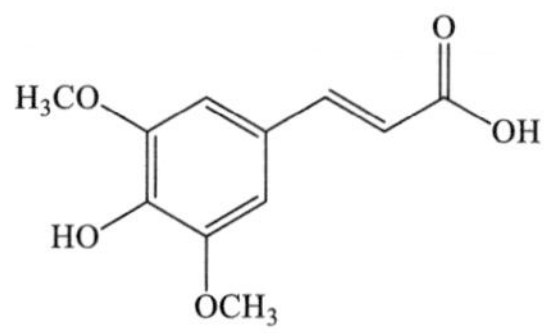

Sinapic acid (150)

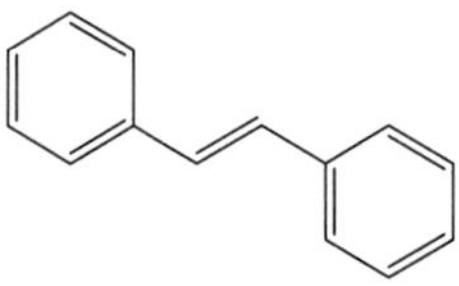

Stilbene (151)

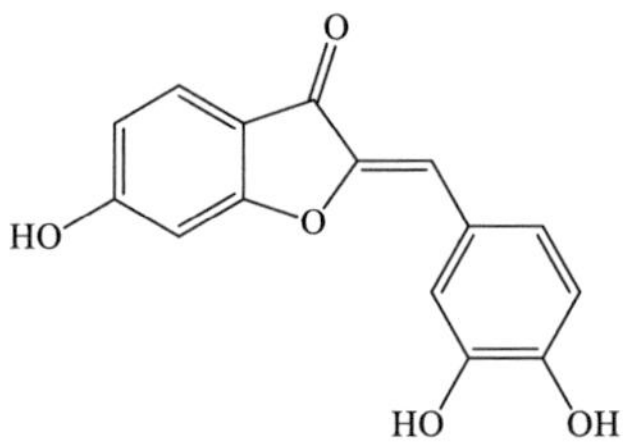

Sulfuretin (152)

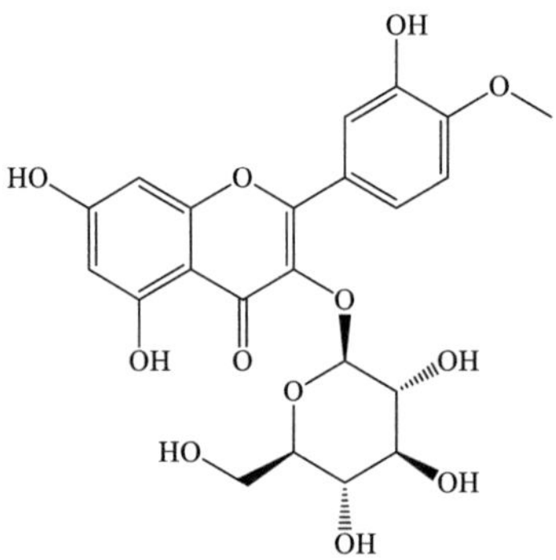

Tamarixin (153)

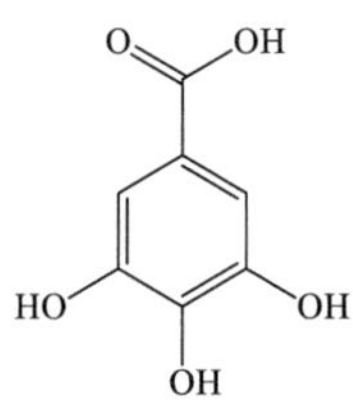

Tannins (154)

Fig. 1.5 (continued)

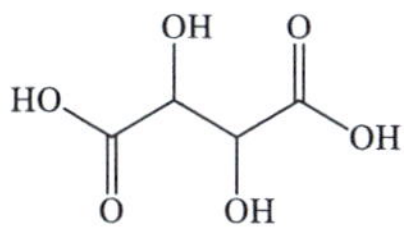

Tartaric acid (155)

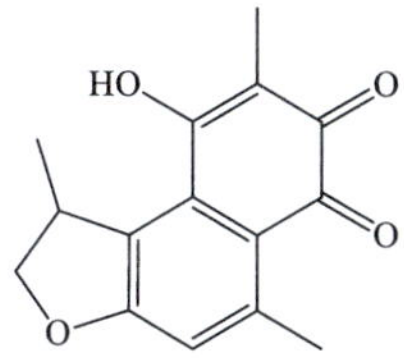

Thespesone (156)

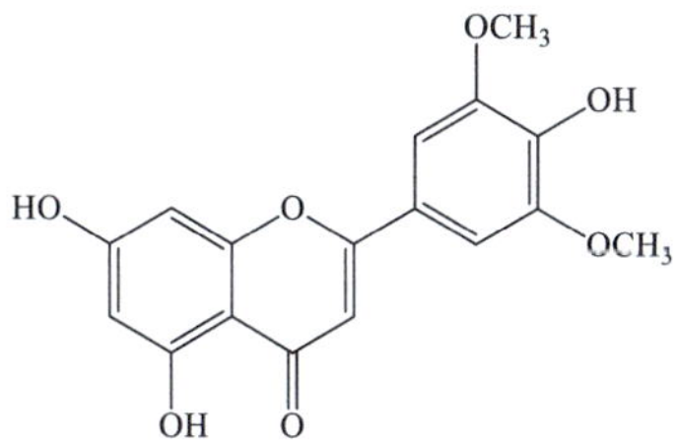

Tricin (157)

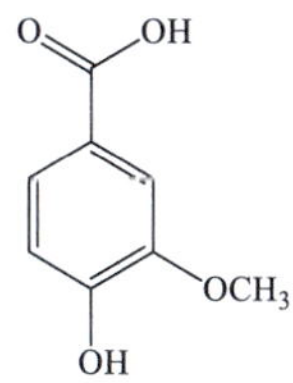

Vanillic acid (158)

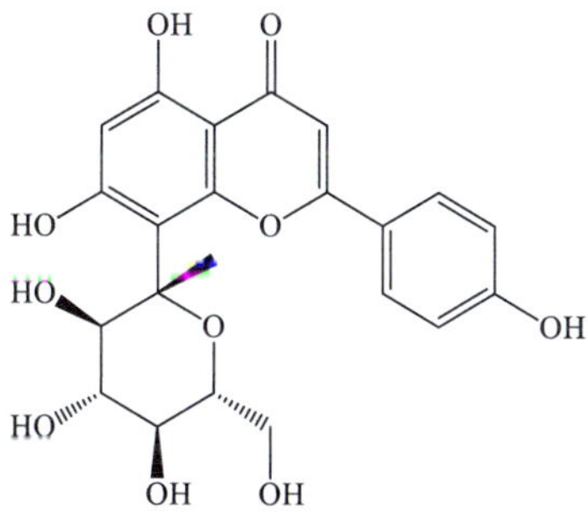

Vitexin (159)

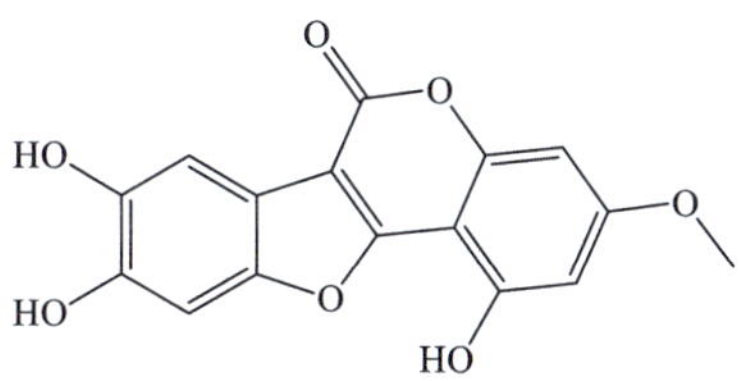

Wedelolactone (160)

Fig. 1.5 (continued)

Pigments obtained from these minerals are widely preferred because of their vibrant colours, opacity and stability. These pigments are also used in paints, ceramics, and cosmetics.

Fig. 1.6 Chemical structures of polysaccharides (161–163)

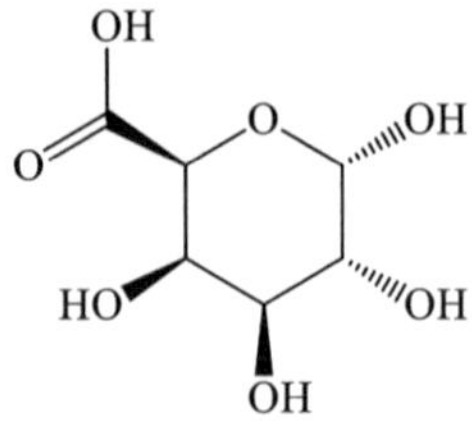

α-D-Galacturonic acid (161)

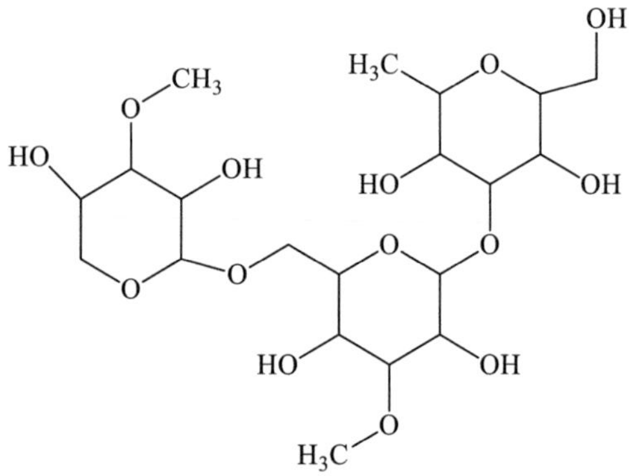

Arabinogalactan (162)

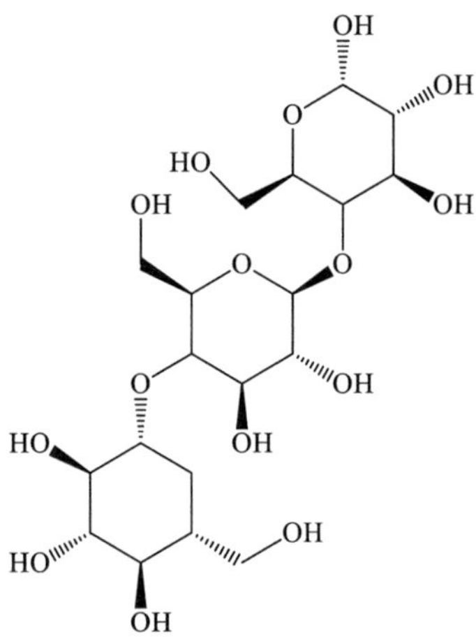

β -Glucan (163)

1.4 Uses of Natural Dyes and Pigments in Food and Medicine

The natural dyes and pigments are widely used in food and medicine. They are used as preservatives and for better appearances due to their appealing colour (Yusuf 2018; Al Fahad et al. 2024). Apart from providing the colour, these dyes and pigments also have therapeutic properties. For example, anthocyanins derived from fruits and red cabbage have vibrant red and purple colours, and they have antioxidant properties with health benefits. Curcumin, a yellow pigment obtained from turmeric, has anti-inflammatory properties and is beneficial for treating skin disorders such as psoriasis and eczema. Similarly, beta-carotene from carrots, which gives yellow-orange-red colours, is rich in vitamin A, which is beneficial for vision, especially in low light (Li et al. 2022).

Fig. 1.7 Chemical structures of quinones (164–188)

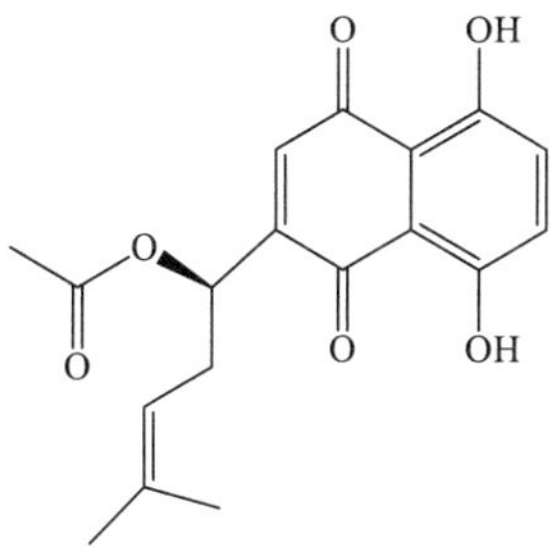

Acetylshikonin (164)

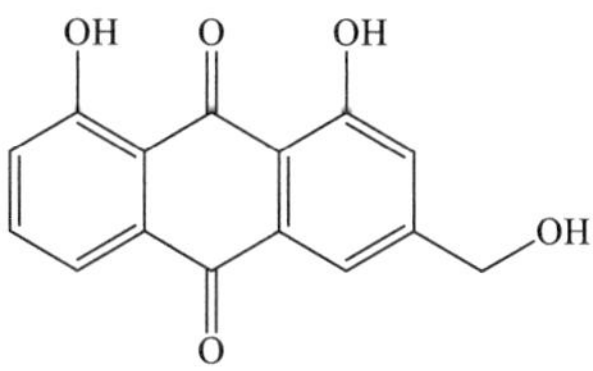

Aloe-emodin (165)

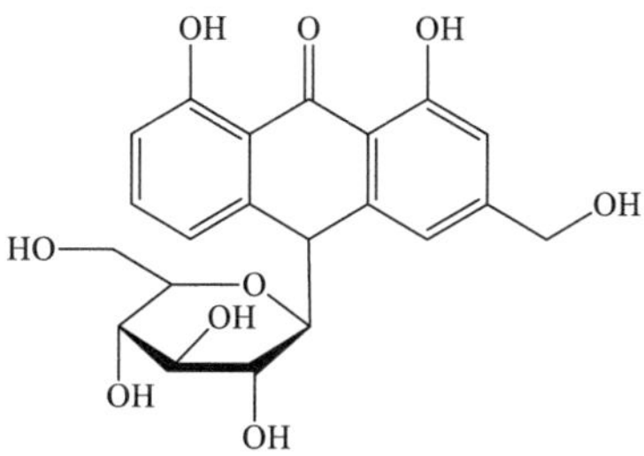

Aloin (166)

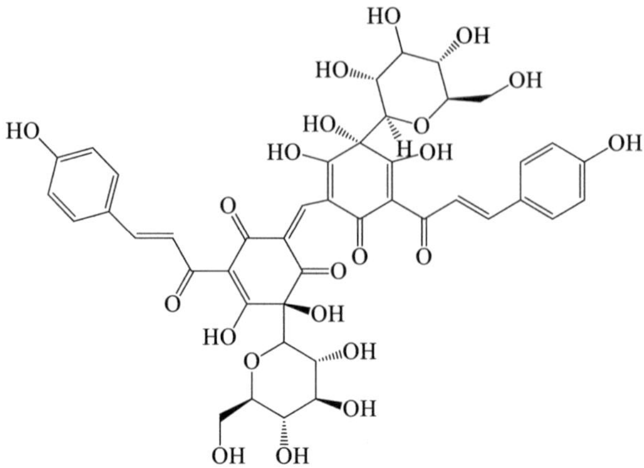

Carthamin (167)

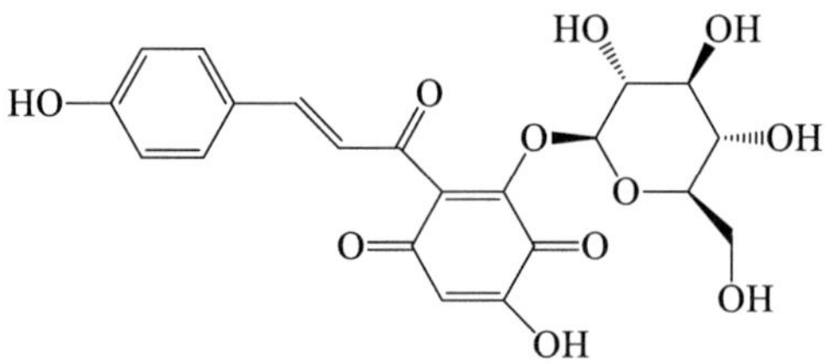

Carthamone (168)

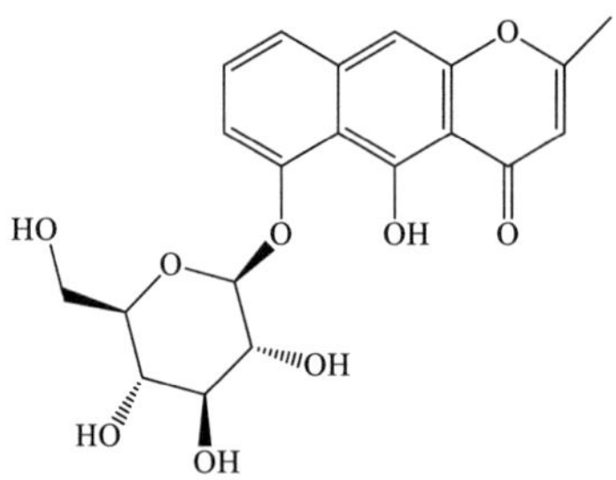

Cassiaside (169)

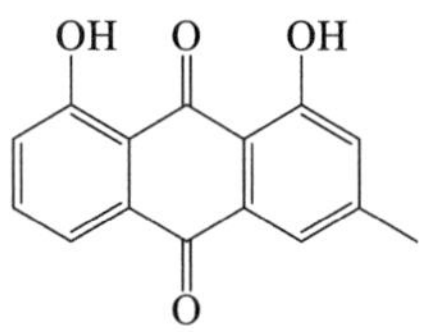

Chrysophanol (170)

Fig. 1.7 (continued)

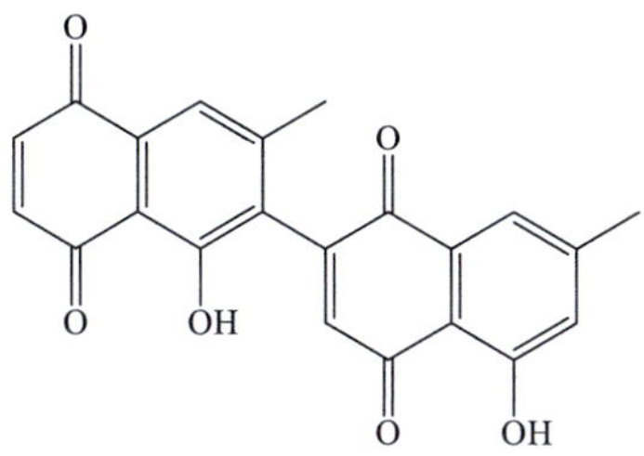

Diospyrin (171)

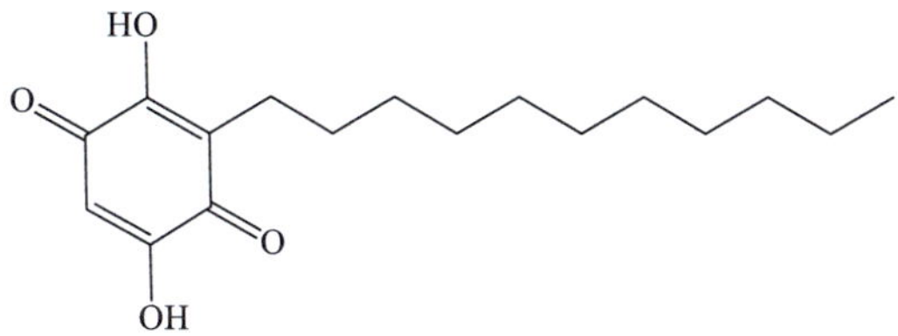

Embelin (172)

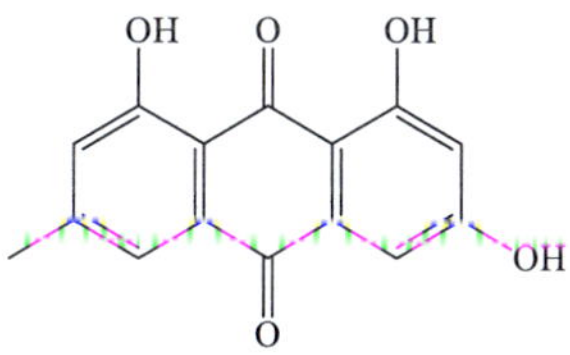

Emodin (173)

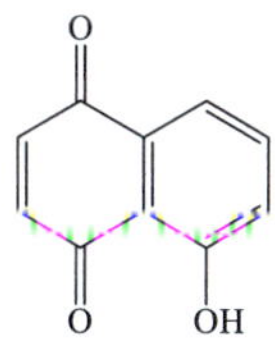

Juglone (174)

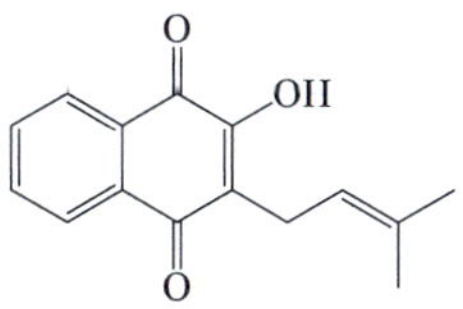

Lapachol (175)

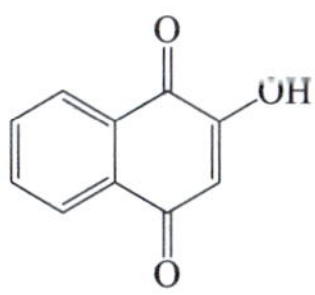

Lawsone (176)

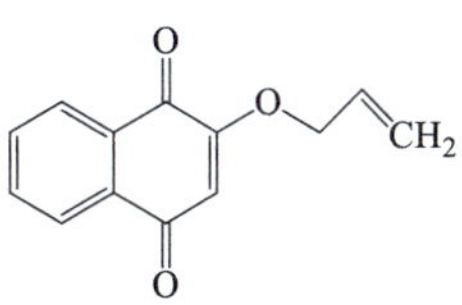

Lawsone and derivatives (177)

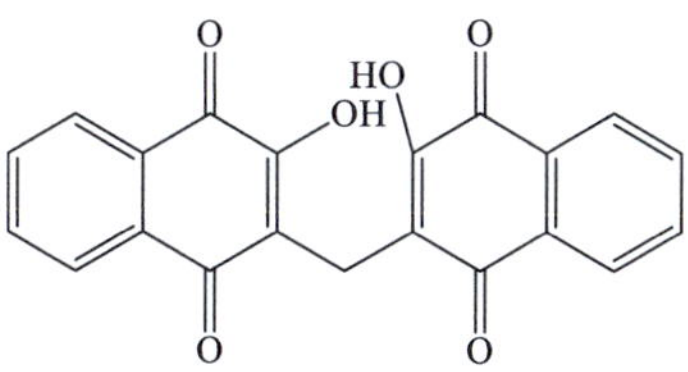

Methylene-3,3'-bilawsone (178)

Fig. 1.7 (continued)

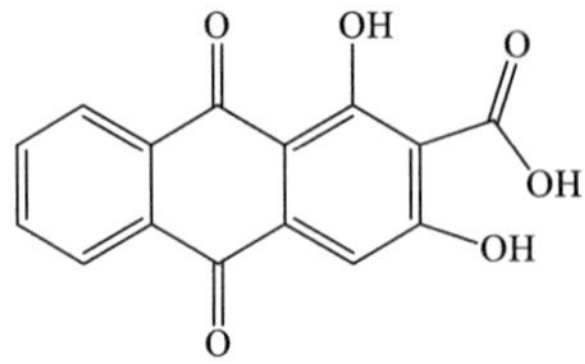

Munjistin (179)

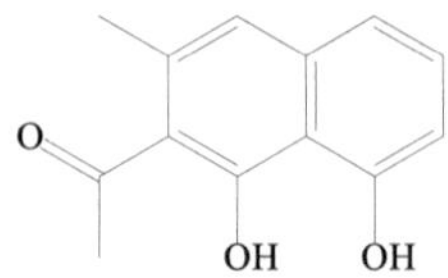

Nepodin (180)

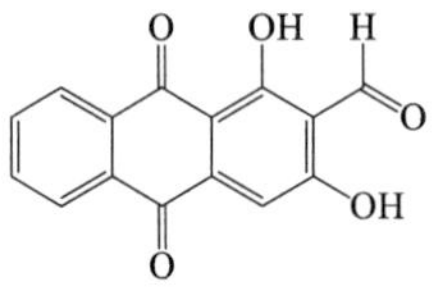

Nordamnacanthal (181)

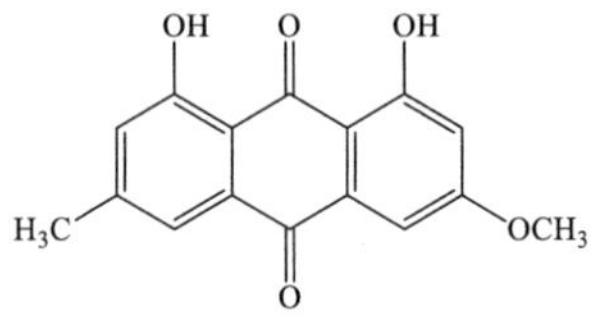

Physcion (182)

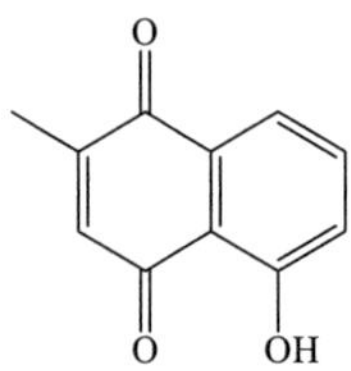

Plumbagin (183)

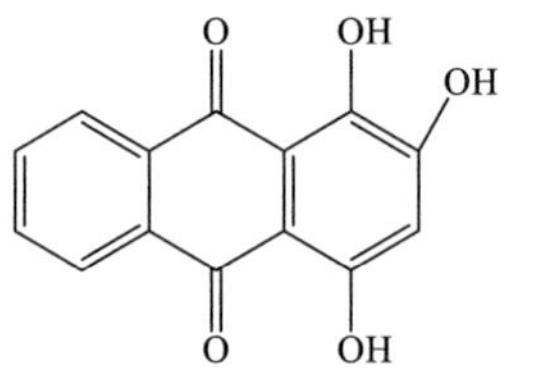

Purpurin (184)

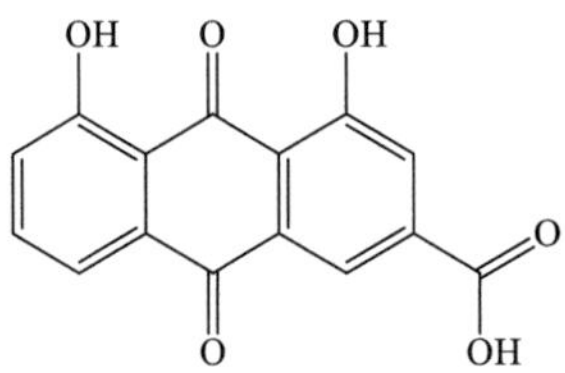

Rhein (185)

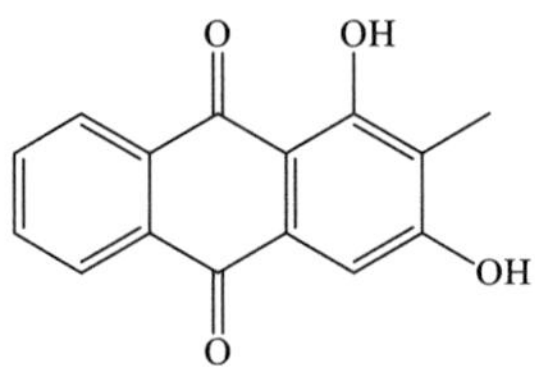

Rubiadin (186)

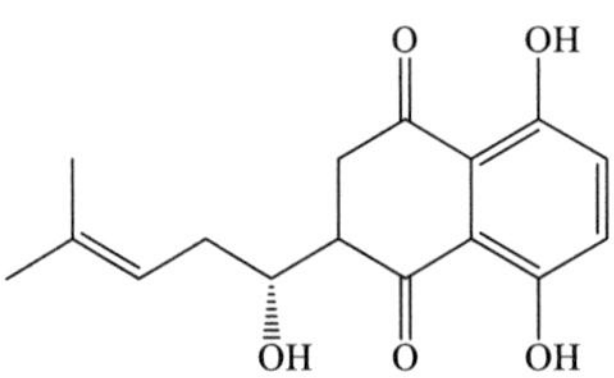

Shikonin (187)

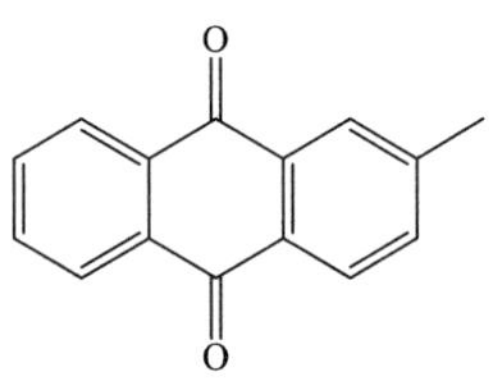

Tectoquinone (188)

Fig. 1.7 (continued)

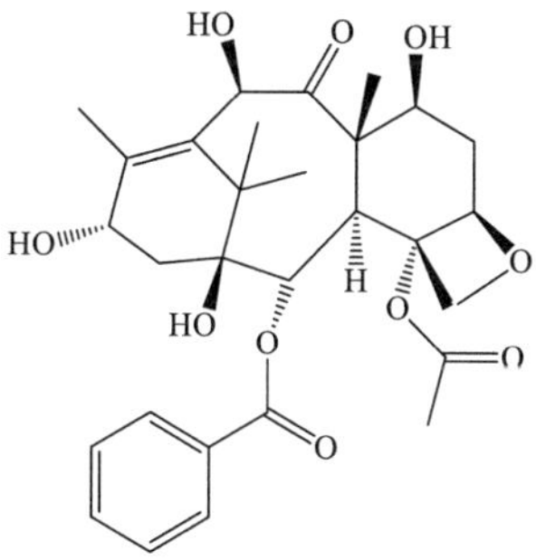

10-deacetylbaccatin III (189)

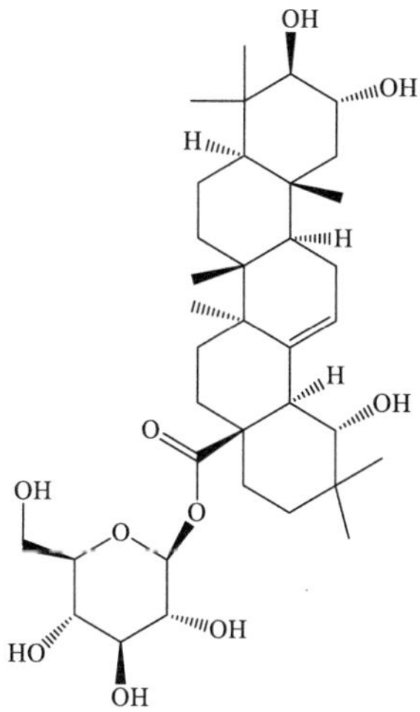

Arjunetin (190)

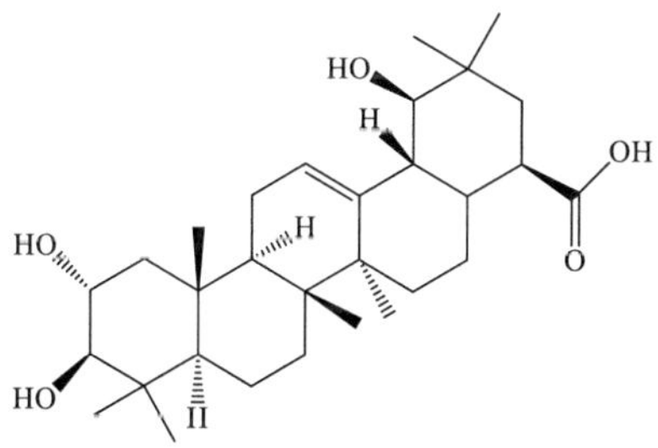

Arjunic acid (191)

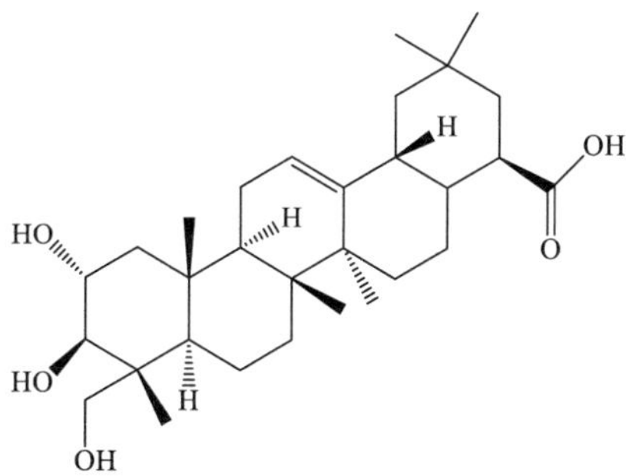

Arjunolic acid (192)

Fig. 1.8 Chemical structures of terpenoids (189–257)

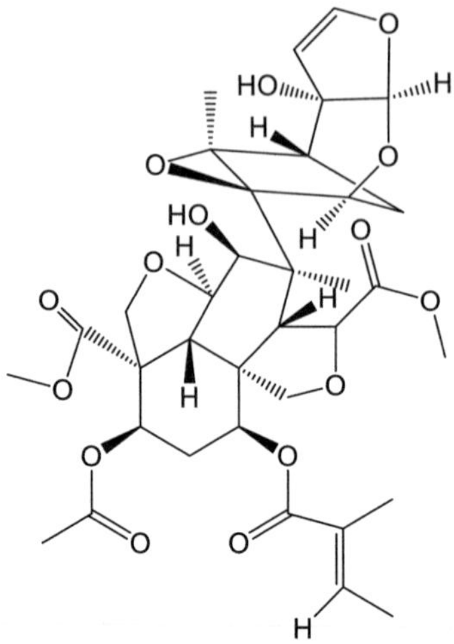

Azadirachtin (193)

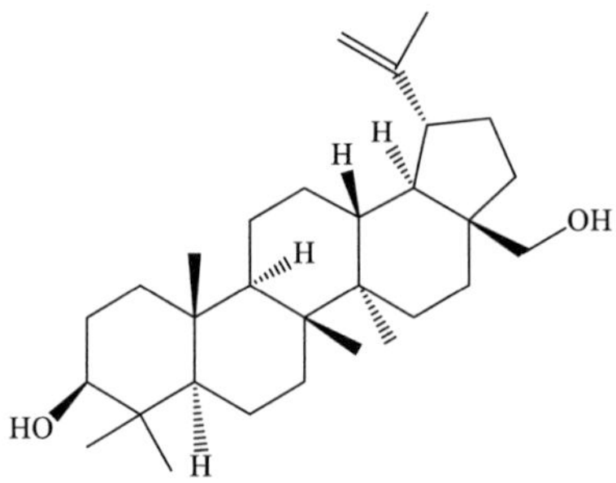

Betulin (194)

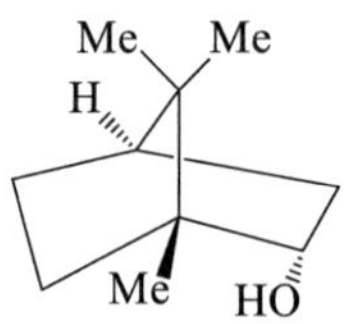

Borneol (195)

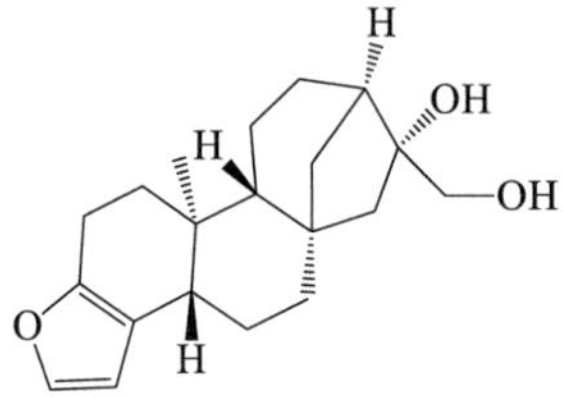

Cafestol (196)

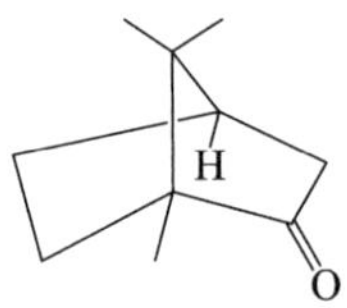

Camphor (197)

Fig. 1.8 (continued)

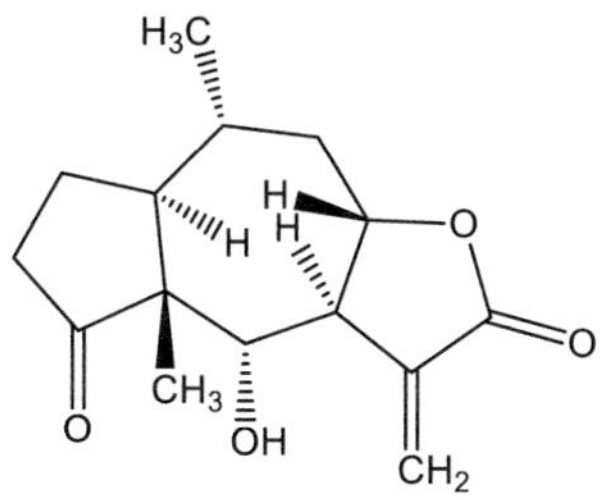

Carpesiolin (198)

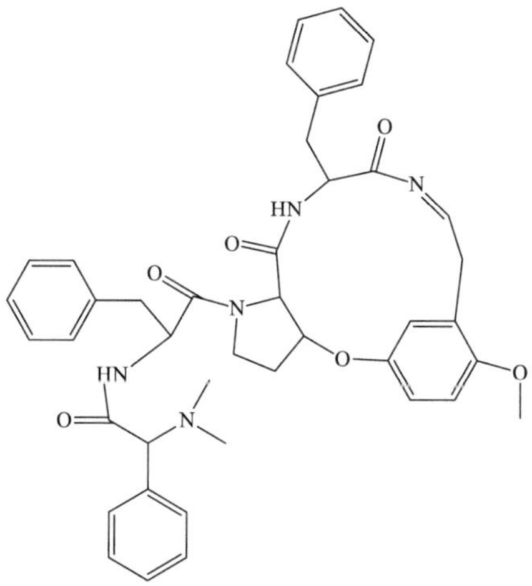

Chukrasins (199)

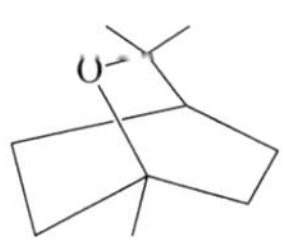

Cineole (200)

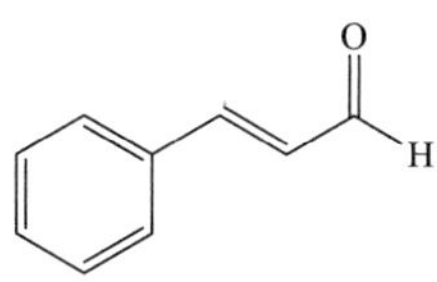

Cinnamaldehyde (201)

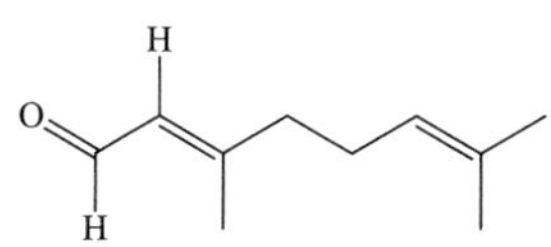

Citral (202)

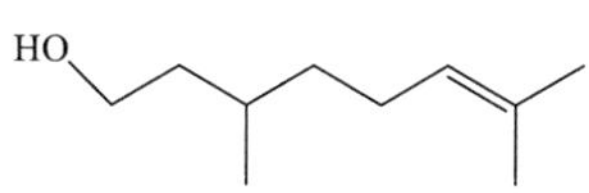

Citronellol (203)

Fig. 1.8 (continued)

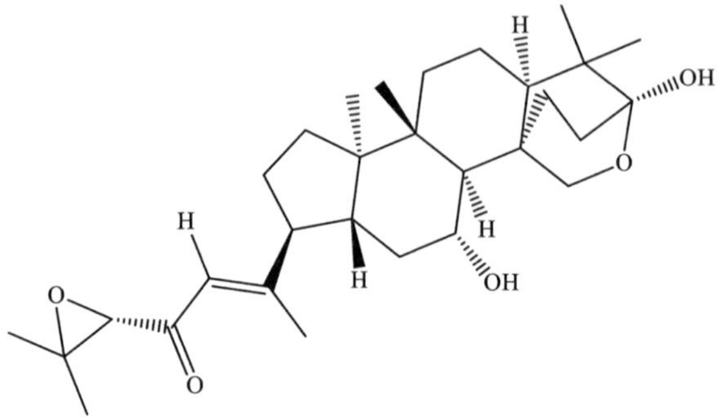

Cordialin A (204)

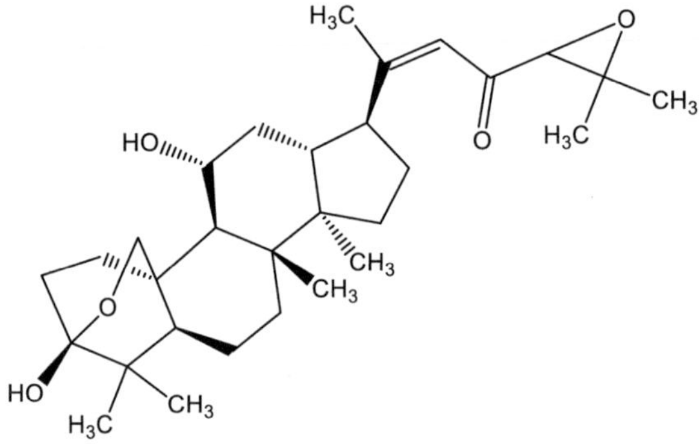

Cordialin B (205)

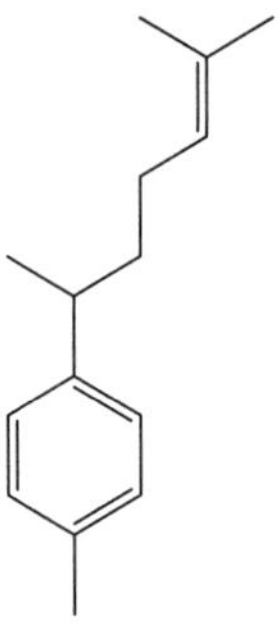

Curcumene (206)

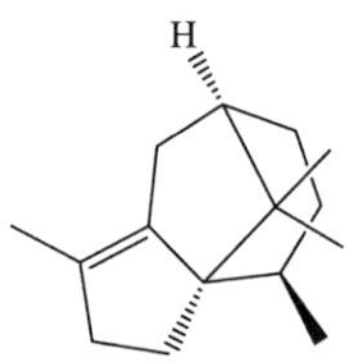

Cyperene (207)

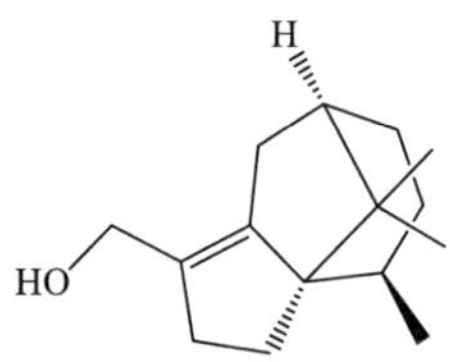

Cyperenol (208)

Fig. 1.8 (continued)

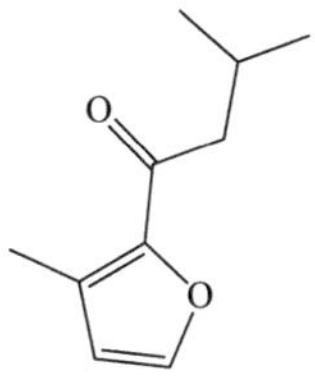

Elsholtzia ketone (209)

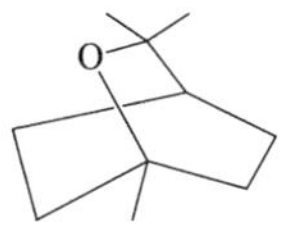

Eucalyptol (1,8-cineole) (210)

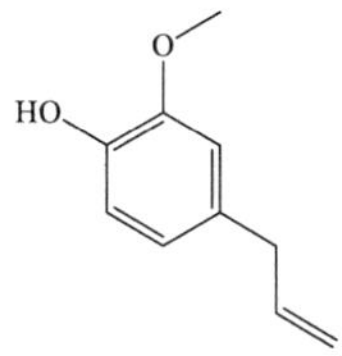

Eugenol (211)

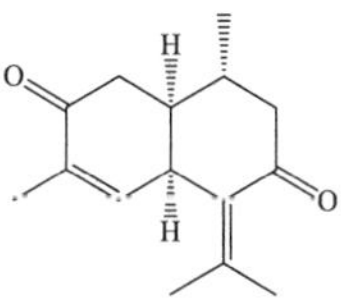

Euptox A (212)

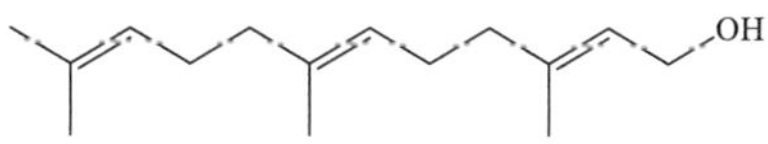

Farnesol (213)

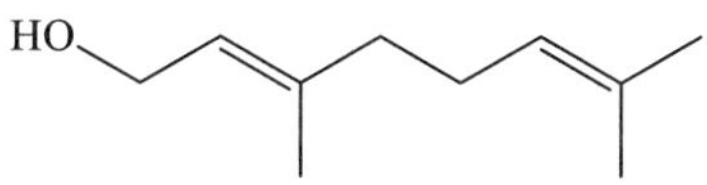

Geraniol (214)

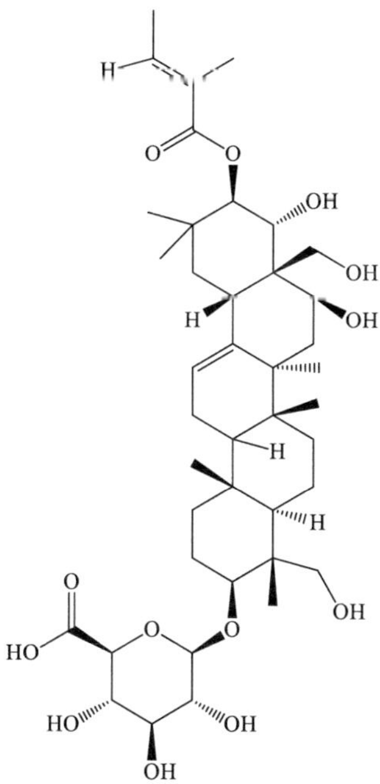

Gymnemic acid (215)

Fig. 1.8 (continued)

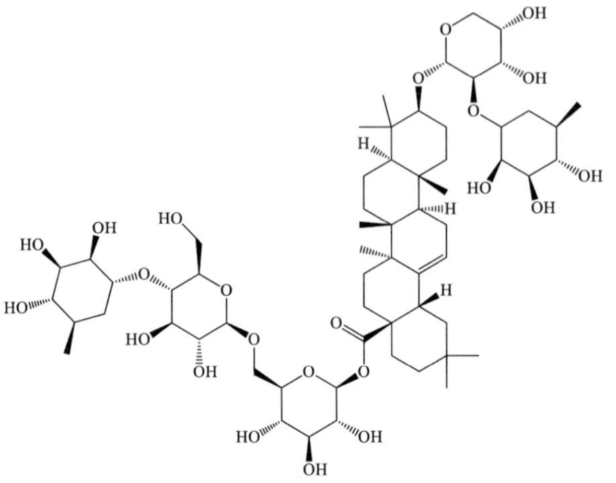

Hederasaponins (216)

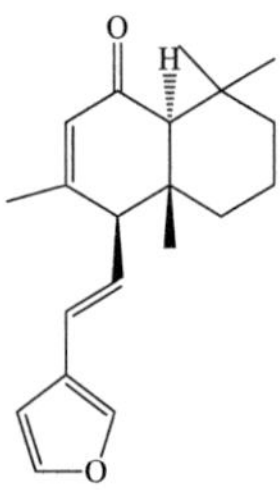

Hedychenone (217)

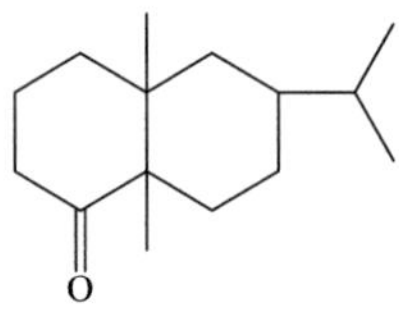

Jatamansone (218)

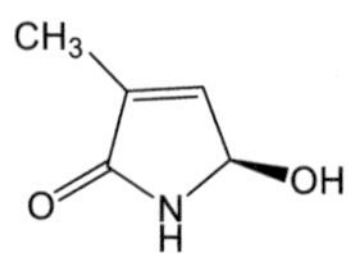

Jatrophine (219)

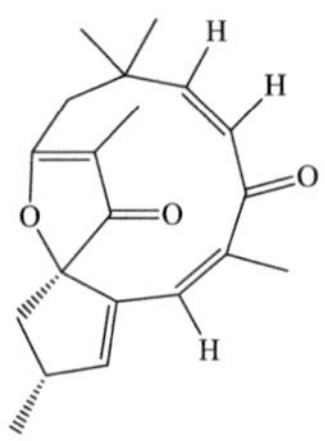

Jatrophone (220)

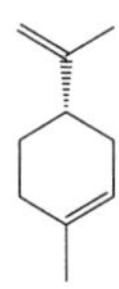

Limonene (221)

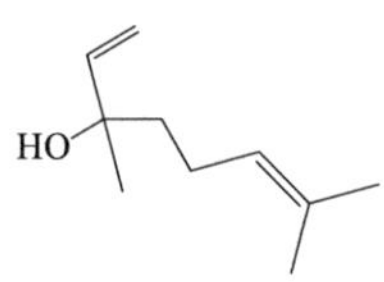

Linalool (222)

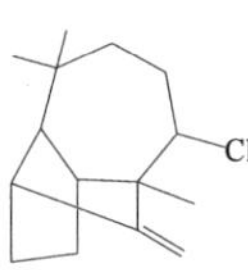

Longifolene (223)

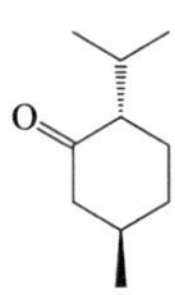

Menthone (224)

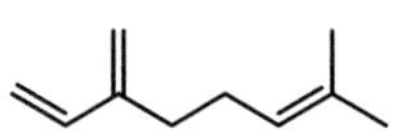

Myrcene (225)

Fig. 1.8 (continued)

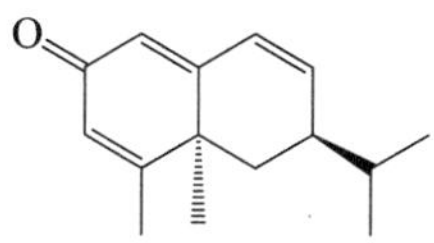

Nardostachone (226)

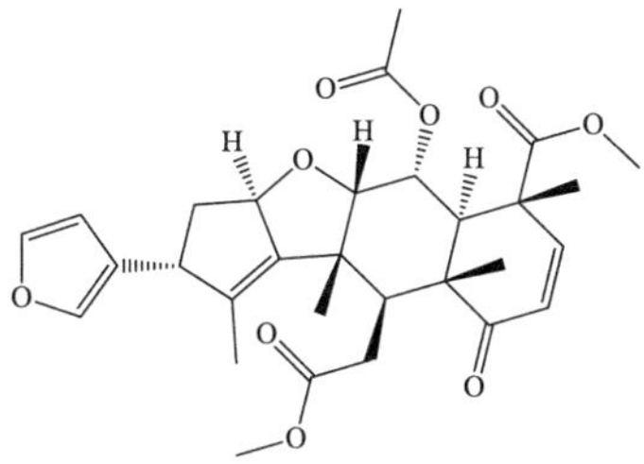

Nimbin (227)

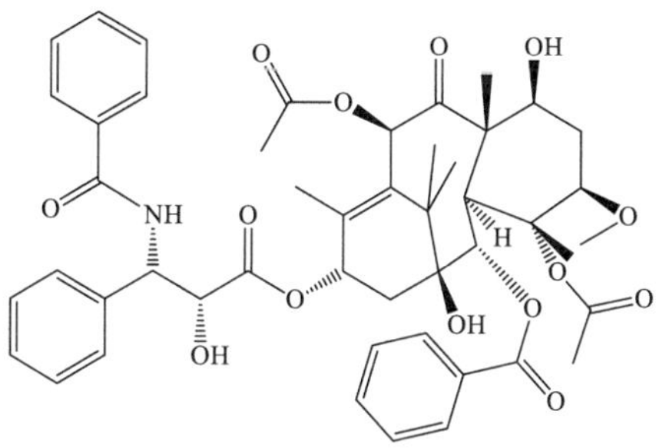

Paclitaxel (228)

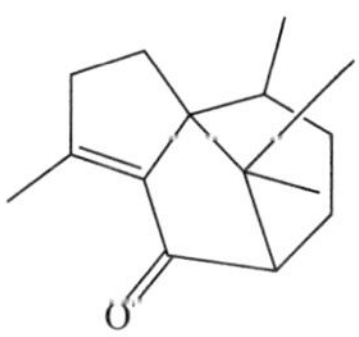

Patchoulenone (229)

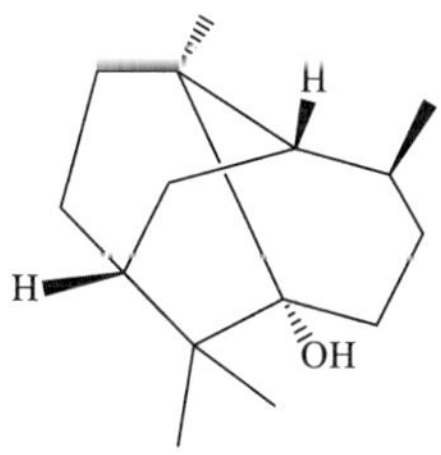

Patchouli alcohol (230)

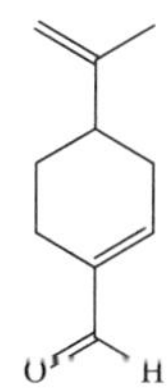

Perillaldehyde (231)

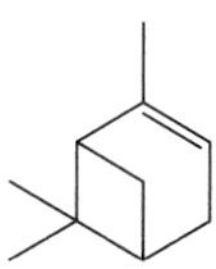

Pinene (232)

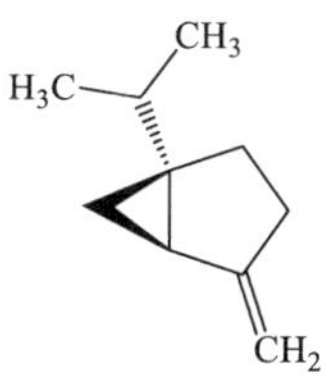

Sabinene (233)

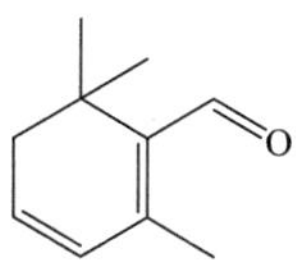

Safranal (234)

Fig. 1.8 (continued)

Fig. 1.8 (continued)

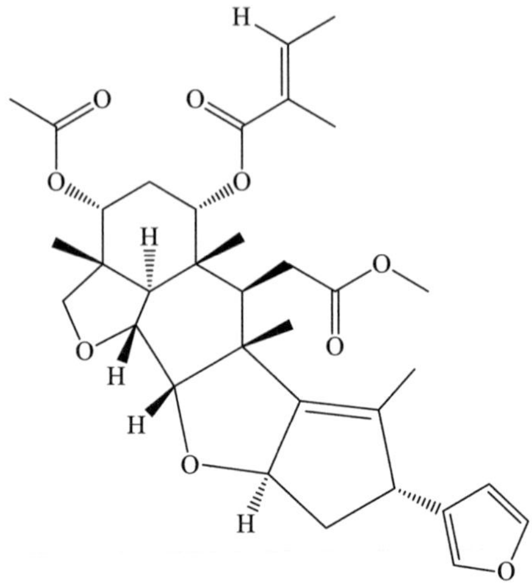

Salannin (235)

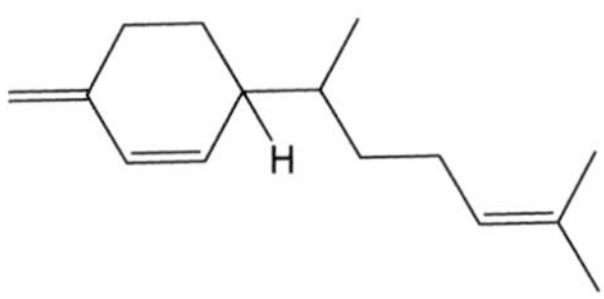

Sesquiphellandrene (236)

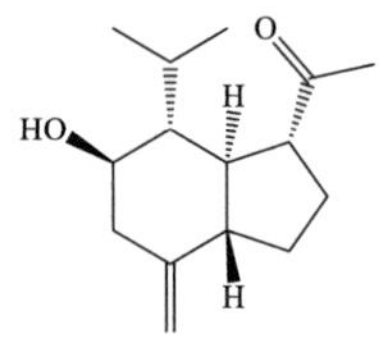

Sesquiterpenes (237)

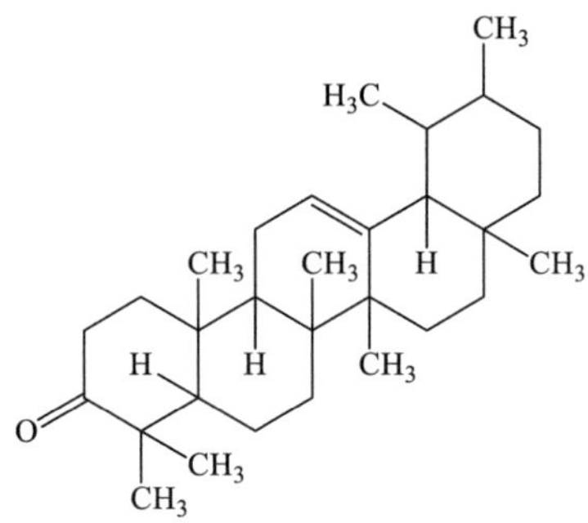

Shorea acid (238)

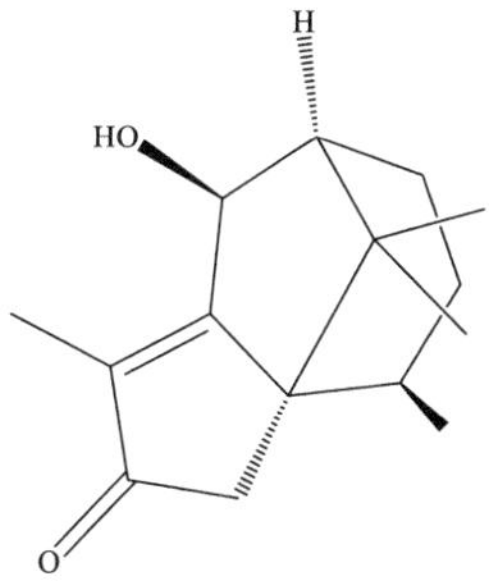

Sugeonol (239)

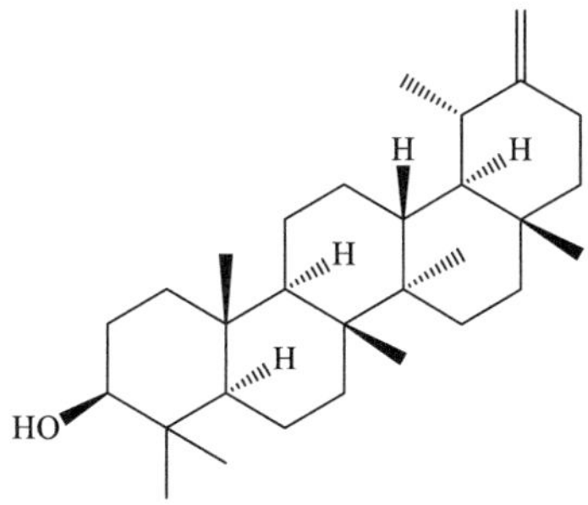

Taraxasterol (240)

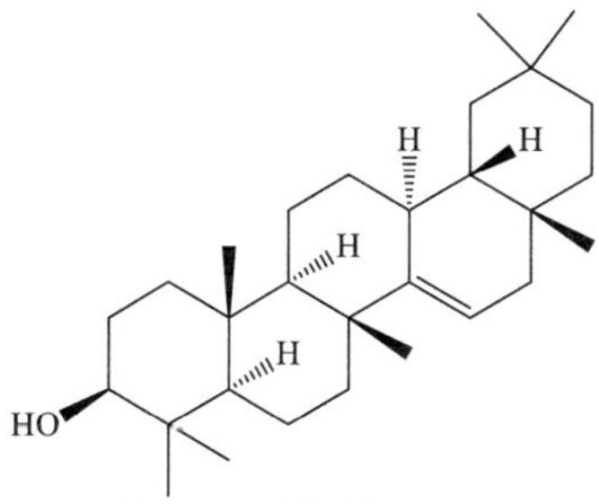

Taraxerol (241)

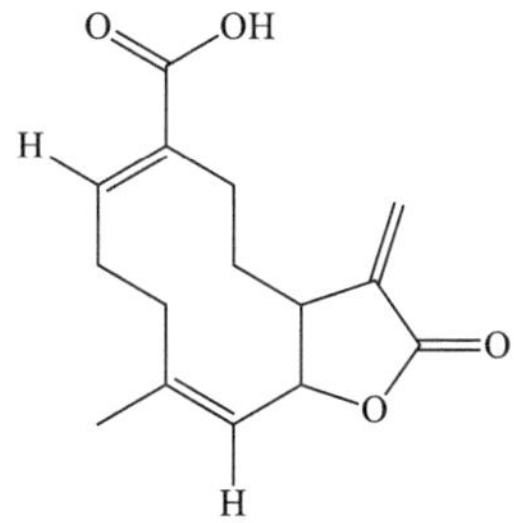

Taraxinic acid (242)

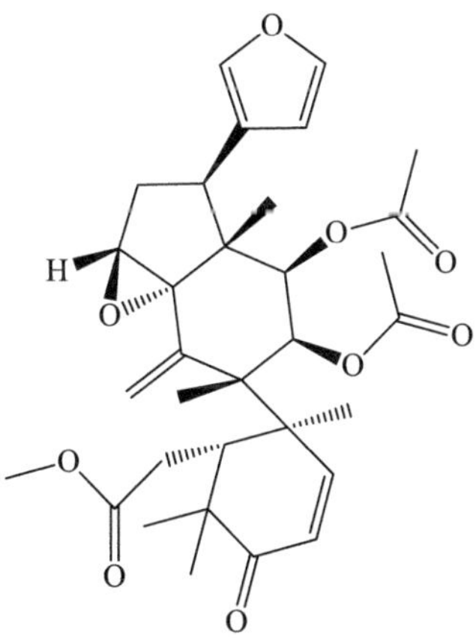

Toonacilin (243)

Fig. 1.8 (continued)

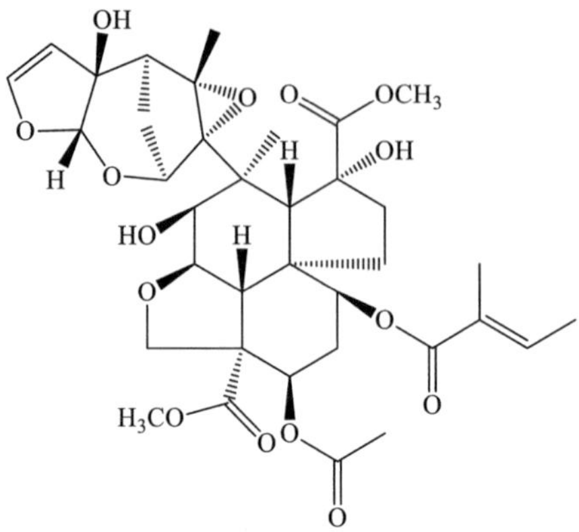

Toonafolin (244)

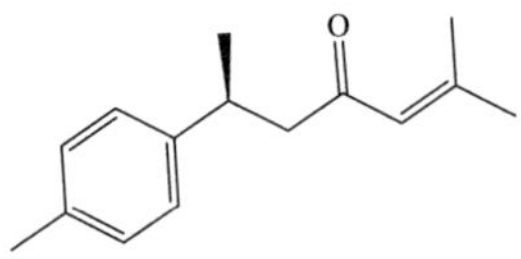

Turmerone (245)

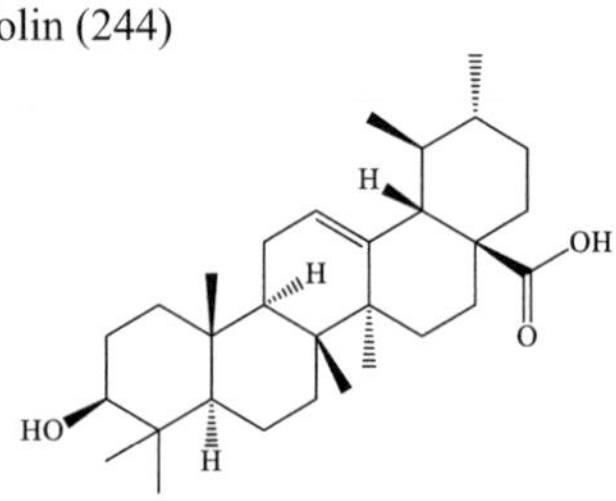

Ursolic acid (246)

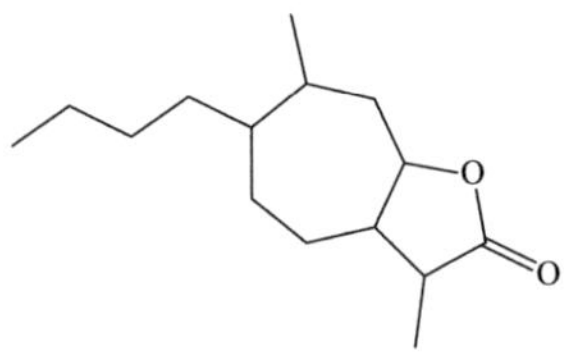

Xanthanolides (247)

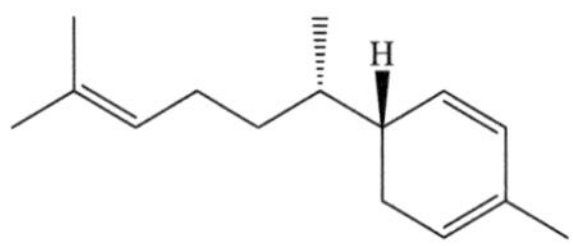

Zingiberene (248)

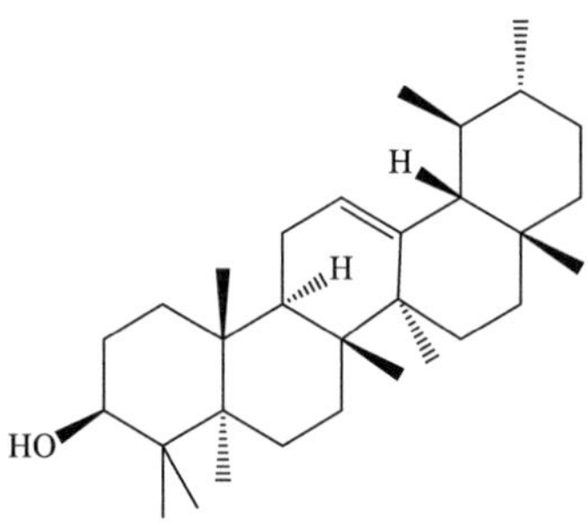

α-amyrin (249)

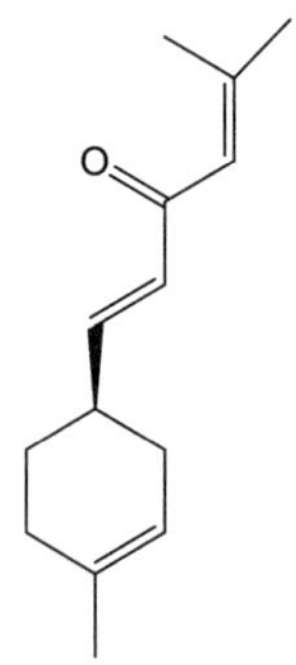

α-Atlantone (250)

Fig. 1.8 (continued)

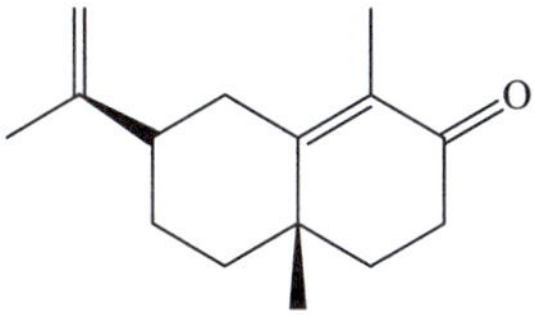

α-cyperone (251)

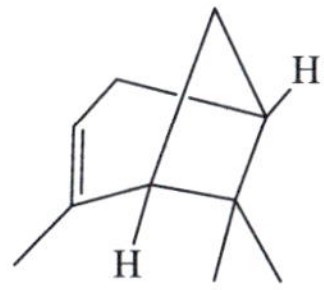

α-pinene (252)

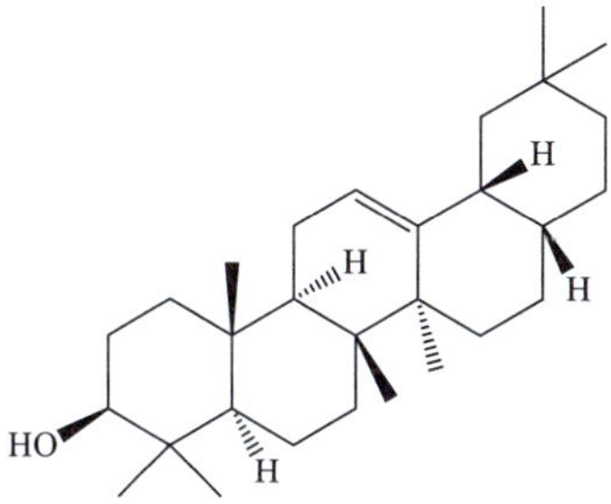

β-amyrin (253)

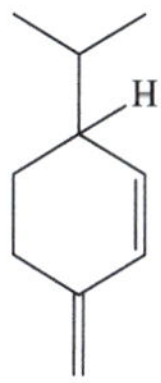

β-phellandrene (254)

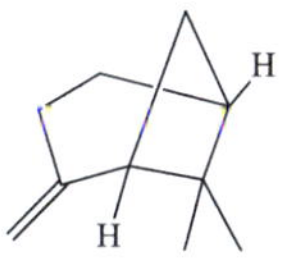

β-pinene (255)

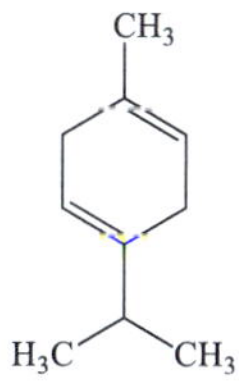

γ-terpinene (256)

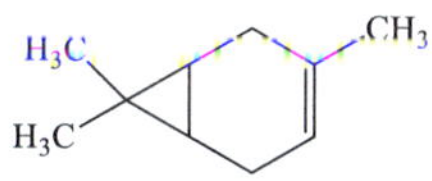

Delta carene (257)

Fig. 1.8 (continued)

Natural dyes and pigments are used for coating medicinal tablets and tinting gelatin capsules. They are used for product identification, to enhance patient adherence, and to conceal unattractive active ingredients (Al Fahad et al. 2024). In addition to these, numerous dyes and pigments provide direct medical benefits and establish a link between the drug's efficacy and appearance.

1.5 Advantages and Disadvantages of Natural Colouring

Natural dyes and pigments are typically obtained from plants, minerals, or animals and are not toxic or hazardous. They are biodegradable and have several benefits for health. However, natural sources are often obtained from agricultural products, and they are limited in supply, so they cannot fulfil high demands, and this makes their supply inconsistent. They are expensive and give less colour brilliance, and their reproducibility is difficult compared to synthetic dyes (Bechtold and Mussak 2009).

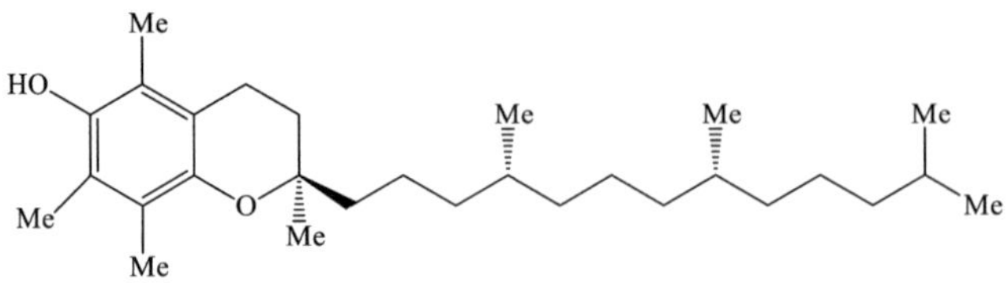

α-Tocopherol (258)

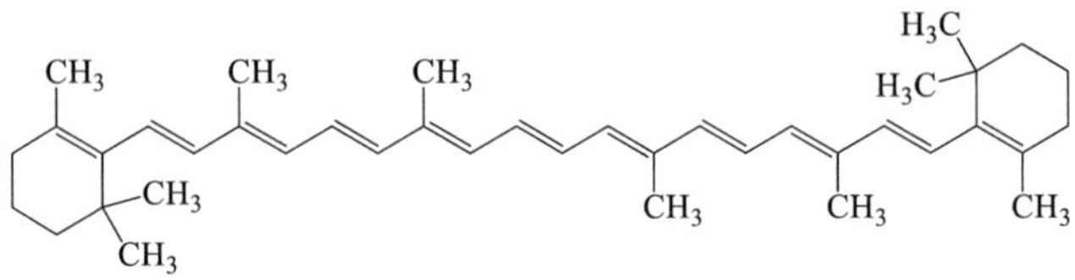

β-carotene (259)

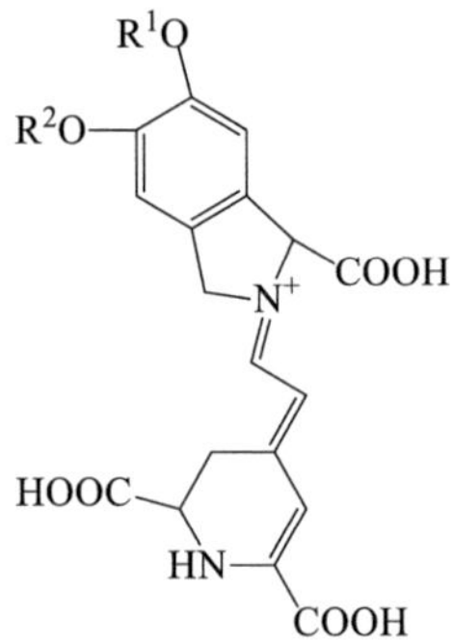

β-cyanins (260)

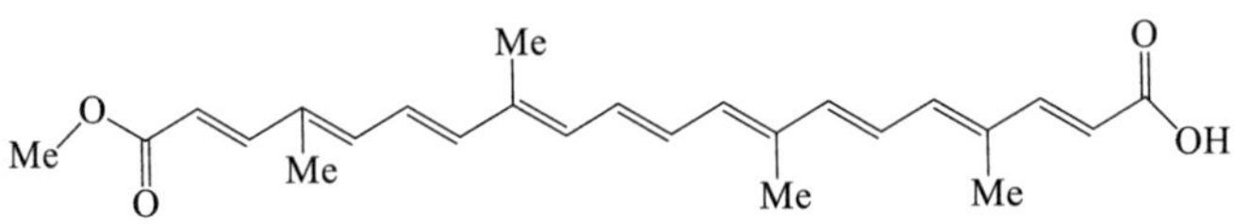

Bixin (261)

Fig. 1.9 Chemical structures of vitamins and pigments (258–274)

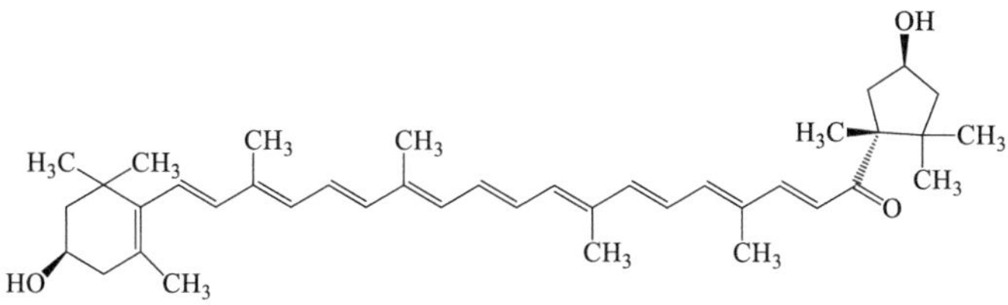

Capsanthin (262)

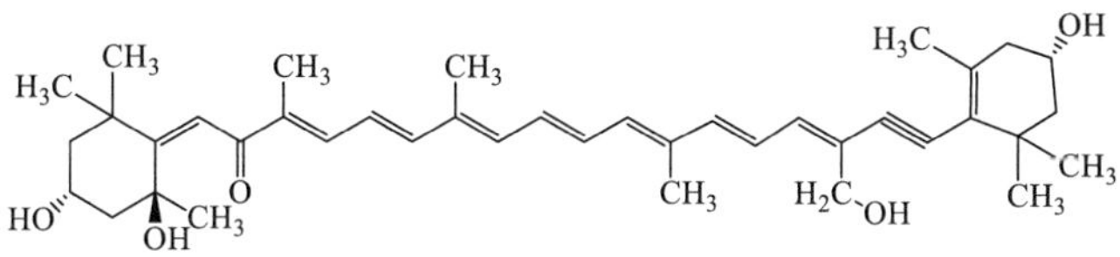

Carotenoids (263)

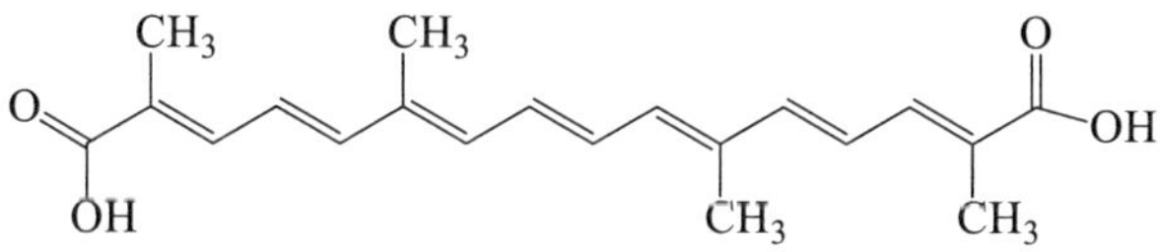

Crocetin (264)

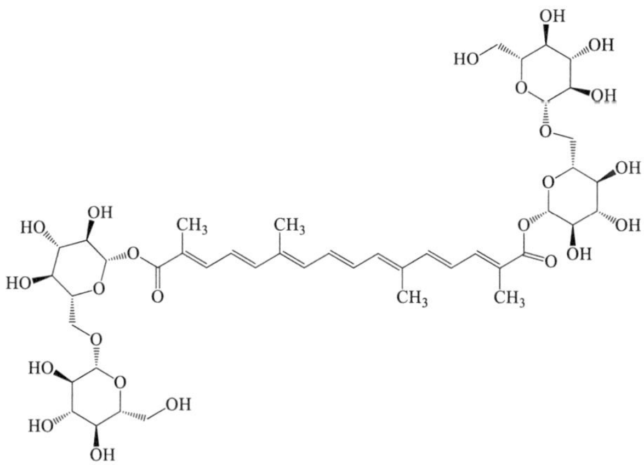

Crocin (265)

Fig. 1.9 (continued)

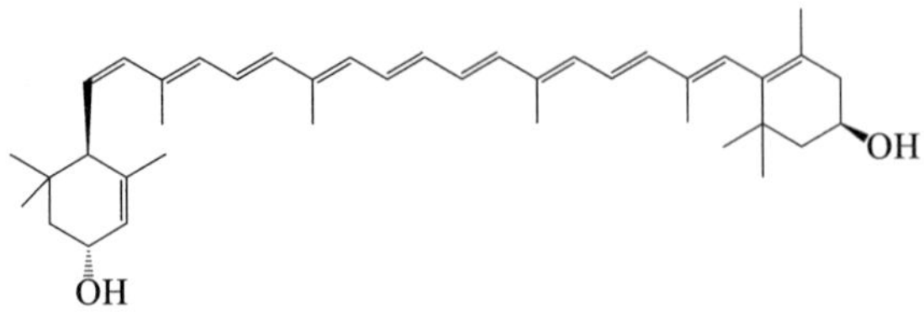

Lutein (266)

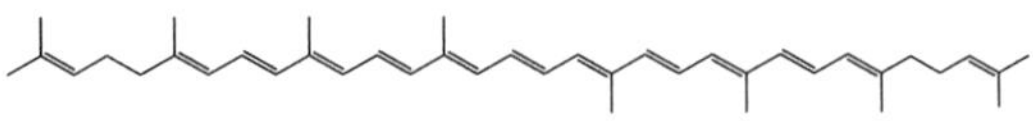

Lycopene (267)

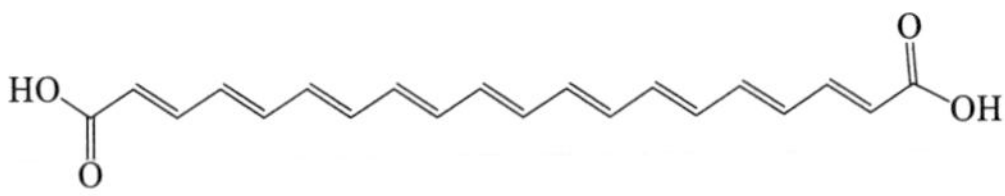

Norbixin (268)

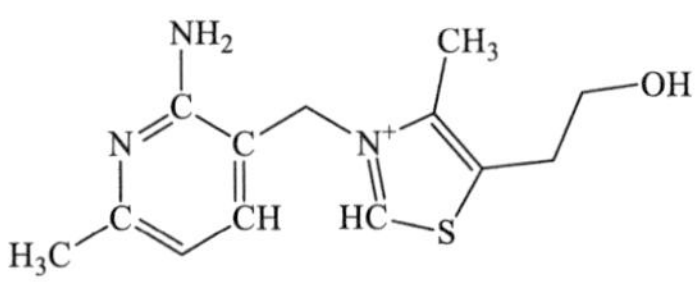

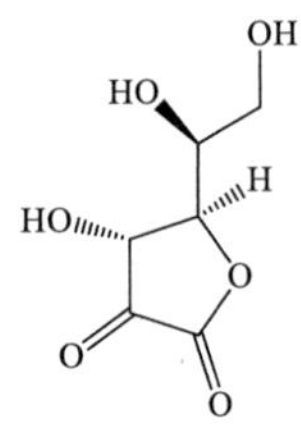

Vitamin B complex (269) Vitamin C or Ascorbic acid (270)

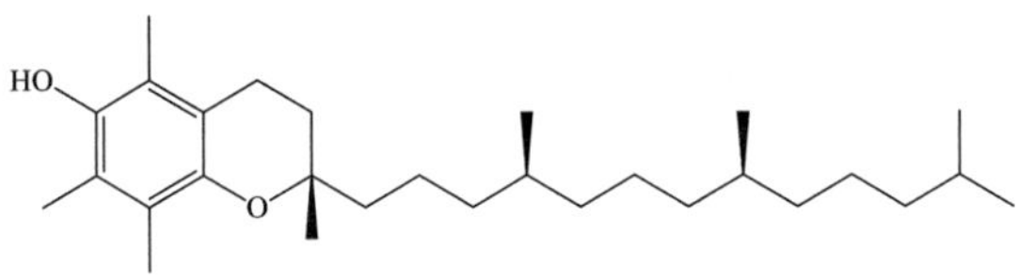

Vitamin E (271)

Fig. 1.9 (continued)

Fig. 1.9 (continued)

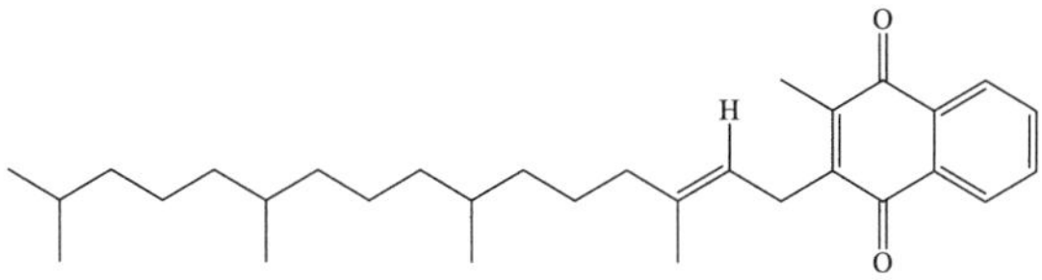

Vitamin K (272)

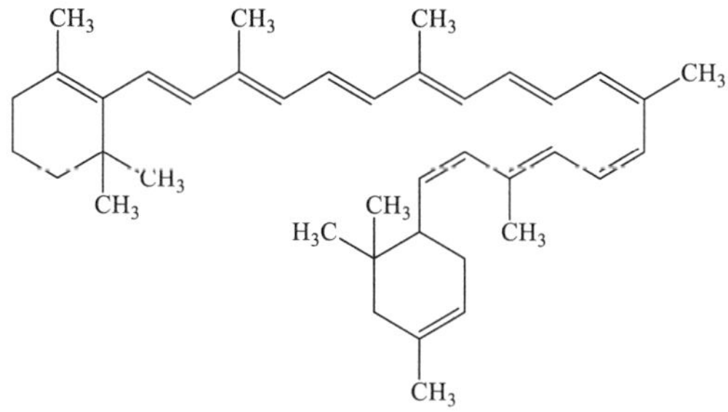

α-carotene (273)

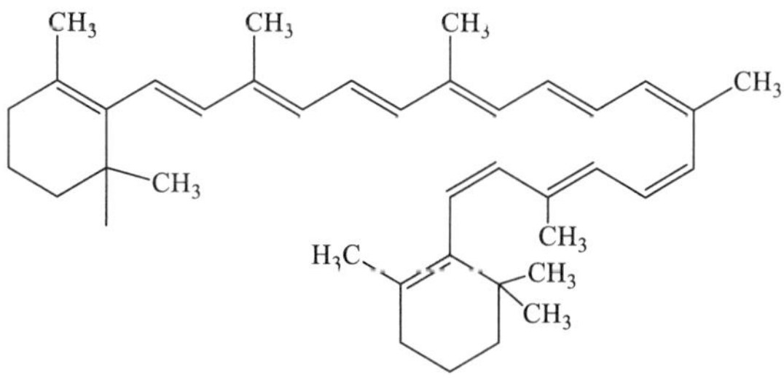

β-carotene (274)

1.6 Conclusions

Natural dyes and pigments are attractive, and their use is found mainly in traditional dyeing industries. They are a good source of income for local farmers and indigenous communities. These natural dyes and pigments are biodegradable, non-toxic, and hypoallergenic. They have medicinal benefits and have a low carbon footprint due to the simple technology used during their production. Thus, natural dyes and pigments are widely used in the food and pharmaceutical industries. However, these natural dyes and pigments are difficult to produce, as they deteriorate due to changes in light, heat, and pH conditions. In addition, their colours are less vibrant than those of synthetic dyes and pigments.

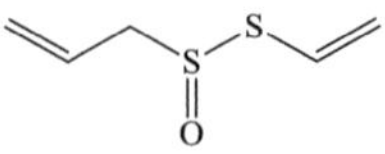

Allicin (275)

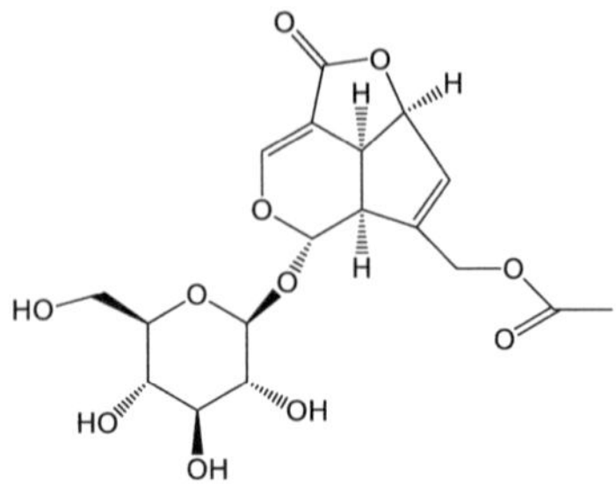

Asperuloside (276)

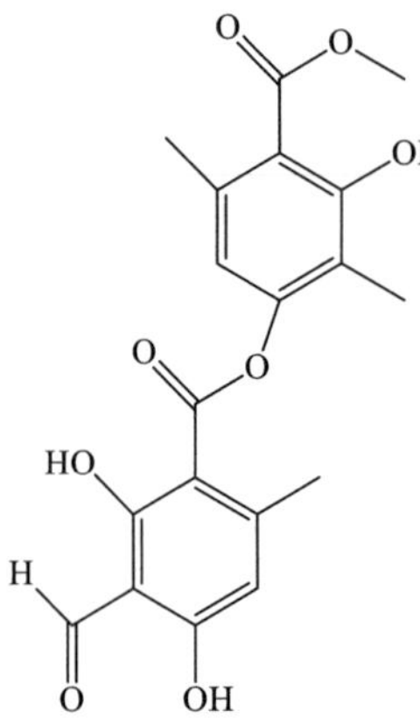

Atranorin (277)

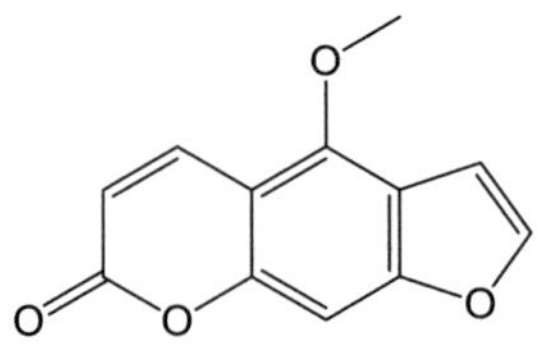

Bergapten (278)

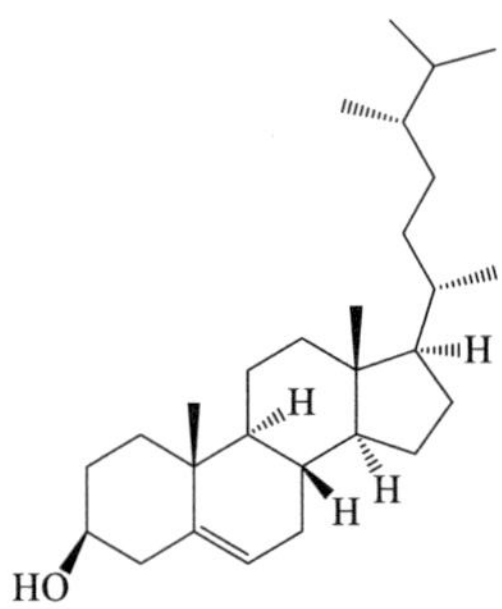

Campesterol (279)

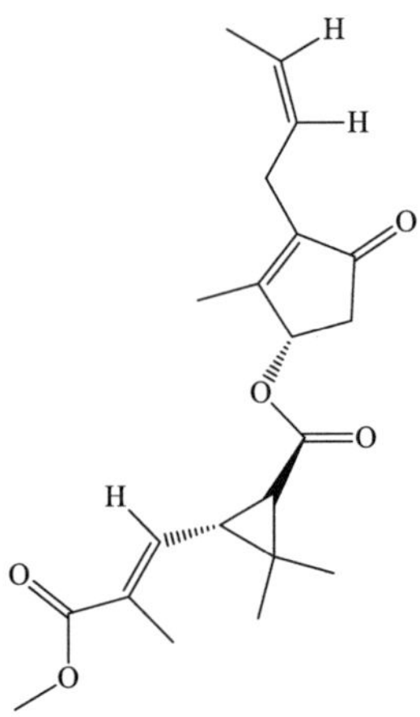

Cinerins (280)

Fig. 1.10 Chemical structures of other compounds (275–316)

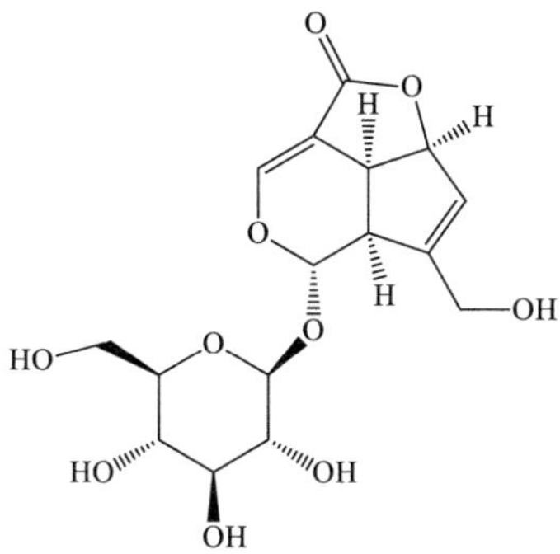

Deacetylasperuloside (281)

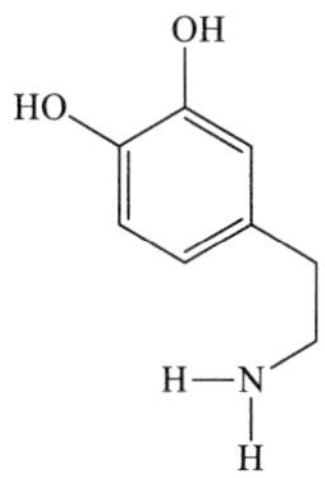

Dopamine (282)

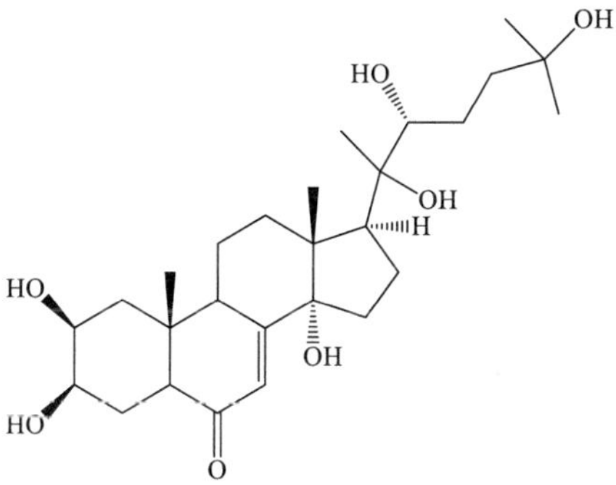

Ecdysterone (283)

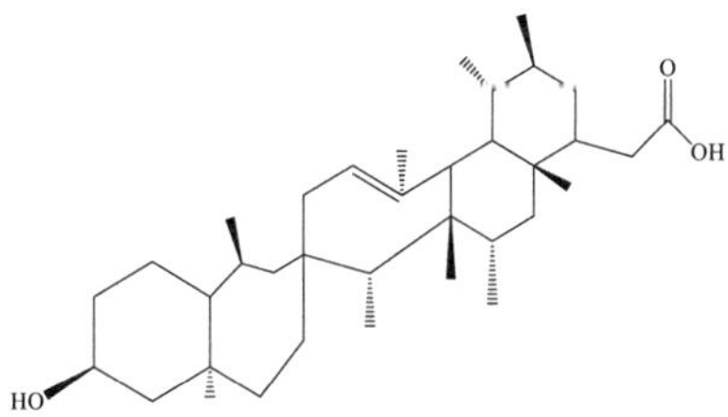

Engelhardic acid (284)

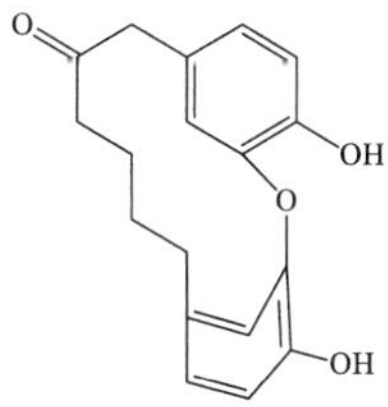

Engelhardione (285)

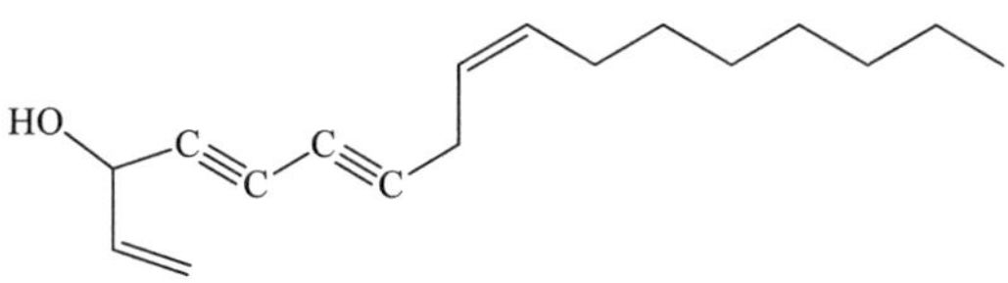

Falcarinol (286)

Fig. 1.10 (continued)

Fig. 1.10 (continued)

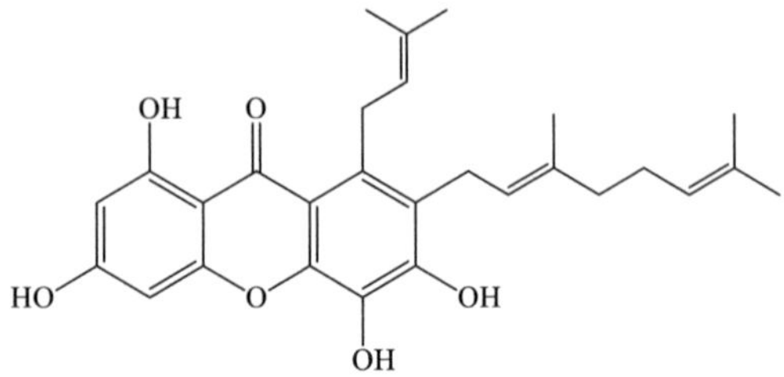

Garciniaxanthone (287)

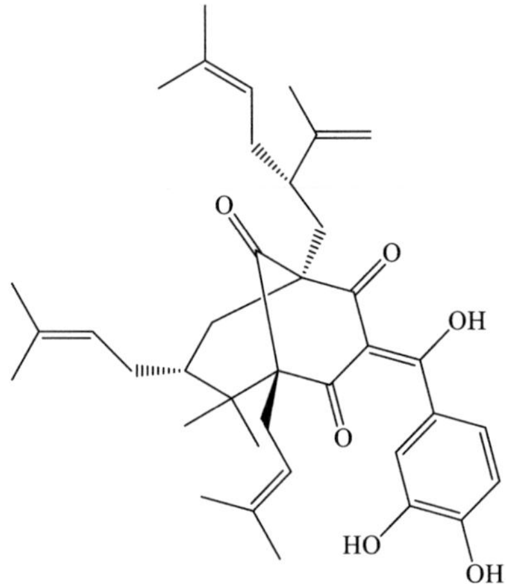

Garcinol (288)

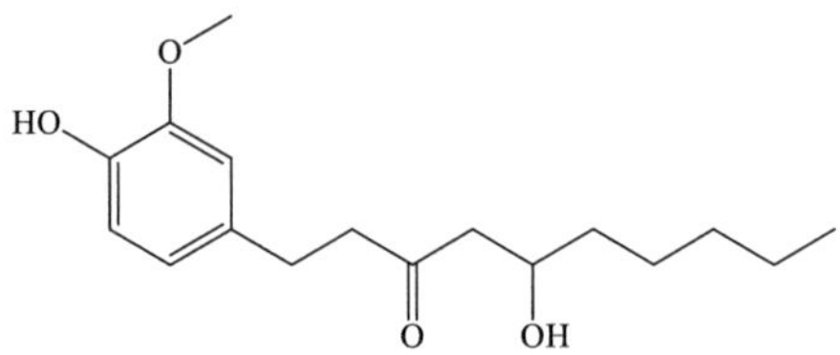

Gingerol (289)

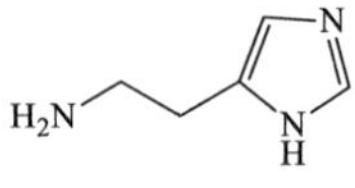

Histamine (290)

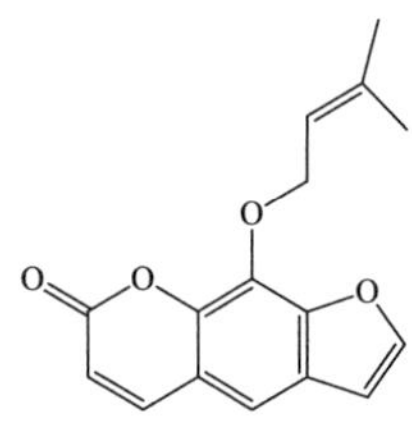

Imperatorin (291)

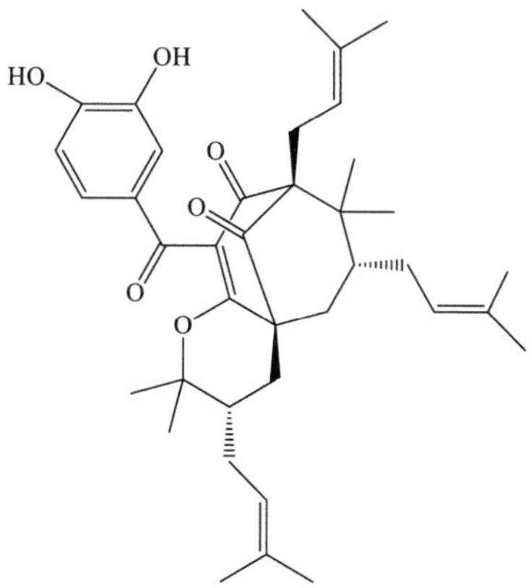

Isogarcinol (292)

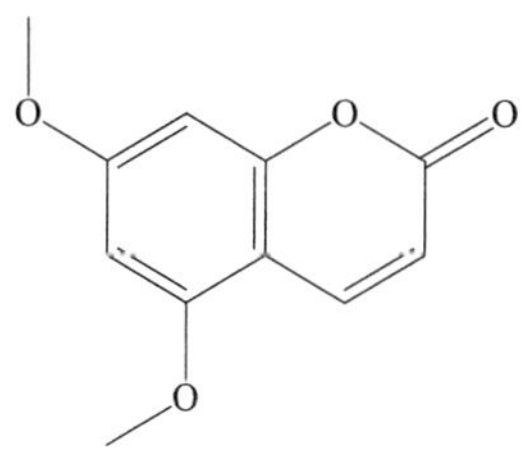

Limettin (293)

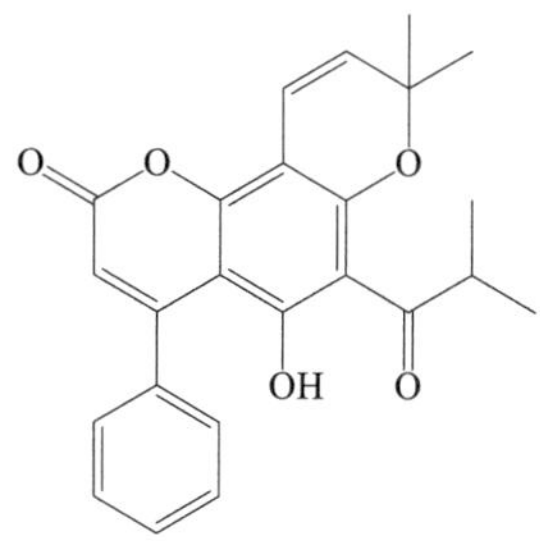

Mammeigin (294)

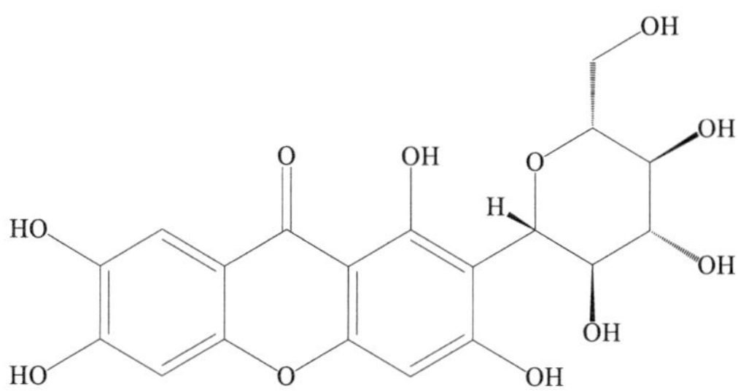

Mangiferin (295)

Fig. 1.10 (continued)

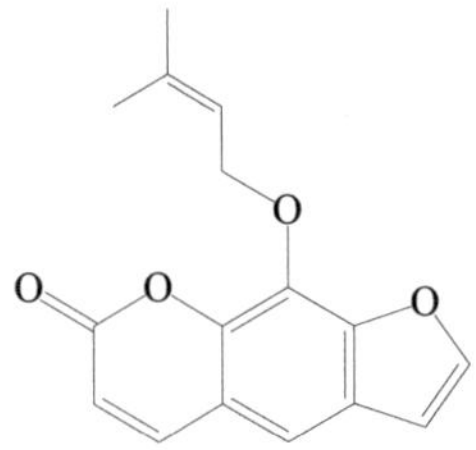

Marmelosin (296)

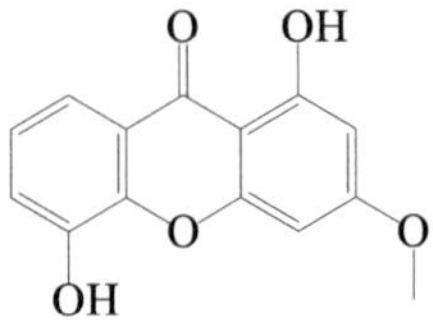

Mesuaxanthone a (297)

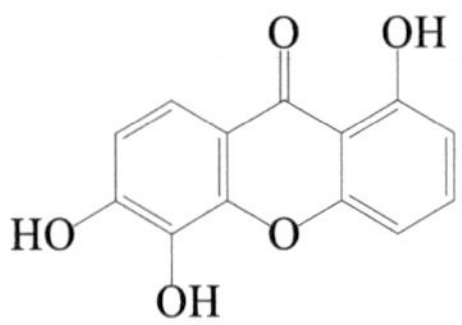

Mesuaxanthone b (298)

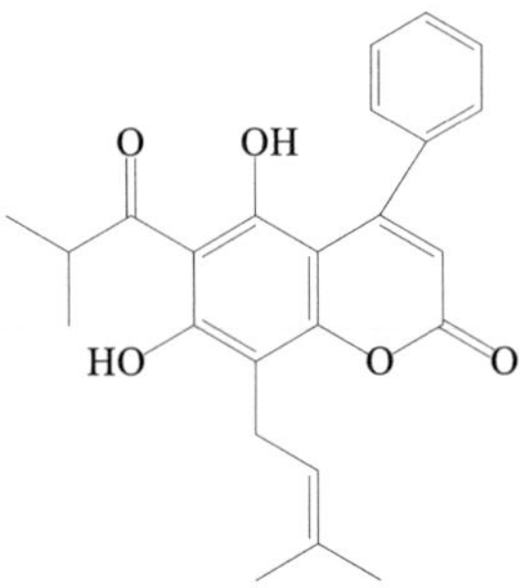

Mesuol (299)

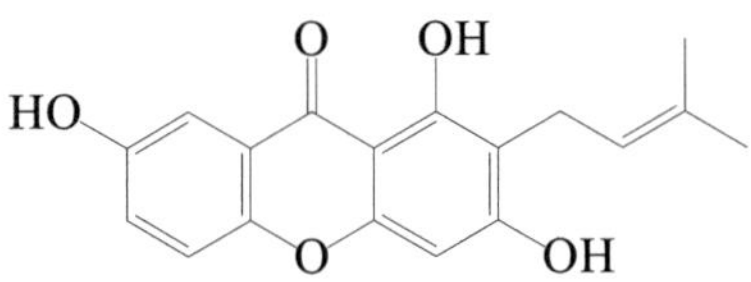

Mesuone [Xanthone derivative: 1,3,7-Trihydroxy-2-(3-methylbut-2-enyl)] (300)

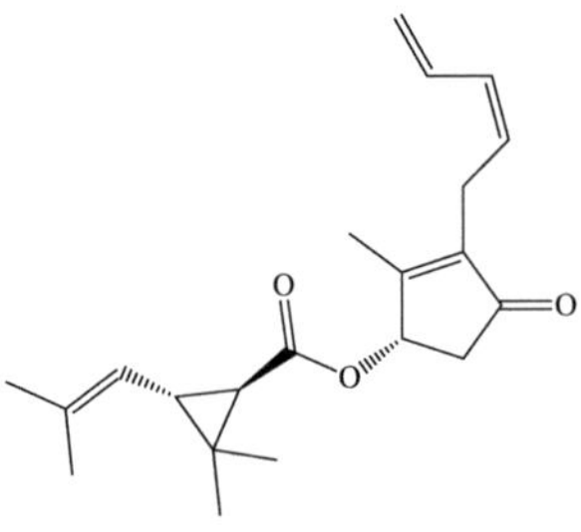

Pyrethrin I (301)

Fig. 1.10 (continued)

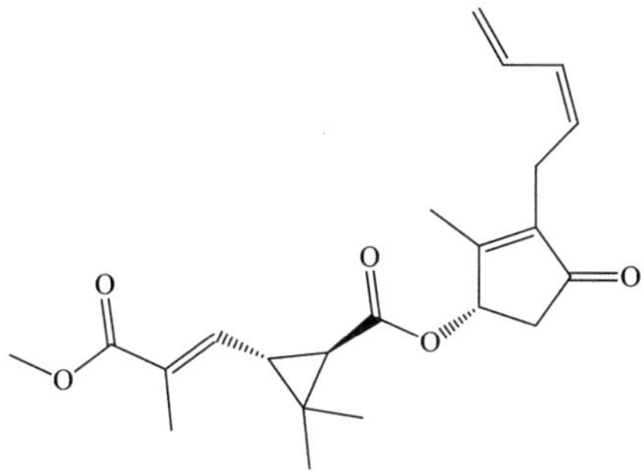

Pyrethrin II (302)

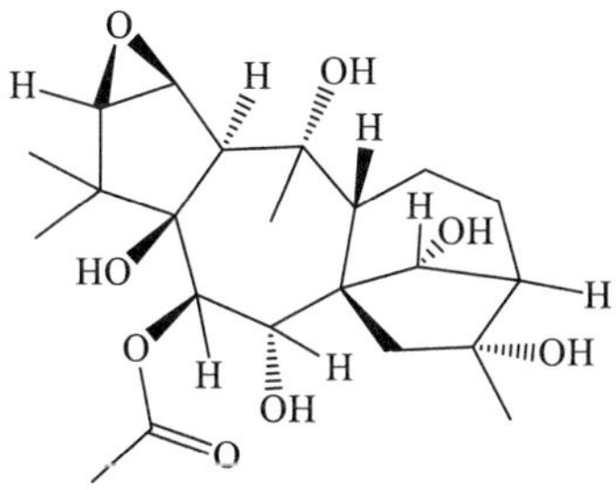

Ricin (303)

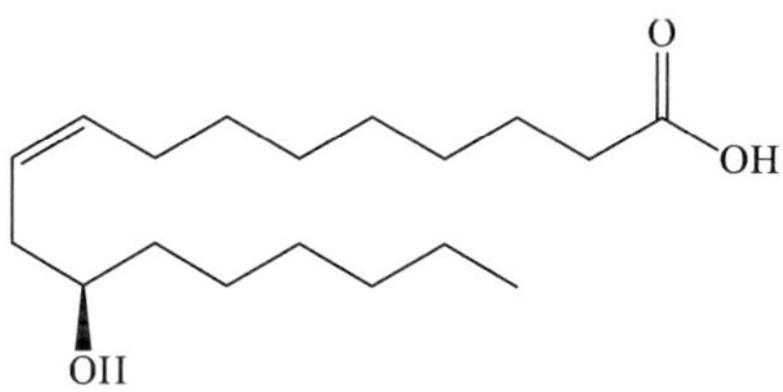

Ricinoleic acid (304)

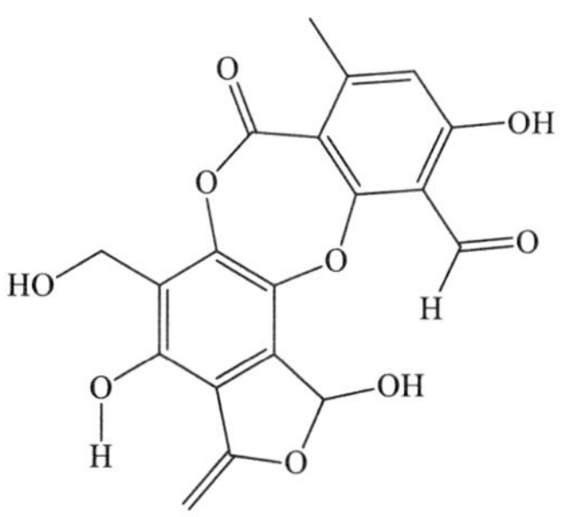

Salazinic acid (305)

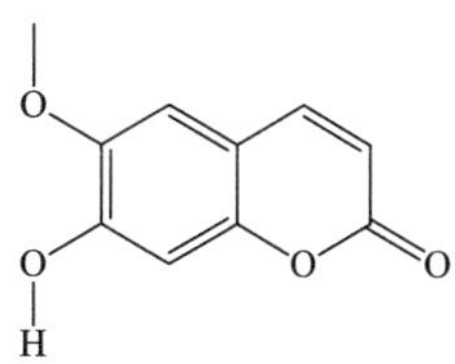

Scopoletin (306)

Fig. 1.10 (continued)

Fig. 1.10 (continued)

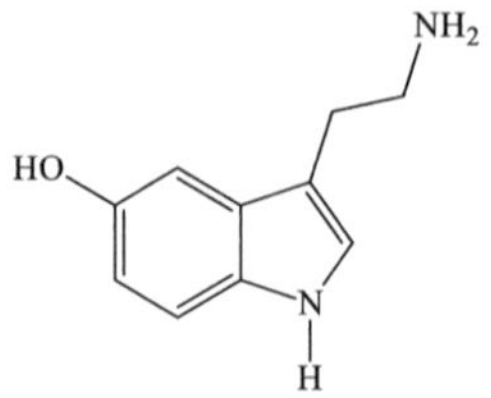

Serotonin (307)

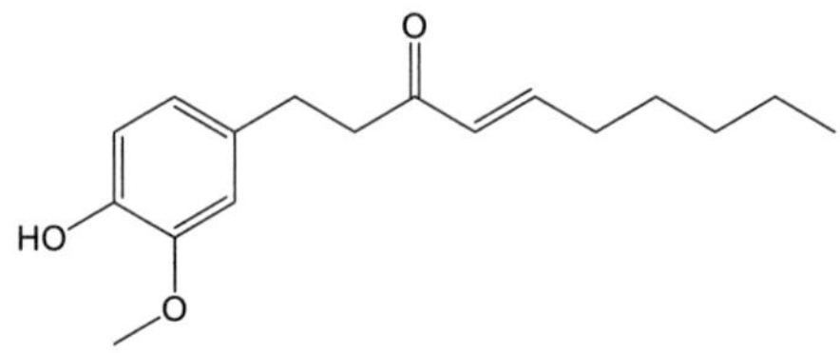

Shogaols (308)

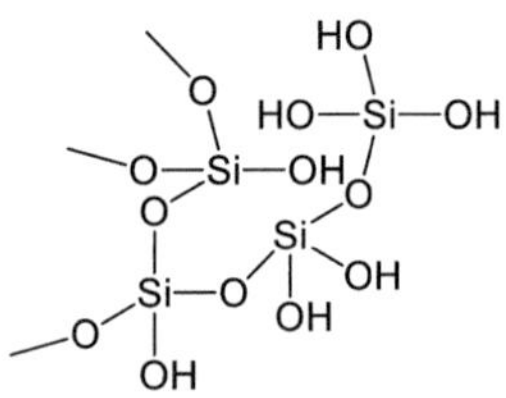

Silica (309)

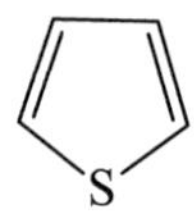

Thiophenes (310)

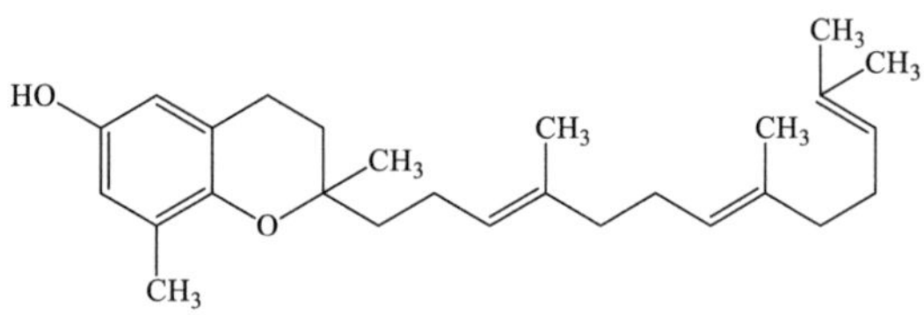

Tocotrienols (311)

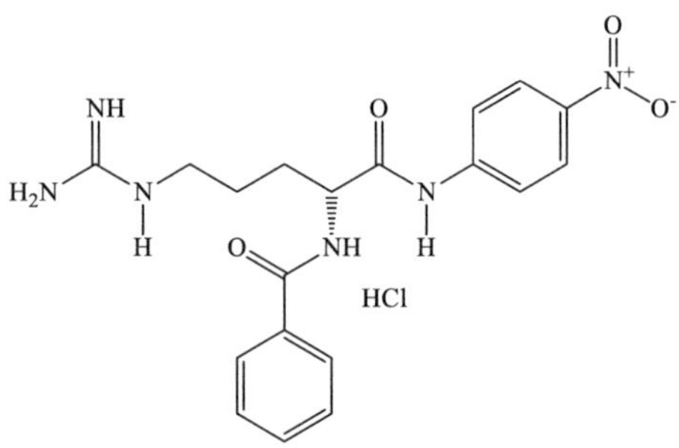

Trypsin inhibitors (312)

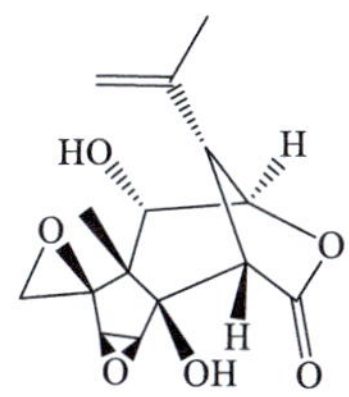

Tutin (313)

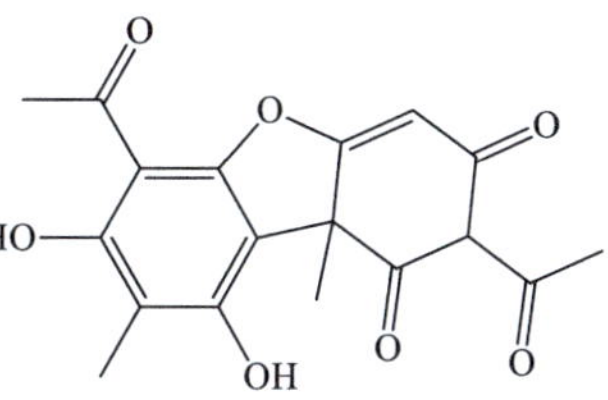

Usnic acid (314)

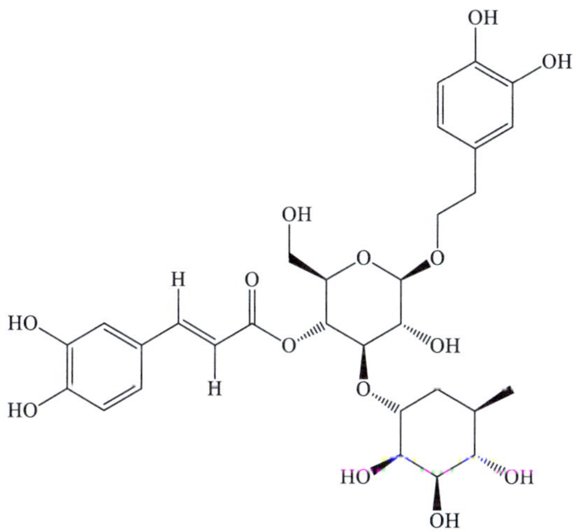

Verbascoside (315)

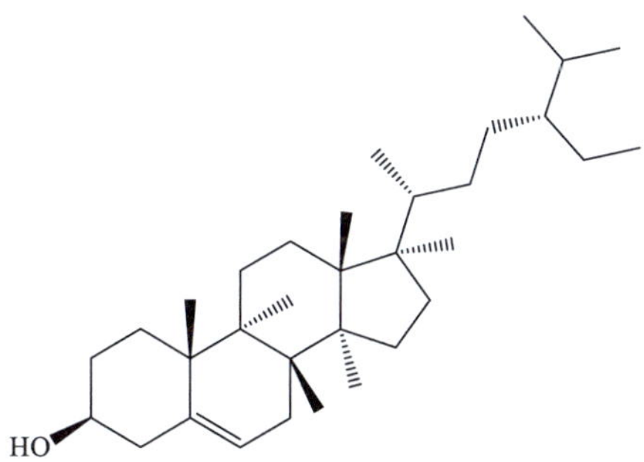

β-Sitosterol (316)

Fig. 1.10 (continued)

References

Abdel-Kareem O (2012) History of dyes used in different historical periods of Egypt. Res J Text Appar 16:79–92. https://doi.org/10.1108/RJTA-16-04-2012-B009

Al Fahad MdA, Ahamed R, Ahmed T et al (2024) Pharmaceutical applications of natural dyes and pigments. In: Ul Islam S (ed) Renewable dyes and pigments. Elsevier, pp 165–175

Alem F-Z, Gita SA, Cougnaud L, et al (2020) Lawsone quantification in *Lawsonia inermis* L. by HPLC-MS: How does the temperature and pluviometry affect lawsone concentration? Industrial Crops and Products 158:112960. https://doi.org/10.1016/j.indcrop.2020.112960

Alemayehu T, Teklemariam Z (2014) Application of natural dyes on textiles: a review. Int J Res 2:61–68. https://doi.org/10.29121/granthaalayah.v2.i2.2014.3067

Anonymous (2025) Definition of DYEING. https://www.merriam-webster.com/dictionary/dyeing. Accessed 8 Oct 2025

Balfour-Paul J (2013) Indigo in South and South-East Asia. Text Hist 98–112. https://doi.org/10.1179/004049699793710688

Baqri SSR, Omkar (2023) Dye-yielding insects. In: Commercial insects. CRC Press

Bechtold T, Mussak R (2009) Handbook of natural colorants. John Wiley & Sons

Bhandari NL, Bist K, Ghimire J et al (2020a) Feasibility study of the Euphorbia pulcherrima plant extract as datural dye with different mordants for fabric dyeing. J Inst Sci Technol 25:30–36. https://doi.org/10.3126/jist.v25i1.29421

Bhandari NL, Pokhrel B, Bhandari U et al (2020b) An overview of research on plant based natural dyes in Nepal: scope and challenges. J Agric Nat Resour 3:45–66. https://doi.org/10.3126/janr.v3i2.32328

Bhandari P, Shrestha K, KC A, Shrestha KK (2020c) Three new records of orchids from Nepal. J Jpn Bot 95:41–46

Chaudhary RP, Nepal M, Gupta VNG, Vetaas OR (2002) Traditional use of plants by the indigenous people of Makalu-Barun Region, Eastern Nepal. In: Chaudhary RP, Subedi BP, Vetaas OR, Aase TH (eds) Vegetation and society: their interaction in the himalayas. Tribhuvan University, Nepal, and the University of Bergen, Norway, pp 83–97

Cheng Y, Li P, Yang Y, et al (2017) Microsatellites for *Phytolacca acinosa* (Phytolaccaceae), a traditional medicinal herb. Applications in Plant Sciences 5:1700028. https://doi.org/10.3732/apps.1700028

Dangol DR (2008) Traditional uses of plants of common land habitats in Western Chitwan Nepal. J Inst Agric Anim Sci 29:71–78

Dangol DR (2002) Economic uses of forest plant resources in western Chitwan Nepal. Banko Janakari 12:56–64

Degano I, Ribechini E, Modugno F, Colombini MP (2009) Analytical methods for the characterization of organic dyes in artworks and in historical textiles. Appl Spectrosc Rev 44:363–410. https://doi.org/10.1080/05704920902937876

Derby CD (2014) Cephalopod Ink: production, chemistry, functions and applications. Mar Drugs 12:2700–2730. https://doi.org/10.3390/md12052700

Feeser A, Goggin MD, Tobin BF (2012) The materiality of color: the production, circulation, and application of dyes and pigments. Ashgate Publishing, Ltd., pp 1400–1800

Fujiwara Y, Kono M, Ito A, Ito M (2018) Anthocyanins in perilla plants and dried leaves. Phytochemistry 147:158–166. https://doi.org/10.1016/j.phytochem.2018.01.003

Gaboriaud-Kolar N, Nam S, Skaltsounis A-L (2014) A colorful history: the evolution of Indigoids. In: Kinghorn AD, Falk H, Kobayashi J (eds) Progress in the chemistry of organic natural products 99. Springer International Publishing, Cham, pp 69–145

Ghimire SK, Lama YC, Tripathi GR et al (2001) Conservation of the plant resources, community development and training in applied ethnobotany at shey phoksundo national park and its bufferzone, Dolpa. WWF Nepal, Kathmandu, Nepal

Gilbert (nee Stoker) KG, Cooke DT (2001) Dyes from plants: Past usage, present understanding and potential. Plant Growth Regul 34:57–69. https://doi.org/10.1023/A:1013374618870

Gleba M, Vanden Berghe I, Aldenderfer M (2016) Textile technology in Nepal in the 5–7th centuries CE: the case of Samdzong. STAR Sci Technol Archaeol Res 2:25–35. https://doi.org/10.1080/20548923.2015.1110421
Grae D (1974) Nature's colors: dyes from plants, First. Macmillan Publishing Co., New York, USA
Gupta VK (2019) Fundamentals of natural dyes and its application on textile substrates. In: Samanta, Awwad N, Algarni HM (eds) Chemistry and technology of natural and synthetic dyes and pigments. IntechOpen, London, United Kingdom, p 04
Gürses A, Açıkyıldız M, Güneş K, Gürses MS (2016) Classification of dye and pigments. In: Gürses A, Açıkyıldız M, Güneş K, Gürses MS (eds) Dyes and pigments. Springer International Publishing, Cham, pp 31–45
Hansen H (2011) The quest for quercitron: revealing the story of a forgotten dye. University of Delaware, USA
Harmon J, Siddika A (2022) A new purple?: Color exploration and extraction from the catkins of cottonwood trees. International Textile and Apparel Association Annual Conference Proceedings 79. https://doi.org/10.31274/itaa.15887
Hassan SHA (2024) An analytical vision of using Indian yellow in the Mughal painting school and Pahari school (11-13AH / 17-19 AD) and its impact on European painting. J Abydos 6:138–173. https://doi.org/10.21608/abidus.2025.410613
Iqbal S, Ansari TN (2021) Extraction and application of natural dyes. In: Sustainable practices in the textile industry. Wiley, pp 1–40
IUCN (2004) National register of medicinal and aromatic plants (Revised & updated). The World Conservation Union (IUCN), Kathmandu, Nepal
Johnston W (2008) The discovery of aniline and the origin of the term "aniline dye." Biotech Histochem 83:83–87. https://doi.org/10.1080/10520290802136793
Kenoyer JM (1984) Shell working industries of the indus civilization: a summary. Paléorient 10:49–63
Kumar A, Dixit U, Singh K et al (2021) Structure and properties of dyes and pigments. In: Papadakis R (ed) Dyes and pigments—novel applications and waste treatment, 1st edn. IntechOpen, London, United Kingdom, p 216
Kunwar RM, Uprety Y, Burlakoti C et al (2009) Indigenous use and ethnopharmacology of medicinal plants in far-west Nepal. Ethnobot Res Appl 7:005–028
Lama YC, Ghimire SK, Aumeeruddy-Thomas Y (2001) Medicinal plants of Dolpo: Amchis' knowledge and conservation. Worldwide Fund for Nature Conservation (WWF) Nepal, Kathmandu, Nepal
Li N, Wang Q, Zhou J et al (2022) Insight into the progress on natural dyes: sources, structural features, health effects, challenges, and potential. Molecules 27:3291. https://doi.org/10.3390/molecules27103291
Luong NT, Hop NV, Quy NV, Hoan VM (2023) Diversity of plant species for food coloring in Vietnam. Nusant Biosci 15:95–105. https://doi.org/10.13057/nusbiosci/n150112
Magray JA, Wani BA, Ganie AH et al (2025) From folklore to pharmacology: a comprehensive review of *Phytolacca acinosa* Roxb. A multipurpose medicinal herb. Fitoterapia 182:106446. https://doi.org/10.1016/j.fitote.2025.106446
Malaisse F, Claus W, Drolkar P et al, Ü Ethnomycology and Ethnobotany (South Central Tibet). Diversity, with emphasis on two underrated targets: plants used for dyeing and incense. Geo-Eco-Trop 36:185–199
Malla L (1993) The natural dyes of Nepal. Research Centre for Applied Science & Technology, Tribhuvan University, Kirtipur, Nepal
Malla SB (1994) Medicinal herbs in the Bagmati zone. International Centre for Integrated Mountain Development (ICIMOD), Kathmandu, Nepal
Manandhar NP (2002) Plants and people of Nepal. Timber Press, Portland, Oregon, USA
Maulik S, Chakraborty L, Pandit P (2021) Evaluation of cellulosic and protein fibers for coloring and functional finishing properties using simultaneous method with *Eucalyptus* bark extract as a natural dye. Fibers Polym 22:711–719. https://doi.org/10.1007/s12221-021-0092-0

Mollon JD (2003) The origins of modern color science. In: Shevell SK (ed) The science of color. Elsevier Science, Second Edition, pp 1–39

Mullen L (2008) Cave art and the origins of typography. Vis Commun Q 15:6–16. https://doi.org/10.1080/15551390801914546

Nambela L (2023) The potentials of plant-based colorants for sustainable textile dyeing industry. Res J Text Appar 29:132–148. https://doi.org/10.1108/RJTA-04-2023-0043

Nautiyal S, Kaul AK (eds) (2003) Non-timber forest products of India. Jyoti Publishers and Distributors, New Delhi, India

Önal A, Özbek O, Tombul KC, Nached S (2021) Investigation of the dyeing properties of cotton fabrics and wool yarns using Prunus persica leaf extract. J Indian Chem Soc 98:100092. https://doi.org/10.1016/j.jics.2021.100092

Pal P (1974) The arts of Nepal II.: painting. Brill, Leiden, The Netherlands

Pandey MR (2006) Use of medicinal plants in traditional Tibetan therapy system in upper Mustang Nepal. Our Nat 4:69–82

Pohle P (1990) Useful plants of manang district: a contribution to the ethnobotany of the Nepal-Himalaya. Franz Steiner Verlag Wisbaden GMBH, Stuttgart

Poudel RC, Gao L-M, Möller M et al (2013) Yews (*Taxus*) along the Hindu Kush-Himalayan region: exploring the ethnopharmacological relevance among communities of Mongol and Caucasian origins. J Ethnopharmacol 147:190–203. https://doi.org/10.1016/j.jep.2013.02.031

Pradhan SP, Adhikari K, Nepal S, Pandey BP (2020) Determination of sun protective factor of selected medicinal plants from western Nepal. J Nepal Chem Soc 41:51–55. https://doi.org/10.3126/jncs.v41i1.30487

Pyakurel D, Baniya A (2011) Impetus for conservation and livelihood support in Nepal. A reference book on ecology, conservation, product development and economic analysis of selected NTFPs of langtang area in the sacred himalayan landscape. WWF Nepal, Kathmandu, Nepal

Pyakurel D, Smith-Hall C, Bhattarai-Sharma I, Ghimire SK (2019) Trade and conservation of Nepalese medicinal plants, fungi, and lichen. Econ Bot 73:505–521. https://doi.org/10.1007/s12231-019-09473-0

Rokaya MB, Maršík P, Münzbergová Z (2012) Active constituents in *Rheum acuminatum* and *Rheum australe* (Polygonaceae) roots: a variation between cultivated and naturally growing plants. Biochem Syst Ecol 41:83–90. https://doi.org/10.1016/j.bse.2011.11.004

Rokaya MB, Münzbergová Z, Timsina B (2010) Ethnobotanical study of medicinal plants from the Humla district of western Nepal. J Ethnopharmacol 130:485–504. https://doi.org/10.1016/j.jep.2010.05.036

Sagiv G (2015) Deep blue: notes on the Jewish Snail fight. Contemp Jew 35:285–313. https://doi.org/10.1007/s12397-015-9138-1

Sanai A, Nachai T, Merat K (2023) Extraction of natural dye from *Oroxylumb indicum* (L.) Kurz by using different solvents for Thai eri silk dyeing. J Appl Res Sci Technol JARST 22:251099. https://doi.org/10.60101/jarst.2023.251099

Sanda B, Liliana I (2021) Natural dye extraction and dyeing of different fibers: a review. In: Innovative and emerging technologies for textile dyeing and finishing. Wiley, pp 113–135

Sarkar PK, Sinha A, Das B, et al (2022) Kusum (*Schleichera oleosa* (Lour.) Oken): A potential multipurpose tree species, it's future perspective and the way forward. Acta Ecologica Sinica 42:565–571. https://doi.org/10.1016/j.chnaes.2021.04.003

Sengupta R (2019) An artisanal history of Kalam? J Text des Res Pract 7:25–37. https://doi.org/10.1080/20511787.2019.1580437

Sharma KK, Chowdhury AR, Srivastava S (2020) Chemistry and applications of Lac and its by-product. In: Kumar D, Shahid M (eds) Natural materials and products from insects: chemistry and applications. Springer International Publishing, Cham, pp 21–37

Shigwan AV, Khade AB, Hatpakki BC, Ghurghure SM (2013) A comprehensive review on *Pogostemon benghalensis* (Burm. F.) O. Kuntze. Res Rev J Pharmacogn Phytochem 1:10–15

Shrestha B (1994) Dye-yielding plants of Nepal. Research Centre for Applied Science & Technology, Tribhuvan University

Sigdel SR, Rokaya MB (2011) Utilization of plant resources in Dang district, West Nepal. Banko Janakari 21:45–49
Singh S (2017) Ethnobotanical study of wild plants of Parsa district Nepal. Ecoprint 24:1–12. https://doi.org/10.3126/eco.v24i0.20641
Siva R (2007) Status of natural dyes and dye-yielding plants in India. Curr Sci 92:916–925
Smith C (2018) Treasures from the Sea. Sea silk and Shellfish purple dye in antiquity. Int J Naut Archaeol 47:237–240. https://doi.org/10.1111/1095-9270.12283
Susmann NM (2020) Tyrian, true, royal, or real: archaeological assumptions about the Roman Murex dye industry. J East Mediterr Archaeol Herit Stud 8:159–173. https://doi.org/10.5325/jeasmedarcherstu.8.2.0159
Svanberg I (1997) Field horsetail (*Equisetum arvense*) as a Food Plant: Kannubjølluvísa til matna. Fróðskaparrit - Faroese Scientific Journal 45:45–55. https://doi.org/10.18602/fsj.v45i.698
Yusuf M (2018) Handbook of renewable materials for coloration and finishing, 1st edn. Wileys, Beverly, MA, USA
Zhao C, Liu J, Zhu P, et al (2014) Research on the dyeing behavior of wool fabric with natural phytolacca berry dyes. Wool Text J 42:12

Chapter 2
Chemistry of Dyes and Pigments

Abstract Colour plays an important role in our daily life as a medium of communication, art, culture, and business. The chemistry of dyes and pigments describes the interaction between molecular structures with visible light through chromophores and auxochromes giving specific types of colours, intensity, and substrate affinity. Natural dyes and pigments are of particular attention, highlighting the Nepal Himalayas as a nexus of rich biodiversity, abundant minerals, and existing indigenous knowledge. This chapter presents the colour chemistry with its chemical principles, sources and applications, together with the roles of mordants, coordination chemistry, and functional groups in colour fixation and fastness. In addition to this, comparisons between natural and synthetic dyes in terms of performance, sustainability, and environmental impact, along with their differences in mechanisms of application, and their industrial significance are also addressed.

2.1 Background

Colours are used in our daily life because they influence our appearance, perception, emotion, and communication. Colours are used in the form of vibrant shades in paintings on textiles, or in the shade of small commercial logos in advertisements for big business purposes. Thus, the practice of using colour for various purposes is found in art, culture, industry, and other fields.

The chemistry of dyes and pigments is essentially about understanding how molecules interact with light to give colour. These chemical compounds in dyes and pigments have chemical structures that are responsible for absorbing specific wavelengths of visible light. The light that is not absorbed is either reflected or transmitted, and this phenomenon is observed as colour by the human eye. The colour and saturation are additionally affected by auxochromes, which are known as functional groups. These functional groups change the chromophore's capacity to absorb light and frequently enable the molecule to attach to a substrate.

The Nepal Himalaya serves as a prime example of the cultural significance of colour chemistry. There are rich resources of natural dyes and pigments due

N. L. Bhandari et al., *Natural Dyes and Pigments of Nepal Himalaya*,
https://doi.org/10.1007/978-3-032-23229-8_2

to existing biodiversity, altitudinal variation, and traditional knowledge systems (Bhandari et al. 2020b). There are many plant species with high percentages of dye-yielding compounds like anthraquinones, flavonoids, and tannins (Malla 1993). In addition to this, there are several minerals such as haematite, limonite, and malachite present in nature that give various colours such as red, yellow, and green pigments (Malla 1993; Bhandari et al. 2020b). The use of natural resources to yield dyes and apply them in fabric colouring, painting, and cultural rituals in Nepal has existed for generations as a major part of long-standing traditional knowledge (Bhandari et al. 2021b).

Different mordants are used in the dyeing process to bond dye or pigment molecules with the substrate and improve colour fastness. Metal cation mordants like aluminium (Al^{3+}) or iron (Fe^{2+}) form complex bonds with anionic functional groups like hydroxyl (–OH) or carbonyl (>C=O) of the dye (Arora et al. 2017; Bhandari et al. 2021a). Pigments are insoluble particles, so they physically disperse and are retained within a binder, such as paints and plastics, to give better colour and are well appreciated for their opacity and stability.

Understanding the chemistry of dyes and pigments is very important because it allows dyes to give the best colour that remains on dyed items and is the backbone of the dyeing industry. Knowing about the functional groups in natural dyes and pigments and different phytoconstituents that have excellent affinity for ultraviolet (UV) light absorption and antimicrobial properties allows them to be more desirable for the textile industry. For example, flavonoids can absorb harmful UV radiation, whereas tannins and anthraquinones can kill microbes, and these are preferred in dyeing many items (Kanchana et al. 2013; Bhandari et al. 2021a).

This chapter deals with the colour chemistry of dyes and pigments. First, it will trace the historical journey from natural dyes and pigments to a vast array of synthetic colourants. Then the section will discuss the basic chemistry of dyes and pigments and how they are differentiated by examining their molecular structures, modes of application, and properties. Understanding the chemistry of dyes and pigments is not only important for creating new colours but also for developing suitable technologies in materials science, medicine, and other manufacturing industries worldwide.

2.2 Natural and Synthetic Dyes

Natural dyes that are mostly extracted from biological substances and inorganic minerals have been in use for hundreds of years (Malla 1993). Although natural dyes are not as widely preferred as synthetic dyes, they are increasingly valued in this modern world because of their eco-friendly, biodegradable, and non-toxic nature (Bhandari et al. 2020a). The isolation and characterisation of active constituents of dyes is done by using several methods that precisely identify specific compounds that give colour and also explain the mechanism of appearance of different colours (Yadav et al. 2023). Understanding the detailed chemistry of dyes and their colours

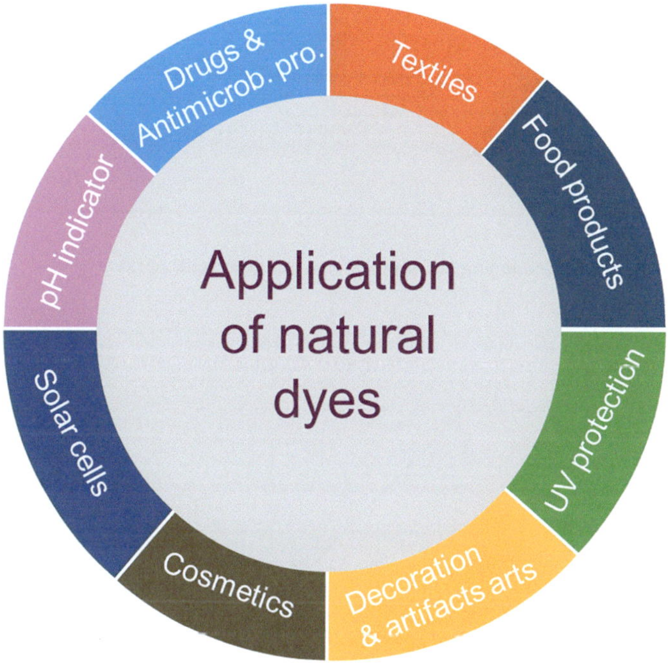

Fig. 2.1 Various applications of natural dyes

allows them to be used in textiles, food, cosmetics, pharmaceutical products, dye-sensitised solar cells, arts, paintings, etc. (Fig. 2.1) (Shahid and Shahid-Ul-Islam 2013; Tayade et al. 2016; Yadav et al. 2023).

A classic example is the behaviour of indigo; a natural dye whose colour formation depends on the chemical transformation. Indigo (indigotin, $C_{16}H_{10}N_2O_2$) is first reduced to its leuco form in an alkaline medium, which is water-soluble. The leuco indigo dye is converted to indigo during the dyeing process. When leuco indigo comes into contact with air during dyeing, it oxidises back to its initial form (indigo), resulting in the visualisation of colour on cotton fabrics (Saikhao et al. 2018). The reaction process during these mechanisms is illustrated in Fig. 2.2. Similarly, madder root is another important source of natural dye, which bears anthraquinone derivatives that convert into alizarin, a historically significant red dye (Blackburn 2017). This transformation suggests how minor structural modifications within natural compounds can yield strong and stable colours.

The colour shades of natural dyes depend on the type of mordant. To explain the chemical interaction and coordination chemistry between natural dyes, substrates and mordants, let's take curcumin as an example, which is mostly found in turmeric. It forms coloured complexes with metal ions such as aluminium (imparting colour yellow–orange shades) or iron (producing darker brown tones) (Fig. 2.3), demonstrating the importance of coordination chemistry in natural dye application (Habib et al. 2024).

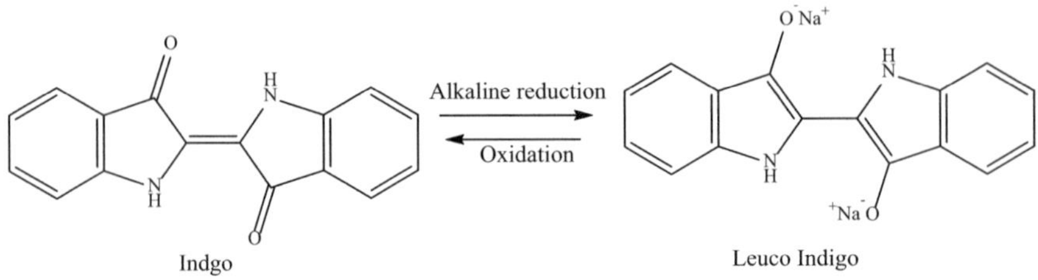

Fig. 2.2 Oxidation–reduction process of indigo dye (Saikhao et al. 2018)

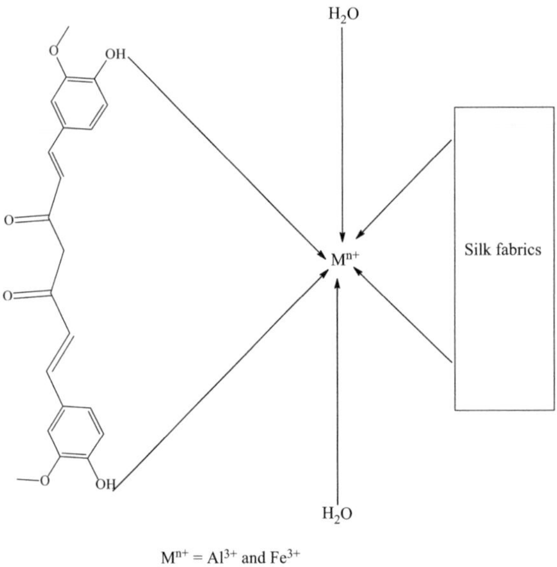

Fig. 2.3 Dye–metal–substrate interactions (Habib et al. 2024)

The use of both natural and synthetic dyes has advantages and disadvantages. Natural dyes, as mentioned above, are mostly safe to use, but some of them are difficult to produce synthetically, such as the complex reds from cochineal insects. Natural dyes generally do not provide a wide range of colours and vary in quality from batch to batch production. If natural dyes are obtained from different places, then achieving superior colour fastness is almost impossible, as plant properties differ from place to place. Thus, at times, natural dyes need different mordants for proper fixation, and this makes them expensive.

Synthetic dyes are produced through controlled chemical synthesis using petrochemicals (Ben et al., 2021; Mabuza et al. 2023). These synthetic dyes are usually cheap, colourfast, and consistent in mass production. Thus, they are highly preferred and widely used in the dyeing industry worldwide. However, the use of synthetic dyes comes at the cost of environmental pollution and health hazards. Numerous synthetic dyes are toxic, carcinogenic, and environmentally persistent (Ali 2024). The discharge of effluent from textile and dyeing industries that use synthetic dyes contains stable compounds that pollute water, and due to pollution, there is a reduction in sunlight penetration, reduced photosynthetic activity in aquatic plants, and the generation of toxic byproducts through degradation that ultimately destroys aquatic ecosystems on a large scale.

Azo dyes, the largest class of synthetic dyes, are synthesised through a two-step chemical process involving diazotisation of an aminobenzene followed by azo coupling with β-napthol. The reaction mechanism is depicted in Fig. 2.4, where the –N=N– azo group is responsible for vivid colours such as those seen in methyl orange and similar dyes (Pramod 2023).

Malachite green is another example of a synthetic dye known as a triphenylmethane dye, which is synthesised by the condensation of benzaldehyde with dimethylaniline. This dye undergoes oxidation, giving rise to an intense green colour (Bello et al. 2017). Similarly, aniline blue is obtained via sulfonation followed by oxidation of aniline derivatives, demonstrating how structural modification leads to diverse colour properties in synthetic dyes (Xu 2024).

From a molecular perspective, the basic mechanism of colour production in both natural and synthetic dyes is possible due to conjugated electron systems, which

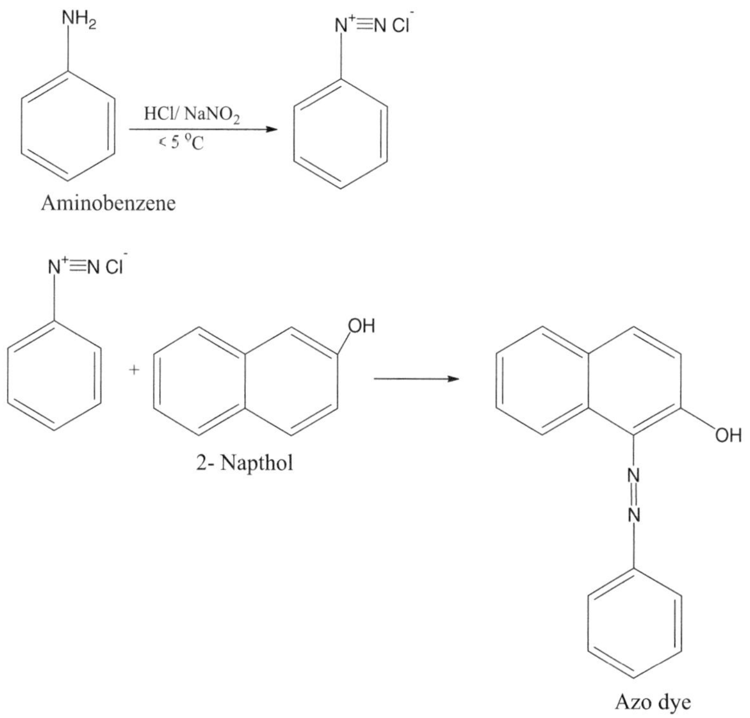

Fig. 2.4 Reaction steps involved in the synthesis of an azo dye (Pramod 2023)

Table 2.1 Colour perception in visible light with wavelengths

Perceived colour	Absorbed colour	Absorbed wavelength (nm)
Yellow-Green	Violet	400–435
Yellow	Blue	435–480
Orange	Green-Blue	480–490
Red	Blue-Green	490–500
Purple	Green	500–560
Violet	Yellow-Green	560–580
Blue	Yellow	580–595
Green-Blue	Orange	595–605
Blue-Green	Red	605–700
Black	Violet-Red	380–720

absorb discrete wavelengths of visible light (Table 2.1) and functional groups such as hydroxyl (–OH) or amino (–NH_2) that help enhance colour intensity and have proper substrate affinity (Gürses et al. 2016). The wavelength of light absorbed is controlled by quantized electronic transitions between molecular orbitals of matching energy differentials, resulting in colour.

The extraction of dyes is carried out using different methods, and these methods have been improving over time to get high yields and also develop novel biodegradable synthetic dyes. An example of biodegradable synthetic dyes is azobenzene derivatives with introduced cleavage points, and the principle is based on the concept of having green chemistry. There has been progress in enhancing the quality of colour and reproducibility of natural dyes through nanotechnology improvements, and there is still ongoing scientific investigation into both natural and synthetic dyes worldwide. These innovative solutions are aimed at balancing aesthetic requirements, optimum technical performance, while minimising environmental impact for both natural and synthetic dyes.

2.3 Dyes and Pigments

There are basic differences between dyes and pigments in terms of their solubility, composition, and application mechanisms (Shindy 2016; Candela et al. 2021). Dyes are organic compounds and water-soluble compounds, but pigments are inorganic compounds and are water-insoluble. Dyes penetrate into the substrate at a molecular level, forming a chemical bond that results in vibrant and intense colouration. In contrast, pigment particles are suspended and distributed within a medium (Shindy 2016; Mansour 2018). Early examples of pigments are ochre, charcoal, and lapis lazuli, and they were used in prehistoric art. Ancient artists used water as a binding medium to apply pigments with their fingers or by spraying from the mouth. Bison cave paintings in Altamira, Spain, and hand stencils in Avignon, France (see Fig. 2.5a, b) are the earliest evidence of symbolic pigment dating back 30,000 years (Barnett et al. 2006). The use of specific minerals is widely observed throughout human civilization. For example, azurite was used in ancient Egypt, while ultramarine became the most important blue pigment of the Middle Ages, and it was made from ground lapis lazuli sourced from Afghanistan (Fig. 2.5c) (Barnett et al. 2006).

Dyes give colour but often exhibit lower lightfastness and can fade over time. Pigments, while generally less brilliant, are applied on surfaces requiring long-lasting colour and superior stability (Gürses et al. 2016; Candela et al. 2021). Dyes have a natural chemical affinity for textiles or paper, allowing them to colour effectively from a solution, but pigments need a binding agent (e.g., a polymer) to attach to the surface because there are no direct interactions between the pigment and its substrate (Gürses et al. 2016).

There is great industrial importance attached to dyes and pigments worldwide. It is estimated that over 10,000 different dyes and pigments are commercially used globally. The annual global production of synthetic dyes alone exceeds 700,000 tonnes, and this high amount highlights the importance of dyes in enhancing the aesthetic appeal of different products (Gürses et al. 2016; Tayade et al. 2016; Bhandari et al. 2021a).

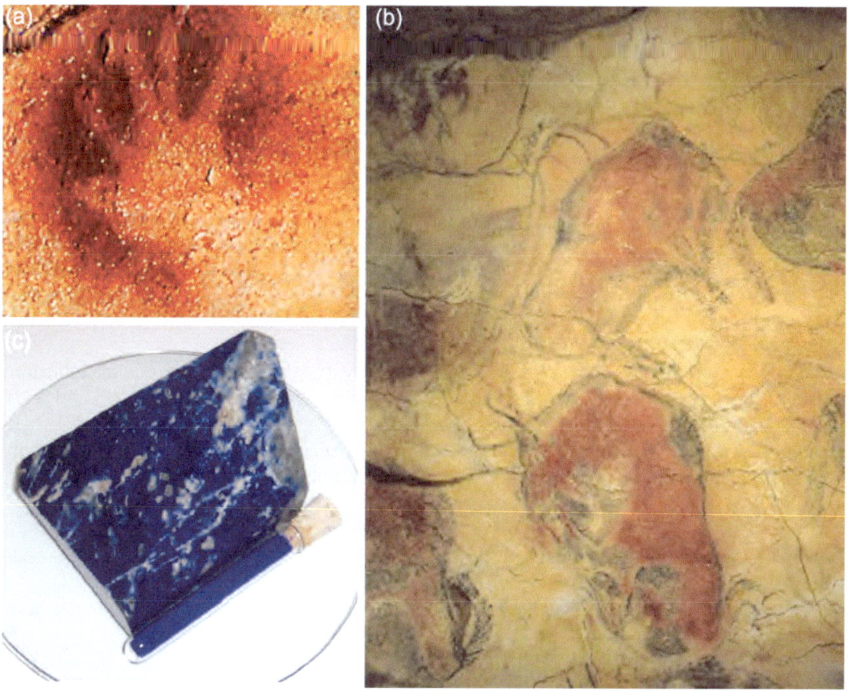

Fig. 2.5 **a** Hand stencil from a cave in Avignon, France; **b** Paintings of bison in the Altamira Caves, Spain; and **c** Lapis rock and lapis pigment (Barnett et al. 2006)

2.4 Conclusions

Dyes and pigments play an important role in daily life, influencing perception, culture, industry, and technology. Understanding the chemistry of dyes and pigments supports the elucidation of molecular structures, how they absorb and reflect light and generate colour. Natural dyes, obtained from plants, animals, and minerals and their traditional usage patterns, are eco-friendly but prone to fading, while synthetic dyes offer consistency and colourfastness but raise environmental concerns. The proper knowledge of chromophores, auxochromes, and the interactions between dye and mordants helps to improve colour shades for advancements in textile colouring, food, pharmaceutical, and other sectors. Thus, dyeing technology helps to demonstrate the scientific and industrial significance of colour chemistry, which improves human life.

References

Ali AN (2024) A comprehensive study of natural and synthetic dyes: their properties, methods of preparation, and uses. Shifaa 2024:1–17. https://doi.org/10.70470/shifaa/2024/001

Arora J, Agarwal P, Gupta G (2017) Rainbow of natural dyes on textiles using plants extracts: sustainable and eco-Friendly processes. Green Sustain Chem 07:35–47. https://doi.org/10.4236/gsc.2017.71003

Barnett JR, Miller S, Pearce E (2006) Colour and art: a brief history of pigments. Opt Laser Technol 38:445–453. https://doi.org/10.1016/j.optlastec.2005.06.005

Bello IA, Bello OS, Adegoke KA (2017) Effects of modifying agents on the dyeability of cotton fabric using Malachite green dye. Ann Sci Technol 2:26–36. https://doi.org/10.2478/ast-2018-0005

Ben SH, Bouket AC, Pourhassan Z et al (2021) Diversity of synthetic dyes from textile industries, discharge impacts and treatment methods. Appl Sci 11:1–21. https://doi.org/10.3390/app11146255

Bhandari NL, Bist K, Ghimire J et al (2020a) Feasibility study of the *Euphorbia pulcherrima* plant extract as natural dye with different mordants for fabric dyeing. J Inst Sci Technol 25:30–36. https://doi.org/10.3126/jist.v25i1.29421

Bhandari NL, Ghimire J, Shrestha S et al (2021a) Green extraction, characterization, applications and antimicrobial analysis of natural dye from *Phyllanthus emblica*. Asian J Chem 33:404–410. https://doi.org/10.14233/ajchem.2021.23008

Bhandari NL, Pokhrel B, Bhandari U et al (2020b) An overview of research on plant based natural dyes in Nepal: scope and challenges. J Agric Nat Resour 3:45–66. https://doi.org/10.3126/janr.v3i2.32328

Bhandari NL, Shrestha S, Bhandari G et al (2021b) Extraction of dye from *Castanopsis indica* for its use in textile dyeing and medicinal purpose with natural mordant. Egypt J Chem 64:6681–6690. https://doi.org/10.21608/EJCHEM.2021.48519.2991

Blackburn RS (2017) Natural dyes in Madder (*Rubia* spp.) and their extraction and analysis in historical textiles. Color Technol 133:449–462. https://doi.org/10.1111/cote.12308

Candela RG, Lazzara G, Piacente S et al (2021) Conversion of organic dyes into pigments: extraction of flavonoids from blackberries (Rubus ulmifolius) and stabilization. Molecules 26. https://doi.org/10.3390/molecules26206278

Gürses A, Açıkyıldız M, Güneş K, Gürses MS (2016) Dyes and pigments: their structure and properties, pp 13–29. https://doi.org/10.1007/978-3-319-33892-7_2

Habib N, Batool F, Adeel S et al (2024) Green extraction and application of yellow natural curcumin colorant from *Curcuma aromatica* rhizomes for silk dyeing. Sci Rep 14:1–11. https://doi.org/10.1038/s41598-024-63927-7
Kanchana R, Fernandes A, Bhat B et al (2013) Dyeing of textiles with natural dyes—an eco-friendly approach. Int J ChemTech Res 5:2102–2109
Mabuza L, Sonnenberg N, Marx-Pienaar N (2023) Natural versus synthetic dyes: Consumers' understanding of apparel coloration and their willingness to adopt sustainable alternatives. Resour Conserv Recycl Adv 18:200146. https://doi.org/10.1016/j.rcradv.2023.200146
Malla L (1993) The natural dyes of Nepal, p 142
Mansour R (2018) Natural dyes and pigments: extraction and applications. Handb Renew Mater Color Finish 75–102. https://doi.org/10.1002/9781119407850.ch5
Pramod DP (2023) Synthesis of azo dye by coupling of diazonium salt with α-naphthol, β-naphthol. Peer Rev Ref J 10:66–71
Saikhao L, Setthayanond J, Karpkird T et al (2018) Green reducing agents for Indigo dyeing on cotton fabrics. J Clean Prod 197:106–113. https://doi.org/10.1016/j.jclepro.2018.06.199
Shahid M, Shahid-Ul-Islam MF (2013) Recent advancements in natural dye applications: a review. J Clean Prod 53:310–331. https://doi.org/10.1016/j.jclepro.2013.03.031
Shindy HA (2016) Basics in colors, dyes and pigments chemistry: a review. Chem Int 2:29–36
Tayade SV, Pandey S, Goswami Giri AS, Giri AG (2016) Chemical science review and letters green extraction of dyes from *Solanum xanthocarpum* leaves. Chem Sci Rev Lett 5:41–45
Xu Y (2024) The Chemical oxidation of aniline: mechanisms, applications, and environmental implications. Adv Mater Sci Res 7:209–210. https://doi.org/10.37532/aaasmr.2024.7(5).209-210
Yadav S, Tiwari KS, Gupta C et al (2023) A brief review on natural dyes, pigments: recent advances and future perspectives. Results Chem 5:100733. https://doi.org/10.1016/j.rechem.2022.100733

Chapter 3
Methods of Extraction of Natural Dyes

Abstract Many natural and biomaterials contain colour pigments. Plant-based natural dyes can be extracted from various plant parts, such as the bark, stems, wood, flowers, fruits, leaves, and roots. Different chromatic pigments, such as chlorophylls, xanthophylls, flavonoids, carotenoids, and anthocyanins, impart colour in plants. Similarly, insects, lichens, fungi, animals, and minerals are other sources for natural pigments for extraction. This chapter discusses the various extraction techniques, from simple traditional to complex modern methods, especially focusing on eco-friendly and greener methods with a high yield of dye pigment. Conventional extraction methods are widely used because of their simplicity and low cost. However, they are lengthy methods and have limited reproducibility with low yield. These involve aqueous extraction, acid, or alkaline extraction processes. Modern approaches include ultrasound- and microwave-assisted extraction, solvent extraction, and fermentation extraction methods. Modern extraction processes have also focused on emerging techniques like enzyme-assisted extraction and supercritical fluid extraction methods, mainly to increase dye yield with minimal energy use without damaging the pigmented compounds. Further, the modern approaches derive maximum proficiency, enhanced dye stability, reduced solvent requirements, with greater expandability. The chapter focuses on operating principles, dye stability, benefits, and drawbacks of such varied extraction methods. It also addresses the influencing factors responsible for the efficacy of extraction. Overall, the study presents multiple synergistic non-thermal, enzyme-assisted, and supercritical extraction methods with maximum yield using greener solvents for the sustainable implementation of natural dye resources.

3.1 Background

The different plant parts, such as fruits, flowers, bark, roots, leaves, and wood, contain colours. Besides the plants, the dyes are also acquired from fungi, lichens, insects, animals, and minerals (Saha and Ayman 2022). The various plant pigments, such as chlorophyll (green), carotenoid (yellow, orange, and red), anthocyanin (red, purple,

N. L. Bhandari et al., *Natural Dyes and Pigments of Nepal Himalaya*,
https://doi.org/10.1007/978-3-032-23229-8_3

and blue), flavonoids (yellow, red, and blue), and betalains (red-violet or yellow-orange), either individually or in combination, are responsible for natural colours or dyes (Swami et al. 2020). The extraction of dyes from plants is a challenging and vital process since these pigments are frequently attached to a complex plant matrix (Vithushika et al. 2024). Hence, the natural dye extraction and dyeing processes required to achieve desired colour shades demand long experience and a perfect hand.

Natural dyes are biocompatible, non-toxic, and rich in bioactive compounds. They are favourable for application in textiles, food, fabrics, cosmetics, and pharmaceuticals. Nowadays, research on the production and use of sustainable dyes has been revitalised due to the growing demand for chemical-free and eco-friendly products (Hamdy and Hassabo 2021). This extends promising potential for local and commercial utilisation in countries like Nepal, which are rich in several dye-yielding plants, fungi and lichens (Rehman et al. 2022). Therefore, natural colours derived from plant species as dyes are used for various purposes and are extracted using both conventional and modern techniques. Conventional techniques are those methods that use aqueous extraction, solvent extraction, and acidic or alkaline extraction (Salauddin et al. 2021). In contrast, modern techniques employ advanced technologies such as ultrasound-assisted, microwave-assisted, enzyme-assisted, and supercritical fluid extraction methods. The use of modern techniques is increasing compared to conventional methods due to their environmental sustainability, economic viability, lower operational costs, and higher yields (Pizzicato et al. 2023).

The process of extracting dyes is quite complex. Dye-bearing plant material usually contains a minor proportion of the actual colouring compound, and the rest of the constituents are non-dye constituents, such as structural components like cellulose, proteins, water-insoluble fibres, carbohydrates, and so on, and chemical compounds like chlorophyll, tannins, terpenes, phenolics, alkaloids, and fatty acids, etc. (Bhandari et al. 2024). During the extraction of dye properties of the target pigment, including its solubility and stability, should be thoroughly studied, and this should include understanding the nature of the target pigment (Bhandari et al. 2021; Chungkrang et al. 2021; Kumar Gupta 2020; Saxena and Raja 2014).

Several factors affect the effective extraction process through any method, such as the solvent, pH, temperature, plant parts, growth, age, and vessel types (Wizi et al. 2018). Because of the complex natural pigments and the need for a stable, eco-friendly procedure, the development and enhancement of the extraction technique are challenging (Ismal and Yildirim 2019). Several extraction processes are mentioned in this chapter, including traditional and modern processes.

3.2 Methods of Extraction

3.2.1 *Traditional Approaches*

Natural dyes have a long-standing history dating back to ancient times. The use of natural dyes, especially colours like green, red, blue, and yellow, has been observed since the Bronze Age to dye archaeological objects and fabric fragments in various parts of the world (Tamburini 2021). China has a 5000-year-old record of the utilisation of natural dyes in its earliest documents (Balachandran and Sabumon 2023).

Traditional methods have a fundamental procedure for extracting colours. At first, the dye sources are collected. These are dried and ground into powder. The pigment-containing powder is soaked in water. Then heated steadily, resulting in the colour. Finally, fabrics are immersed in it and heated until the colour is absorbed by the fabric (Kumar Gupta 2020). Cold methods are also used for sensitive dye sources. During this process, the plant source is immersed in water for days or weeks to make the dyeing material softer and less harsh, allowing the desired colour through this process (Saxena and Raja 2014). Regardless of its simple procedure and profound cultural background, current research focuses on the limitations of traditional dye extraction. One of the major issues with natural dyes is that they are prone to discolour or fading very easily because of heat or light. The growing condition, harvest time, and age of the plant vary the concentration and composition of natural dyes, causing a challenge to standardise dye for homogeneous colour (Zaky et al. 2024).

Several mordants like alum ($KAl(SO_4)\ 12H_2O$), ferrous sulphate ($FeSO_4$), and salts of tin and copper are used for dyeing as a molecular glue to set the dye on fabric and to increase the colour intensity and duration. These mordants permanently attach the dye molecule to the fibre during the dyeing process to form long-lasting colours (Jabar et al. 2021). Traditional dyers generally achieve an astonishing range of rich, beautiful, and durable colours by continuously experimenting with different mordants and combining various natural sources, leading to the transformation of simple cloth into a lasting work of art that becomes a piece of history and culture. Figure 3.1 presents an example of traditional dye extraction from Papaya *(Carica papaya)* leaves and Turmeric *(Curcuma longa)* tubers, demonstrating the percolation method, which is commonly used in traditional and cultural practices.

3.2.2 *Aqueous Extraction*

Aqueous extraction is a traditional and essential method for extracting dyes, and this method begins by breaking down the raw materials into small pieces, grinding them into a powder, and sieving them to increase the surface area for more effective colour release (Salauddin et al. 2021). Subsequently, either the cold or hot percolation method is used for the aqueous extraction process. In cold percolation, the powdered

Fig. 3.1 Traditional method of dye extraction from **a** papaya (*Carica papaya*) leaves, and **b** turmeric (*Curcuma longa*) tubers in water using the percolation method, and colouring on cotton fabrics and threads, respectively

material is soaked in cold water for a long period of time, often overnight or longer, using vessels that are made of soil, wood, or stainless steel or copper and generally non-reactive in nature (Bhandari et al. 2021; Chungkrang et al. 2021; Saxena and Raja 2014). This process allows water to slowly penetrate the dye source and facilitate colour release. On the other hand, the hot percolation process involves boiling of the source material in water to quicken the breaking down of tissues and to allow the release of dyes into the solution. Finally, the solution is filtered to remove solid residues (Periyasamy 2024). To maximise yields, boiling and filtration are often repeated with stainless steel vessels. The aqueous extraction of dyes from walnut (*Juglans regia*) powder is shown in Fig. 3.2.

The yields in aqueous extraction depend on the type and properties of the source material (Saxena and Raja 2014). The cold extraction method is used to extract limited natural sources, while hot extraction is used for a wide range of colourant sources, which is done by Soxhlet steam extraction. In the traditional method, the resulting yield of dye is altered by various factors like water quality, pH level, and the nature of the extraction vessel, such as iron or copper, that can modify the hue during dyeing.

Aqueous extraction is cheap, sustainable, uses existing water, and is applied instantly to fabrics, making it highly advantageous (Salauddin et al. 2021). However, it operates at high temperature, needs a large amount of water, is tedious, obtained in a smaller amount. Furthermore, aqueous extract is prone to fungal contamination, thus cannot be preserved for a long duration (Sen and Dawood 2014). The aqueous extraction of a red-coloured dye from Mojito (*Rubia cordifolia* and *Rubia manjith*) is illustrated in Fig. 3.3.

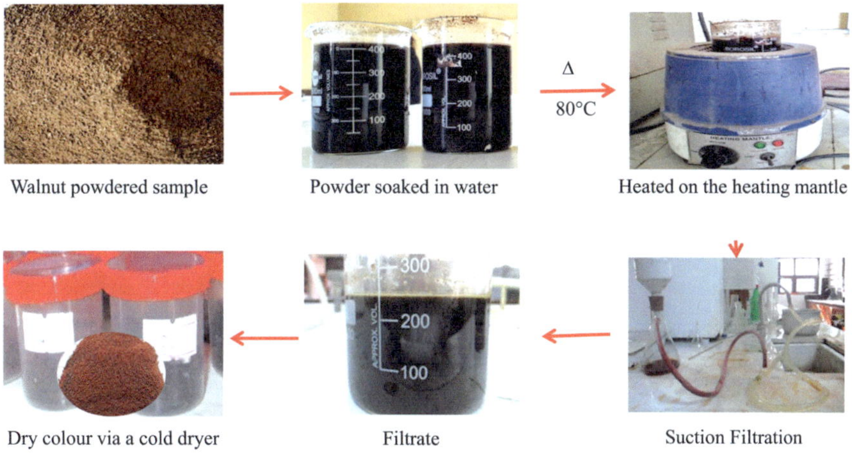

Fig. 3.2 A simple schematic showing the aqueous extraction of dye from walnut (*Juglans regia*) powder

Fig. 3.3 Schematic showing the aqueous extraction of red-coloured dye from Mojito (*Rubia manjith*)

3.2.3 Acidic and Alkaline Extraction

Many natural dyes are present in the form of glycosides. Acidic or alkaline extraction methods are mainly used to break down glycosides using hydrolysis. In this method, the dye is released during hydrolysis, which significantly increases yields (Shrivastava and Daniel 2003). There is no need to dilute the acidic or alkaline extraction medium in this method.

The choice between an acid and a base depends on the nature of the dye. Alkaline solvents are used for dyes that contain phenolic groups, and these dyes become soluble in an alkaline solution. Acidic solvents are used for pH-sensitive dyes, such as flavones, and acids prevent oxidation during the extraction process, which generally degrades the quality of dyes (Bhandari et al. 2021; Kumar Gupta 2020). This method is mainly used to extract red dyes from safflower (*Carthamus tinctorius*) petals and lac dye secreted by lac insects (Islam 2023). During the alkaline extraction method, acid is often added back so that the purified dye can undergo proper precipitation and separation. It is recommended not to use concentrated alkaline or acidic solutions as they may decompose the original colouring compound, leading to faded or unintended colours. Alteration of pH can also result in a loss of colour yield if conditions are not well controlled (Khattab et al. 2020).

The general procedure of acid or alkaline extraction of natural dyes is illustrated in Fig. 3.4. It demonstrates the utilisation of an acidic or alkaline extraction medium for plant material that contains dye-bearing glycoside. The glycosidic bond breaks during hydrolysis and forms a free dye molecule in the solvent. The procedure starts by taking clean, dry raw plant material, and then its size is reduced to ease effective solvent interaction (Islam 2023). After that, it is treated with an acidic or alkaline solution according to the chemical nature and stability of the needed dye. To get the dye molecule, glycosidic bond is broken under controlled heat. Finally, the liquid mixture is separated from the mixture through filtration. For proper precipitation and purification of colourant in alkaline extraction, additional acidification is done. This process is consistent and broadly accepted for the successful formation and retrieval of dye compounds (Shrivastava and Daniel 2003).

3.2.4 Ultrasonic Microwave Extraction

This is a modern extraction method, which employs both ultrasound-assisted extraction (UAE) and microwave-assisted extraction (MAE). These methods have remarkable benefits over conventional approaches. Due to improved results and high effectiveness in the utilisation of solvent, energy, and time (Laura et al. 2020). In this process, plant material is passed through a solvent, and high-frequency sound waves called ultrasonic waves are passed through it, which breaks the plant source and facilitates getting the dye quickly. The method is deployed at low temperatures and is highly relevant to extract heat-sensitive dyes (Nonglait and Gokhale 2024).

Fig. 3.4 Schematic diagram of acid/alkaline extraction

Microwave energy is also applied to heat the material in place of high-frequency sound, to release a larger amount of dye quickly. MAE is used to extract yellow to red–orange dye from annatto (*Bixa orellana*) and the blue pigment from butterfly pea flowers, *Clitoria ternatea* (Sinha et al. 2012, 2013). Both UAE and MAE are known as green extraction technologies as they use less energy, solvent, and the overall process to extract dyes is relatively short (Saxena and Raja 2014). The ultrasonic and microwave method of dye extraction is represented in Fig. 3.5.

3.2.5 *Fermentation*

The fermentation method uses microbial enzymes present in natural resources for dye extraction. This method is generally used for the extraction of indigo from the indigo plant (*Indigofera tinctoria*). Figure 3.6 illustrates the sequential steps involved in microbial extraction. It illustrates the procedure for indigotin dye formation by enzymatic hydrolysis of *indigo*. The leaves and twigs are immersed in water at about 32 °C to form indigotin dye by enzymatic hydrolysis of *indigo.*

In the fermentation process, the glucoside indican existing in the leaves or twigs is hydrolysed by natural enzyme indimulsin to obtain glucose and indoxyl (Kumar Gupta 2020). After a fermentation period of 10–15 h, the resulting yellow indoxyl solution, when exposed to air, produces a blue colour. This blue colour that precipitates upon drying is known as indigotin, which is normally washed and compressed into cakes for future use. Fermentation-based extraction is similar

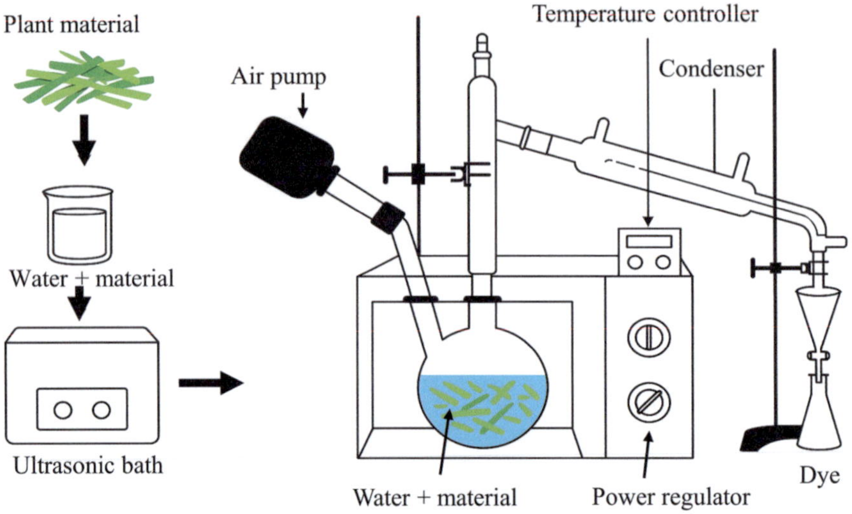

Fig. 3.5 Simplified representation of the ultrasonic–microwave method of dye extraction

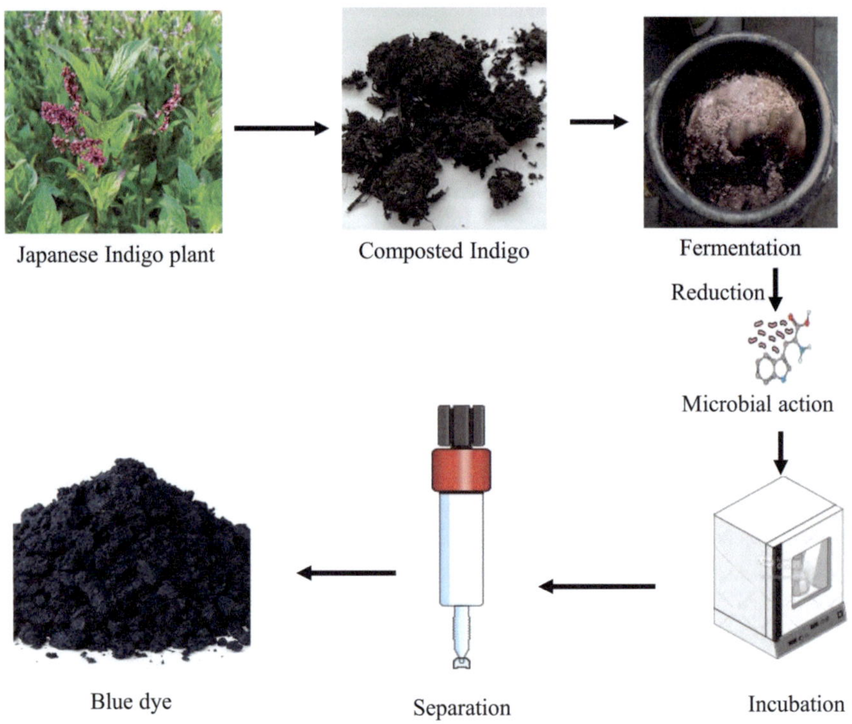

Fig. 3.6 Fermentation process for indigo dye production

to aqueous extraction but operates without using high temperatures, and it relies on microorganisms to degrade the dye source. However, this method requires a long extraction time and produces unpleasant odours due to microbial activity (Kumar Gupta 2020; Saxena and Raja 2014).

Solid-state fermentation is also used for extracting natural colours from seaweed biomass. In a 2017 study, *Chaetomorpha antennina*, *Gracilaria corticata*, and *Ulva fasciata* were processed using this method with *Aspergillus niger* in a 1% sucrose solution as the fermenter, and the highest yield of chlorophyll a concentration (20.625 mg/L) was obtained in *U. fasciata*, followed by chlorophyll b (12.486 mg/L) in *G. corticata*, and carotene content (14.701 mg/L) in *C. antennina* (Sharmila et al. 2017).

3.2.6 Solvent Extraction

The solvent extraction method is also known as liquid–liquid extraction or the partitioning method. It is a basic separation method that uses two immiscible liquids (usually water and an organic solvent) with different solubilities. The dye is separated from a mixture into one of the liquid phases in a lab setting using a separating funnel. This method is also used for analytical purposes, such as concentrating an analyte or systematically separating complex mixtures of dyes (Salauddin et al. 2021).

In the process, organic solvents such as ethanol, methanol, acetone, ether, and chloroform are used, and extraction is carried out at low temperatures, which lowers the risk of destroying heat-sensitive dyes. This extraction process gives a high yield and uses less water, and can extract both water-soluble and water-insoluble dyes at the same time (Krgovic et al. 2025; Khan et al. 2018). Often acids or alkalis are added to alcoholic solvents to increase their effectiveness and selectivity, and due to the use of low temperatures, dyed fabrics are less likely to fade. However, there is the potential for toxic solvent residue used during the extraction process, which may negatively impact the environment (Bhandari et al. 2024; Kumar Gupta 2020). Another disadvantage is the frequent use of aqueous solutions during the dyeing process, and obtaining perfect colour using this process is unlikely (Saxena and Raja 2014). The alcoholic extraction of red coloured dye from *Rhododendron arboreum* flowers is shown in Fig. 3.7. It is also found that the alcoholic extraction is better than aqueous extraction in the rhododendron flower.

3.2.7 Enzymatic and Supercritical Fluid Extraction

Enzymatic extraction is an advanced biotechnological method, and this method, and uses hydrolytic enzymes such as cellulase, pectinase, and amylase to decompose the structural polymers (e.g., cellulose, pectin, and starch) in plant tissues under remarkably mild temperature and pH conditions. This method is efficient mainly for

Fig. 3.7 Schematic showing alcoholic extraction of dye from *Rhododendron arboreum* flowers in ethanol using a Soxhlet extractor

extracting dyes from bark and roots (Saxena and Raja 2014). Enzymatic extraction was first used commercially in Japan in the 1980s to remove caffeine from coffee and extract hops. The use of this method is well preferred as there is high enzymatic accuracy, leading to the hydrolysis of specific binding materials while preserving dye molecules with high purity and high yield (Streimikyte et al. 2022). The use of enzymes is increasing in the textile industry. For example, cellulases are used for the biofinishing of denim, and laccases are used for the treatment of bleaching effluents. Apart from extracting dyes, enzymes are used as textile preservatives, for removing adhesive residues, and as stain removers (Martins et al. 2023).

Several enzymes, like cellulose, hemicellulose, pectinase, protease, and amylase, are applied to hinder the structural stability of the plant's cell walls. As a result, it increases the extraction of bioactive molecules, that is, complex pigment molecules. The cell wall permeability is increased due to the hydrolysis of the cell wall by enzymes, and bioactive compounds are obtained from natural sources in high yield (Hien et al. 2019). In this process, source material is extracted by pretreating with an enzyme at controlled pH conditions, enzyme concentration, time, and temperature (Hien et al 2019). Then the mixture is incubated for several hours with continuous stirring. After homogeneous enzyme contact, it is centrifuged, and the dye-acquiring liquid supernatant is collected for colourant (Martins et al. 2023). This description is schemed in the following Fig. 3.8.

Supercritical fluid extraction (SFE) is a rising extraction and purification method that uses non-toxic solvents, which can be removed easily at a low temperature. It broadens the industrial process as it effectively recovers bioactivity, transfers mass quickly, and continuously transfers fresh fluid (Abou Elmaaty et al. 2022). Supercritical fluid works as a gas above the critical temperature and pressure. They interact with substrate effectively because of low surface tension, low viscosity, and good diffusivity (Saxena and Raja 2014). Carbon dioxide (CO_2) is applied as a solvent to SFE as it is cheap, accessible, and nontoxic. Critical temperature of CO_2 is 31.4 °C,

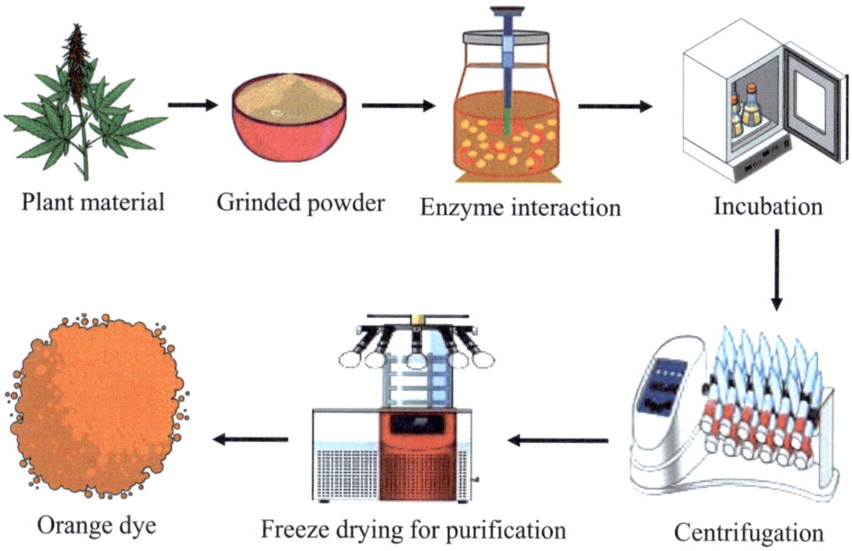

Fig. 3.8 Schematic diagram of the enzymatic extraction method

and pressure is 1,070 pounds per square inch (Psi) or 73.8 bars. The temperature and pressure required to function CO_2 containing SFE are 32–49 °C and 1070–3500 Psi, respectively (Pourmortazavi et al. 2013; Saxena and Raja 2014). As CO_2 is non-polar molecule, for slightly polar molecules, a cosolvent or modifier may be added to enhance solubility. SFE results in a residual-free extract, hence highly beneficial in the extraction of purified natural products (Saxena and Raja 2014).

Figure 3.9 shows the scheme of the laboratory scale SFE instrument, which contains essential components that are the CO_2 reservoir, the co-solvent reservoir, chiller unit, CO_2 pump, co-solvent pump, valve, heater controller, extractor, heating jacket, back pressure regulator vessel, equilibrium coil, pressure vessel, and coil. The SFE process can be done in two modes: static mode and dynamic mode. At first, the plant materials are susceptible to continuous supercritical CO_2 for a period of time in static extraction. The static mode is used to permeate the plant matrix with fluid. However, in dynamic mode, CO_2 is constantly passed to the plant matrix. In most SFE procedures, both modes are combined, where the static mode preceding the dynamic. The plant fluid matrix then passes to the reservoir. After that, analytes from the extraction vessel are separated in a separator. The final dye extract is collected in highly pure form either by recycling CO_2 or releasing (Abou Elmaaty et al. 2022).

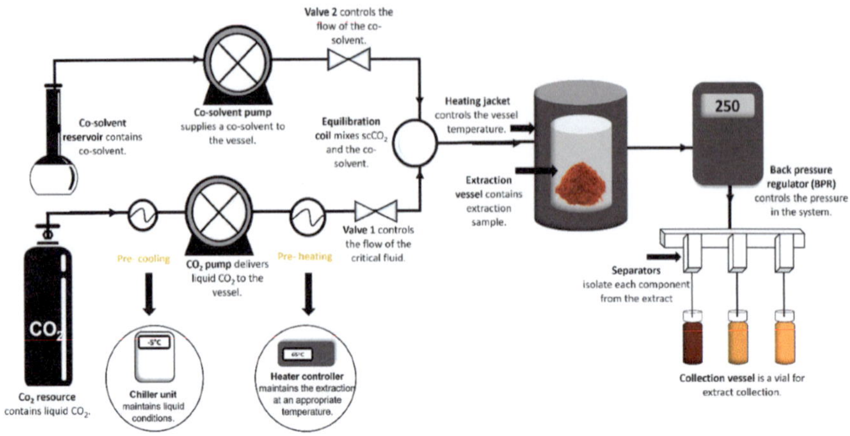

Fig. 3.9 Schematic of a laboratory-scale supercritical fluid extractor (Abou Elmaaty et al. 2022)

3.3 Conclusions

Recent research work has widely focused on sustainable natural dyes to develop multiple extraction techniques. Various extraction methods are investigated in this chapter, such as traditional aqueous methods, which are still used as they are affordable and easy to apply. However, these methods need high temperature that shows the possibility of degradation of thermolabile pigments.

To overcome such mitigations, advanced techniques like enzyme-based extraction, ultrasound-assisted extraction, and fermentation-based extraction have been in progress. They elevate pigment discharge, strengthen cell-wall disruption, less time consumption and solvent usage (Masyita et al. 2025). Overall, the advanced techniques give a transition from conventional. Energy-intensive methods to recent techniques. Modern techniques are highly effective, eco-friendly, and sustain industrial dependence on natural dyes.

In order to increase the yield while reducing environmental hazard, future research should focus on the addition of non-thermal technologies with clean solvents like deep-eutectic solvents (Plyduang et al. 2025). More research should be oriented toward multi-enzyme systems designed specifically for lignocellulosic and pectin-rich plant materials (Kumar et al. 2023).

The extraction of natural dyes is a challenging method. However, the extraction methods are continuously being enhanced to obtain a high yield at a lower price. The extraction of natural dyes has been practised for a long time throughout various civilisations, societies, cultures, and religious ceremonies. In Nepal, there is widespread use of colours, but methods for the extraction of dyes are still conventional, and the use of modern techniques is necessary for the commercialisation of natural dyes.

References

Abou Elmaaty T, Sayed-Ahmed K, Elsisi H, Magdi M (2022) Optimization of extraction of natural antimicrobial pigments using supercritical fluids: a review. Processes 10:2111. https://doi.org/10.3390/pr10102111

Balachandran B, Sabumon PC (2023) Scope of natural dyes and biomordants in the textile industry for cleaner production. In: Muthu SS (ed) Novel sustainable process alternatives for the textiles and fashion industry. Sustainable textiles: production, processing, manufacturing and chemistry. Springer Nature, Cham, pp 73–106. https://doi.org/10.1007/978-3-031-35451-9_4

Bhandari N, Shrestha S, Bhandari G et al (2021) Extraction of dye from *Castanopsis* indica for its use in textile dyeing and medicinal purpose with natural mordant. Egypt J Chem 64(11):6681–6690. https://doi.org/10.21608/ejchem.2021.48519.2991

Bhandari V, Badanayak P, Jose S (2024) Plant-based colorants: isolation and application. In: Thomas S, Jose S, Mathew SS, Samant L (eds) Plant biomass-derived materials. Wiley, pp 159–188

Chungkrang L, Bhuyan S, Phukan AR (2021) Natural dyes: extraction and applications. Int J Curr Microbiol Appl Sci 10(1):1669–1677. https://doi.org/10.20546/ijcmas.2021.1001.195

Hamdy D, Hassabo A (2021) Various natural dyes using plant palette in the colration of natural fabrics. J Text Color Polym Sci 18(2):121–141

Hien NT, Hoa HD, Tu NTM (2019) Study of ultrasound and enzyme-assisted extraction of tannins from mangosteen peel in Vietnam. Food Sci Appl Biotechnol 2:130–139. https://doi.org/10.30721/fsab2019.v2.i2.72

Islam SU (2023) Renewable dyes and pigments. Elsevier, Amsterdam

İşmal ÖE, Yıldırım L (2019) Metal mordants and biomordants. In: The impact and prospects of green chemistry for textile technology. Elsevier, pp 57–82

Jabar JM, Adedayo TE, Odusote YA (2021) Current research in green and sustainable chemistry. Curr Res Green Sustain Chem 4:100151. https://doi.org/10.1016/j.crgsc.2021.100151

Khan B, Sindhyan R, Divan A, Rathod S (2018) Extraction, characterization and applications of natural dyes. AnnPlSci 7:2463. https://doi.org/10.21746/aps.2018.7.11.4

Khattab T, Abdelrahman MS, Rehan M (2020) Textile dyeing industry: environmental impacts and remediation. Environ Sci Pollut Res 27(4):1–16. https://doi.org/10.1007/s11356-019-07137-z

Krgović N, Jovanović MS, Nedeljković SK, Šavikin K, Lješković NJ, Ilić M, Živković J, Menković N (2025) Natural deep eutectic solvents extraction of anthocyanins—effective method for valorisation of black raspberry (*Rubus occidentalis* L.) pomace. Ind Crops and Prod 223:120237. https://doi.org/10.1016/j.indcrop.2024.120237

Kumar A et al (2023) Advances in enzymatic extraction of plant bioactives. Ind Crops and Prod 195:116–128

Kumar Gupta V (2020) Fundamentals of natural dyes and its application on textile substrates. In: Kumar Samanta A, Awwad NS, Majdooa Algarni H (eds) Chemistry and technology of natural and synthetic dyes and pigments. IntechOpen

Laura G, Tiwari B, Garcia-Vaquero M (2020) Emerging extraction techniques: microwave-assisted extraction. In SST. Elsevier, pp 207–24

Martins R, Mouro C, Pontes R et al (2023) Natural deep eutectic solvent extraction of bioactive pigments from spirulina platensis and electrospinning ability assessment. Polymers 15:1574. https://doi.org/10.3390/polym15061574

Masyita A, Hardinasinta G, Astuti AD et al (2025) Natural pigments: innovative extraction technologies and their potential application in health and food industries. Front Pharmacol 15:1507108. https://doi.org/10.3389/fphar.2024.1507108

Nonglait D, Gokhale JS (2024) Review insights on the demand for natural pigments and their recovery by emerging microwave-assisted extraction (MAE). Food Bioprocess Technol 17(7):1681–1705. https://doi.org/10.1007/s11947-023-03192-0

Periyasamy AP (2024) Recent advances in the remediation of textile-dye-containing wastewater: prioritizing human health and sustainable wastewater treatment. Sustainability 16:495. https://doi.org/10.3390/su16020495

Pizzicato B, Pacifico S, Cayuela D et al (2023) Advancements in sustainable natural dyes for textile applications: a review. Molecules 28:5954. https://doi.org/10.3390/molecules28165954

Plyduang T, Monton C, Maneewattanapinyo P, Suksaeree J (2025) Advancing sustainable phytochemical extraction through design of experiments: a data-driven pathway toward low-emission natural product processing. Sustain Chem Clim Act 7:100137. https://doi.org/10.1016/j.scca.2025.100137

Pourmortazavi S, Rahimi-Nasrabadi M, Hajimirsadeghi S (2013) Supercritical fluid technology in analytical chemistry–review. Curr Anal Chem 10(1):3–28. https://doi.org/10.2174/1573411011410010004

Rehman FU, Adeel S, Haddar W et al (2022) Microwave-assisted exploration of yellow natural dyes for nylon fabric. Sustainability 14:5599. https://doi.org/10.3390/su14095599

Saha A, Ayman H (2022) A review on the extraction methods of food colours from natural sources. Ann Food Sci Technol 23(2):274–286

Salauddin Sk, Mia R, Haque MA, Al Mojnun S (2021) Review on extraction and application of natural dyes. TLR 4(4):218–223. https://doi.org/10.31881/TLR.2021.09

Saxena S, Raja ASM (2014) Natural dyes: sources, chemistry, application and sustainability issues. In: Muthu SS (ed) Roadmap to sustainable textiles and clothing. Springer, Singapore, pp 37–80

Sen TK, Dawood S (2014) Review on dye removal from its aqueous solution into alternative cost-effective and non-conventional adsorbents. J Chem Proc Eng 1:1–11

Sharmila S, Kowsalya E, Kamalambigeswari R et al (2017) Production of natural dye by solid state fermentation. Int J Chemtech Res 10(4):219–225

Shrivastava A, Daniel M (2003) Extraction methods for natural dyes. In: Daniel M, Bhattacharya SD, Arya A, Raole VM (eds) Natural dyes: scope and challenges, pp 67–80

Sinha K, Chowdary S, Das Saha P, Datta S (2013) Modelling of microwave-assisted extraction of natural dye from seeds of *Bixa orella* (Annatto) using response surface methodology (RSM) and artificial neural network (ANN). Ind Crops Prod 41:165–171

Sinha K, Saha PD, Ramya V, Datta S (2012) Improved extraction of natural blue dye from butterfly pea using microwave-assisted methodology to reduce the effect of synthetic blue dye. Int J Chem Technol 4(2):57–65

Streimikyte P, Viskelis P, Viskelis J (2022) Enzymes-assisted extraction of plants for sustainable and functional applications. IJMS 23:2359. https://doi.org/10.3390/ijms23042359

Swami SB, Ghgare SN, Swami SS et al (2020) Natural pigments from plant sources: a review. PIJ 9(10):566–574

Tamburini D (2021) Colour analysis: an introduction to the power of studying pigments and dyes in archaeological and historical objects. Heritage 4:4366–4371

Vithushika K, Selvaskanthan S, Kirija P (2024) Extraction of natural dye from flowers and dyeing cotton. SLJoAS 3(1):2024. https://orcid.org/0009-0008-5352-8323

Wizi J, Wang L, Hou X et al (2018) Ultrasound-microwave assisted extraction of natural colorants from sorghum husk with different solvents. Ind Crops Prod 120:203–213. https://doi.org/10.1016/j.indcrop.2018.04.068

Zaky AA, Akram MU, Rybak K et al (2024) Bioactive compounds from plants and by-products: novel extraction methods, applications, and limitations. AIMSMOLES 11:150–188. https://doi.org/10.3934/molsci.2024010

Chapter 4
Characterisation of Natural Dyes

Abstract Natural dye characterisation is important for the identification of chemical constituents and understanding the colour performance, stability, and industrial suitability. This chapter outlines the principles and methods used for the characterisation of natural dyes and pigments using physical, spectroscopic, chromatographic, microscopic, and thermal techniques. Chromophores and auxochromes are responsible for colour. Fastness properties, such as light, wash, and rub are affected by mordants, substrates, and environmental conditions. Advanced analytical techniques, such as Fourier transform infrared (FTIR) spectroscopy, UV–visible spectroscopy, GC–MS, microscopy, and thermal analyses (TGA and DSC), are essential for the identification of functional groups, optical properties, molecular composition, morphology, and thermal stability of natural dyes. Photostabilisation, metal coordination, antioxidants, and hybrid inorganic frameworks enhance the stability of dyes. The characterisation of natural dyes gives an idea of understanding the active colouring components in substrates with their effective utilisation, thereby enhancing long-term stability.

4.1 Background

The characterisation of natural dyes means the process of identifying and analysing the properties of constituent compounds found within a raw extract. It is an essential prerequisite for their application in various items, such as textiles or cosmetics. To achieve better dyes, using suitable analytical techniques (such as physical, spectroscopic, chromatographic, microscopic, and thermal methods) is necessary so that there is also a comprehensive profile of the dye intended for use (Cook 1982; Ahn and Obendorf 2006; Chang et al. 2013). The data obtained from the characterisation of dyes helps in improving dye stability and longevity, so that economic benefits can be achieved by using natural dyes. Certain measures are necessary to avoid the degradation of natural dyes, as they are easily destroyed by light, heat, oxygen, water, etc. For example, it is easy to use spectroscopic and chromatographic characterisation techniques to observe the breakdown of anthocyanin under UV light or the oxidation of curcumin in an alkaline environment. Understanding at the molecular level is

N. L. Bhandari et al., *Natural Dyes and Pigments of Nepal Himalaya*,
https://doi.org/10.1007/978-3-032-23229-8_4

highly important, as it allows for the development of intelligent stabilisation strategies designed to protect vulnerable molecules, such as the use of specific mordants and the selection of types of antioxidants or protective coatings. Thus, to predict an accurate shelf life, key component identification of dyes and pigments through characterisation is very important so that stability is established and the chemical composition of dyes and pigments functions properly.

4.2 Methods of Characterisation

4.2.1 Physical Techniques

The primary goal of physical characterisation of natural dyes and pigments is to assess the type of colour and fastness properties. Assessing colour fastness properties that are affected by washing, light, and rubbing is also necessary to find the durability of natural dyes and pigments used for different purposes. In addition, several factors, like types of mordant used, maintaining pH and temperature, proper use of substrate, and optimising other environmental conditions, must be considered, as they directly affect the durability of the colour obtained from natural dyes and pigments (Cook 1982; Ali et al. 2009).

4.2.1.1 Colour

Chromophores (molecules that absorb light) and auxochromes (–OH, $-NH_2$, –COOH, etc.) are responsible for the observation of colour, as different wavelengths of light are absorbed for the visualisation of colour. Natural dyes and pigments contain various chromophores, which are responsible for their colour by absorbing specific wavelengths of light. Some of the chromophores are diazinyl (N=N) groups, keto groups (C=O), ethylenic groups (C=C), cyano groups (–CN), etc.

Auxochromes are also responsible for modifying colour, and they are functional groups that modify the chromophore's ability to absorb light, shifting the colour's wavelength and/or intensity. Examples of different auxochromes are amino groups ($-NH_2$), hydroxyl groups (–OH), and alkoxyl groups (-OR) (Bhandari et al. 2020b). When dye or pigment molecules absorb visible light at certain wavelengths, electronic transitions occur commonly from bonding to antibonding orbitals, or from non-bonding to antibonding orbitals. This leads to the condition of having an energy gap and creating a transition state that is responsible for colour appearance (Bhandari et al. 2021b; Harsito et al. 2021; Yadav et al. 2023).

4.2.1.2 Fastness Properties

The fastness properties of dyes or pigments are as important as the colour quality on substrates/fabrics, and the performance is affected by the use of different mordants and the type of substrate used (Mongkholrattanasit et al. 2010). The fastness properties of colour are also affected by other additional factors, such as water, chemicals, temperature, humidity, light exposure, pre- and post-treatment processes, dispersion, and dye fixation (Mongkholrattanasit et al. 2010; Zarkogianni et al. 2011). Natural dyes and pigments are very sensitive in terms of retaining colour and durability. Therefore, special care is necessary when selecting the source and the proper use of the dyeing method (Zarkogianni et al. 2011; Okonkwo et al. 2019; Kumar Gupta 2020). However, the use of natural dyes and pigments has seriously declined; thus, different measures should be implemented to improve the quality and promote the wider use of natural dyes and pigments (Okonkwo et al. 2019; Kumar Gupta 2020; Mandal and Venkatramani 2023). The different types of fastness properties are described in the following section.

(a) ***Light fastness***

Natural dyes and pigments usually show low to moderate lightfastness because their organic structures are prone to breakdown by light energy. When molecules of natural dyes and pigments are exposed to light, chromophores may undergo quick structural changes and can easily fade in colour. Generally, the chromophores in natural dyes and pigments are not strong enough and cannot efficiently release absorbed energy through the resonance process. This process leads to structural degradation (Cristea and Vilarem 2006; Kumar Gupta 2020). For the enhancement of fade resistance of bio-dyes and pigments upon fabric, it is necessary to know the mechanisms underlying light fading. For example, using tannin-based mordants during cotton/fabric dyeing is one of the best methods that can improve not only light and wash fastness but also enhance the overall durability of natural dyes or pigments on fabrics. However, compared to synthetic dyes, natural dyes and pigments have lower light stability (Cook 1982).

A study related to the lightfastness properties of cotton fabrics using the dye extracted from *Whitfieldia lateritia* upon exposure to daylight for 72 h retained effective colour compared to unexposed fabrics, showing moderate daylight stability (Okonkwo et al. 2019). The colour endurance of dyed wool with different natural dyes were also assessed by Padfield and Landi (1966), and they reported that the light fastness grades ranged from 1–2 for yellow dyes (ancient fustic and Persian berries), 3–4 for red dyes (cochineal with tin mordant, alizarin with alum mordant, and lac with tin mordant), 4–6 for blue (indigo), and 4–5 for logwood (black). This study explains that lightfastness varies based on the dye type and fixing agent used to dye the material (Padfield and Landi 1966).

The appropriate lightfastness in natural dyes and pigments is determined by the type of mordant used. Bio-dyes used with conventional fixing agents, such as alum and tin, show poorer lightfastness properties than chromium mordant on the same colourant. The natural dyes derived from turmeric, *Berberis asiatica*, *Maclura tinctoria*, and marigold, applied along with mordants like iron and copper, show a weaker shade compared to other yellow colourants (Umbreen et al. 2008; Semwal et al. 2012; Kumar Gupta 2020). Thus, to strengthen dye durability, a suitable fixing agent should be chosen. Despite this, the light resistance property is a more difficult factor than others for enhancing natural dye fastness (Cristea and Vilarem 2006). That means the utilisation of appropriate mordants results in enhanced colour (Umbreen et al. 2008; Bhandari et al. 2021a, b). So, light resistance is improved by using textile additives such as UV absorbers and singlet oxygen quenchers (Oda 2001; Ahn et al. 2014; Kumar Gupta 2020).

To improve the lightfastness of natural dyes and pigments, several methods are used, and one approach involves studying the photofading process, which examines how a specific dye, like carthamin, fades on a substrate and evaluates the protective effects of various additives during dyeing. The application of specific compounds, such as nickel hydroxylaryl sulphonate, has been shown to decrease the rate of light-induced fading. Alternatively, UV absorbers can be added directly to the dye bath. However, this method offers only limited protection against photodegradation. Thus, there is no universal method that has been developed for all natural dyes or pigments to improve lightfastness.

(b) ***Wash and rub fastness***

The wash fastness properties of natural dyes and pigments are tested to preserve their original colour after multiple washings of fabrics in the textile industry. The colour retention can vary, and it is evaluated by subjecting dyed fabrics to multiple washings and rating the extent of colour change from excellent to poor (Alegbe and Uthman 2024). Because natural dyes and pigments make weak chemical bonds with cotton fabrics, they exhibit weak to moderate wash fastness in them (Hassan et al. 2024). Thus, they tend to fade after washing, mainly in the presence of detergent solutions (Zarkogianni et al. 2011).

The study on *Santalum album* with dye obtained from natural sources to dye jute cloth showed that the jute fabrics dyed with red sandalwood, either alone or in combination with 2% hexadecyltrimethylammonium bromide (CTAB), commonly known as Sandofix-HCF, improved wash fastness almost to the first grade. This shows the effectiveness of cationic surfactants in making a bond between natural dyes and fibres during washing. The use of such a process is necessary for post-dye treatments to make perfect bonds between natural dye molecules and textile fibres to prevent colour loss during washing (Samanta et al. 2009) and enhance long-term wash durability (Samanta et al. 2009; Datta et al. 2013).

The stability of the dye to friction is evaluated by applying crocking tests (Geelani et al. 2017). It helps to measure the durability and quality of dyed textile and other materials. These tests are commonly used in routine quality control (Mongkholrattanasit et al. 2010; Geelani et al. 2017). Cotton fabric dyed with marine algae dyes (*Chaetomorpha antennia, Gracilaria corticota,* and *Ulva fasciata*) has exhibited exceptional rubbing fastness (Sharmila et al. 2017). Similarly, the extract of eucalyptus leaves revealed excellent wash fastness. This extract also stays for a long period on silk and wool fabrics (Mongkholrattanasit et al. 2010).

4.2.2 *Spectroscopic Techniques*

Nowadays, modern approaches like spectroscopic techniques have been applied to identify natural dyes in artworks. But this method is often ineffective and presents notable issues in the dye industry. The problem arises due to the limited availability of samples, low concentration of chromophore, decomposition of products, and unidentified formulations. The process of recognising suitable dyes for specific textiles and other substances is challenging (Degano et al. 2009; Bhandari et al. 2020a). Under such conditions, analysis of natural dyes is done by using spectroscopic techniques. These techniques identify dye-sensitive molecules without destroying them. Some of the spectroscopic techniques are explained below (Samanta et al. 2009).

4.2.2.1 Fourier Transform Infrared Spectroscopy (FTIR)

This technique is frequently used to identify the functional groups of natural dyes and pigments. During this process, the infrared radiation interacts with a dye or pigment sample, where certain wavelengths are absorbed, and the remaining are transmitted through, and the absorbed spectrum shows the vibrational frequency of dye molecules (Hassan et al. 2024). These spectra act as molecular fingerprints of functional groups, which are used in the identification of the bonding structure of natural dyes or pigments. The functional groups, such as amines ($R\text{-}NH_2$), amides ($RCONH_2$), carbonyls (C=O), carboxylic acids (RCOOH), esters (RCOOR'), and alcohols (R-OH), are identified, and the fingerprint is obtained based on complex vibration modes (Chang et al. 2013; Bhandari et al. 2021a).

The intensity of the absorption peaks correlates with dye concentration, and the spectrum is typically plotted as absorbance or transmittance versus wavelength or wavenumber. According to the electromagnetic radiation energy (E), its frequency (υ), wavelength (λ), and wave number ($\bar{\upsilon}$), the spectrum is obtained. It is described in Eqs. 4.1 and 4.2. Radiations that perfectly match the vibrational energy levels of dye molecules are absorbed and detected in FTIR analysis.

$$E = h\upsilon \tag{4.1}$$

$$\overline{\upsilon} = hc/\lambda, \tag{4.2}$$

where, λ (lambda) is the light's wavelength; $\overline{\upsilon}$ (nu) represents the energy of one photon; h is Planck's constant, which relates frequency to photon energy (approximately 6.626×10^{-34} J-seconds); and c is the speed of electromagnetic radiation.

Using this principle, studies have demonstrated the effectiveness of FTIR in identifying and characterising the functional components of natural dyes (Bhandari et al. 2021a; Aluko 2024; Bekele et al. 2024). The FTIR absorption of *Phyllanthus emblica* dye extracts showed broad and strong peaks in 3325 cm^{-1} along 3332 cm^{-1} (Fig. 4.1a), assigning O–H stretching vibrations of phenolic compounds like tannins and flavonoids. Aromatic C=C stretching appeared at 1608 cm^{-1} and 1613 cm^{-1}. At 1443 cm^{-1} and 1449 cm^{-1}, the bending and stretching modes of vibration of CH_3, CH_2, and aromatic rings were observed. The phenolic C–O stretching was present between 1200 and 1182 cm^{-1}, whereas peaks at 1033 cm^{-1} and 1028 cm^{-1} indicated the C–O–C stretching (ether groups and pyranose rings). Similarly, aromatic C–H vibrations and O–H wagging intensified at 882 cm^{-1} and in the 720–590 cm^{-1} range, respectively (Bhandari et al. 2021a).

Bixin and norbixin dyes extracted from *Bixa orellana L.* showed similar characteristic peaks to *Phyllanthus emblica.* In the analysis, stretching of –OH (3331 cm^{-1}), –CH_2 bending (2974 cm^{-1}), carboxylic group (1651 cm^{-1}), C=C group (1450 cm^{-1}), –CH bending of methyl groups (1381 cm^{-1}), C=O stretching (1274 cm^{-1}), and ester show vibrations at 1085 and 1043 cm^{-1} were observed. The

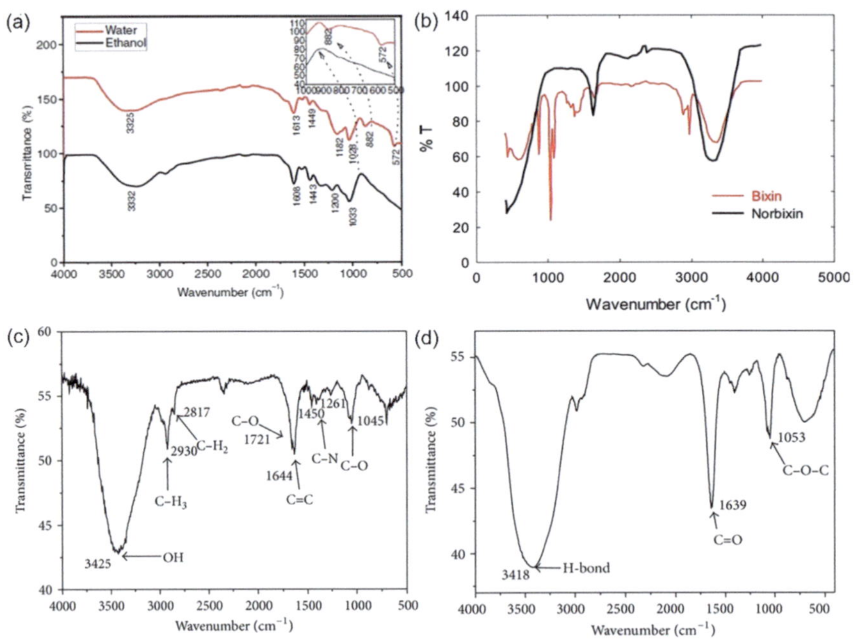

Fig. 4.1 FTIR spectra of **a** dye extract from *Phyllanthus emblica* (Bhandari et al. 2021a); **b** bixin and norbixin dye extract from *Bixa orellana* L. (Aluko 2024); **c, d**chlorophyll and anthocyanin dye extracts from wormwood and purple cabbage, respectively (Chang et al. 2013)

methylene rocking vibrations of trans- and cis-carotenoids appeared at 879, 846, and 798 cm^{-1} (Fig. 4.1b). Norbixin showed O–H stretching (3302 cm^{-1}), bending of $-CH_2$ (2391 and 2193 cm^{-1}), along with carboxylic C=O at (1635 cm^{-1}). These functional groups are commonly found in flavonoids, anthocyanins, and polyphenols and are major colour sources in natural dyes. Wormwood and purple cabbage are used to extract chlorophyll and anthocyanin natural pigment, when analysed using FTIR spectra (Fig. 4.1c, d), showed CH_3 and CH_2 stretching at 2930 cm^{-1} and 2817 cm^{-1}, –CO vibration at 1721 cm^{-1}, –CN vibration at 1644 cm^{-1} (porphyrin structure), and C–O about 1045 cm^{-1} in chlorophyll dye. Anthocyanin dye showed H-bond-related O–H stretching at 3418 cm^{-1}, C=O at 1639 cm^{-1}, and ester stretching at 1053 cm^{-1} (Chang et al. 2013).

All the above data indicate that natural dyes contain similar functional groups like hydroxyl (3200–3600 cm^{-1}), methyl/methylene (2850–2950 cm^{-1}), carbonyl (1650–1750 cm^{-1}), aromatic rings (1450–1600 cm^{-1}), ethers and esters (1000–1300 cm^{-1}), and either hydroxyl groups or transverse bending vibrations of aromatics (600–900 cm^{-1}) (Chang et al. 2013; Bhandari et al. 2021a; Aluko 2024; Bekele et al. 2024).

4.2.2.2 UV–Visible (UV–Vis) Spectroscopy

The analysis of radiation absorption within the ultraviolet and visible areas of the radiation spectrum is the UV–Visible spectroscopy technique. For the identification of optical properties of natural dyes, this method is broadly utilised (Degano et al. 2009), and in the analysis, this technique shows two regions, which are the UV region (wavelengths ranging from 10 to 400 nm) and the visible region, with wavelengths ranging from 400 to 800 nm. Within this region, if a wavelength exists between 200 and 400 nm, then the region is known as the near ultraviolet region, and if the wavelength is from 10 to 200 nm, then the region is known as the far ultraviolet or vacuum UV region. However, the visible range is from 400 to 800 nm, and absorption of light in these regions gives rise to the colour of dye molecules (Asen et al. 1972; Cristea and Vilarem 2006; Samanta and Agarwal 2009; Samanta et al. 2009; Skoog et al. 2017).

In UV–Vis, the amount of UV or visible light absorbed or transmitted by a dye sample is compared to a reference, and it is useful for the identification of the molecular structure of the dye (Samanta et al. 2009; Chang et al. 2013). Thus, the wavelength at which a natural dye exhibits its maximum absorption (λ_{max}) identifies its chemical nature and structure (Alegbe and Uthman 2024). If the molecular environment, functional groups, or solvents are changed, then variations in the peak position and intensity of UV–Vis spectra are also observed (Fig. 4.2). These changes help to explain the mechanisms and interactions in dye molecules.

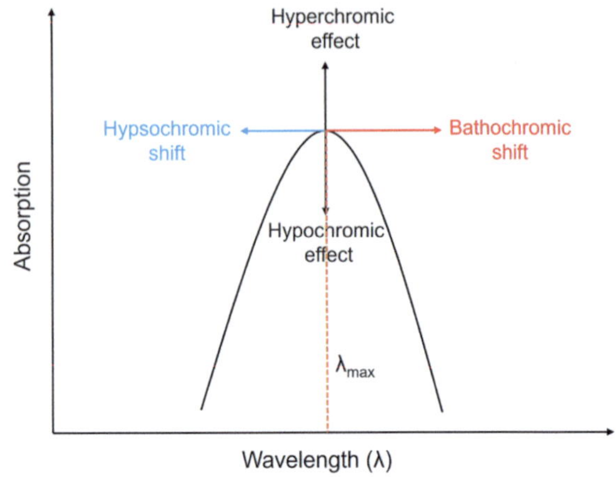

Fig. 4.2 Bathochromic, hypsochromic, hyperchromic, and hypochromic shifts in absorption position and intensity in UV–Vis spectroscopy

Bathochromic Shift (Red Shift)

It appears due to a change in the peak of absorption to a lower energy. Such an effect arises due to the presence of various substituent groups or solvents, and the introduction of an –OH and $-NH_2$ group in flavonoid dyes enhances conjugation and shifts absorption toward the red end of the spectrum, which is an example of red shift (Asen et al. 1972; Morales et al. 2012).

Hypsochromic Shift (Blue Shift)

Hypsochromic or blue shift occurs when the maximum absorption peak moves towards shorter wavelengths. This effect is influenced by solvent polarity or by the reduction of the conjugation system, such as through the protonation of an electron-donating group (Kemp 1991; Li et al. 2023). For example, protonation of $-NH_2$ groups in anthocyanins can cause a loss of electron delocalisation in acidic media, resulting in a spectral shift into shorter wavelengths (Asen et al. 1972).

Hyperchromic Effect

This effect is an intensification (ε_{max}) of the maximum absorbance caused by solvents and substitutions. Auxochromes (such as –OH and $-NH_2$) can boost the absorption intensity of dyes by increasing the electron density and conjugated system of the solution. For example, if the amino group is added to a phenolic solution of natural dyes, there is a significant increase in the absorption intensity because of an improvement in electronic transitions (Kemp 1991; Morales et al. 2012; Bhandari et al. 2021a).

Hypochromic Effect

The hypochromic effect is known as a decrease in the absorption intensity, and it is also caused by the solvent's effect. In natural dyes, this effect may result from structural distortion or steric hindrance that disrupts planarity and the conjugation of the chromophore. An example of the hypochromic effect is that methyl substitution at ortho positions can twist aromatic rings out of coplanarity, reducing conjugation and diminishing absorbance (Kemp 1991; Estiningtyas et al. 2023; Pal et al. 2025).

Chromophore

A chromophore is an unsaturated group bonded covalently that absorbs light in the region of the UV or visible spectrum. Depending on the absorption of visible or UV rays, a chromophore provides a specific colour to the dye molecule. There are no defined standardised procedures for identifying a chromophore because the wavelength and intensity of absorption vary depending on the physical conditions (Kemp 1991; Skoog et al. 2017).

Auxochrome

An auxochrome is a covalently bound saturated group that modifies the maximum absorbance wavelength and intensity. Extending the conjugation through resonance, auxochromes often increase the value of the maximum (λ_{max}) and minimum (ε_{max}) wavelengths (Skoog et al. 2017). Auxochromes do not absorb light themselves at 200 nm, but chromophores and auxochromes in combination behave as a new chromophore differing in λ_{max} and ε_{max} values. For example, benzene (C_6H_6) has a λ_{max} of 255 nm and an ε_{max} of 204 nm, and aniline ($C_6H_5NH_2$) has a λ_{max} of 280 nm and an ε_{max} of 1430 nm (Kemp 1991).

The UV–Vis spectra of wormwood extract show a strong peak in the blue range (~400–500 nm), whereas the purple cabbage shows a shift is at the green range (530–550 nm). Their mixed dye extracts are shown in Fig. 4.3a (Chang et al. 2013). Similarly, a broad absorption band (from 450 to 700 nm) with a maximum peak at 538 nm for purple cabbage extract indicates the presence of anthocyanin as a major component (Ainfatin et al. 2020). The extract from madder and weld contains many active colouring agents that give distinct absorption peaks, as shown in Fig. 4.3b (Hosseinnezhad et al. 2021). Extracts of *Phyllanthus emblica* using water and ethanol as solvents have shown two distinct peaks for ethanol (278 and 380 nm) and one peak for water (274 nm) (Fig. 4.3c) that correspond to flavonoid-based phenolic compounds (Bhandari et al. 2021a). Additionally, components of *Castanopsis indica* dye extracts are shown in Fig. 4.3d (Bhandari et al. 2021b).

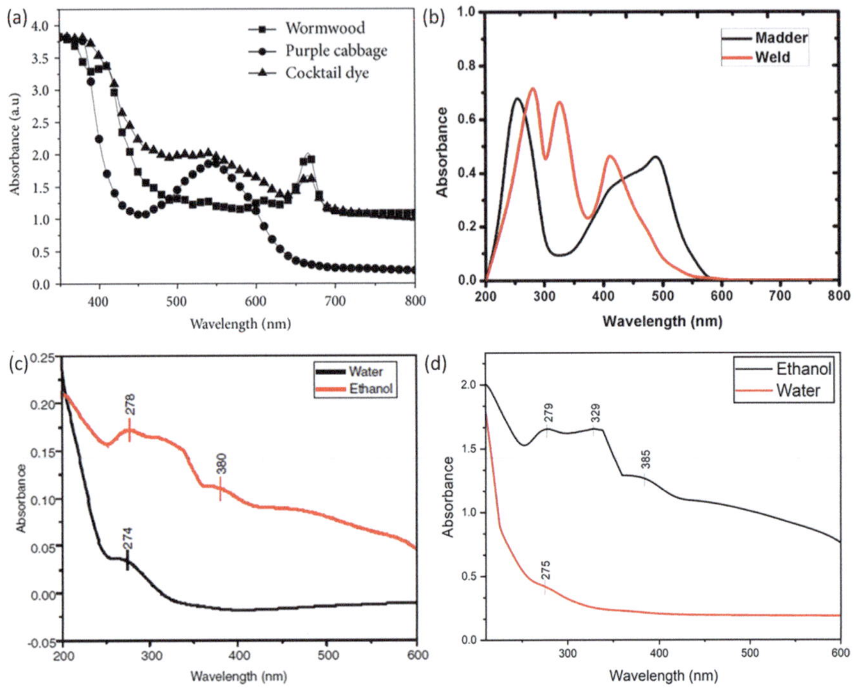

Fig. 4.3 UV–Vis spectra of **a** chlorophyll, anthocyanin, and mixed (cocktail) dyes (Chang et al. 2013); **b** madder and weld dyes (Hosseinnezhad et al. 2021); **c** *Phyllanthus emblica* dye (Bhandari et al. 2021a); and **d** *Castanopsis indica* dye (Bhandari et al. 2021b)

4.2.3 Chromatographic Techniques

Chromatography is the method that enables the separation and identification of individual components of a complex natural dye. In this method, a complex dye is subjected to a mobile phase, and then transported through a stationary phase (Degano et al. 2009). Separation of the required analyte is done on the principle of different affinities of each analyte, and resolution is optimised by modifying the mobile or stationary phase properties, temperature, or pressure during the separation process (Deveoğlu et al. 2012; Kaur 2020).

It is a remarkable hyphenated analytical technique to examine the natural dyes. It serves both for separation as well as comprehensive identification of the colourants or pigments. It is an effective analytical approach. It functions by connecting two essential steps to study complicated dye mixtures. Initially, the mixture is separated into its specific chemical components by gas chromatography. Then, mass spectroscopy characterises every individual component according to its molecular mass. The molecules that occur in a sample are recognised distinctly by using this dual method. It results in high sensitivity and precision.

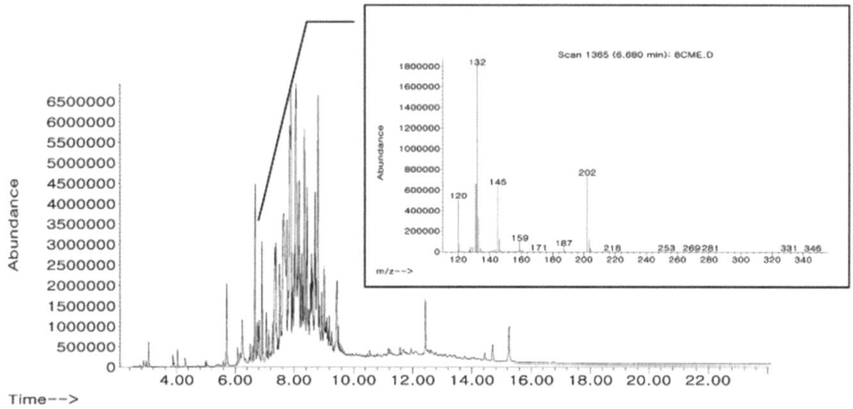

Fig. 4.4 GC analysis of natural dye extracted from turmeric (inset: mass spectrum of curcumene) (Ahn and Obendorf 2006)

For the identification of dyeing properties, several researchers have used GC–MS. Colombini and collaborators recognised varied dye-based compounds by using this technique. For example, they extracted the dye solution from *Rhamnus* species, which contains pigments like morin, genistein, fisetin, kaempferol, apigenin, rhamnetin, quercetin, myricetin, and luteolin (Colombini et al. 2007). Likewise, turmeric dye extract was identified using GC–MS investigation, and different compounds were identified, including curcumene, feruloylmethane, coumaran, vanillin, and zingiberene as the predominant analytes. Figure 4.4 provides the GC chromatogram of the turmeric extract together with the MS profile of curcumin (Ahn and Obendorf 2006). Similarly, using GC–MS analysis on various natural dyes identified flavonoid, quinone, and indigoid compounds. Various pigments, such as alizarin, emodin, apigenin, genistein, kaempferol, fisetin, quercetin, and maclurin (Poulin 2018) were also identified. As illustrated in Fig. 4.5, the natural dye extracts from *Acacia nilotica* pods are composed of organic compounds such as pyrogallol, 6-octadecanoic acid, benzenediol, pyrocatechol, pentadecanoic acid, octadecanoic acid, dibutyl phthalate, benzene, palmitoyl chloride, squalene, and hexadecanoic acid (Jamadar and Sannapapamma 2018).

4.2.4 Microscopic Techniques

Microscopic techniques are used in the characterisation of natural dyes, providing a detailed morphological structure and surface features. This technique also helps to visualise dye particles at the micro- and nanoscale and provides important information associated with the dye extraction process. Different microscopic tools like optical microscopy (OM), scanning electron microscopy (SEM), transmission electron microscopy (TEM), and atomic force microscopy (AFM) are used in the study.

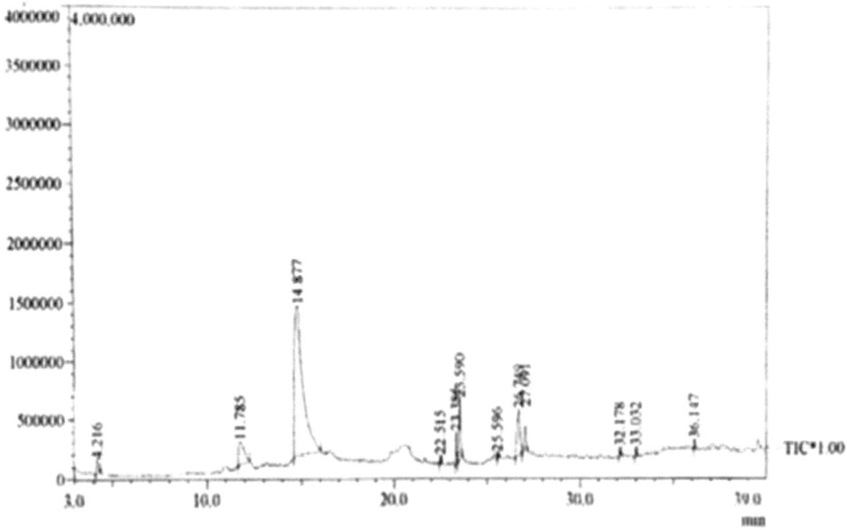

Fig. 4.5 GC–MS spectrum of solvent extracts of *Acacia nilotica* pod dye (Jamadar and Sannapapamma 2018)

The microscopic characterisation technique is utilised to upgrade the dyeing process. Moreover, they facilitate enhanced intensity and strength of bio-colourants (Luo and Zhang 2021; Chajii et al. 2025). Figure 4.6 explains the surface configuration of *Phyllanthus emblica* dye extracted using water and ethanol solvents. The dye extracted from the aqueous solvent shows nonhomogeneous, irregular particles with a curved structure. On the other hand, the ethanolic extract indicates comparable uneven particles. The difference is observed due to their distinct molecular composition (Bhandari et al. 2021a).

a. Water extract of Amala b. Ethanol extract of Amala

Fig. 4.6 Surface morphology of *Phyllanthus emblica* (Amala) dyes: **a** water extract; **b** ethanol extract (Bhandari et al. 2021a)

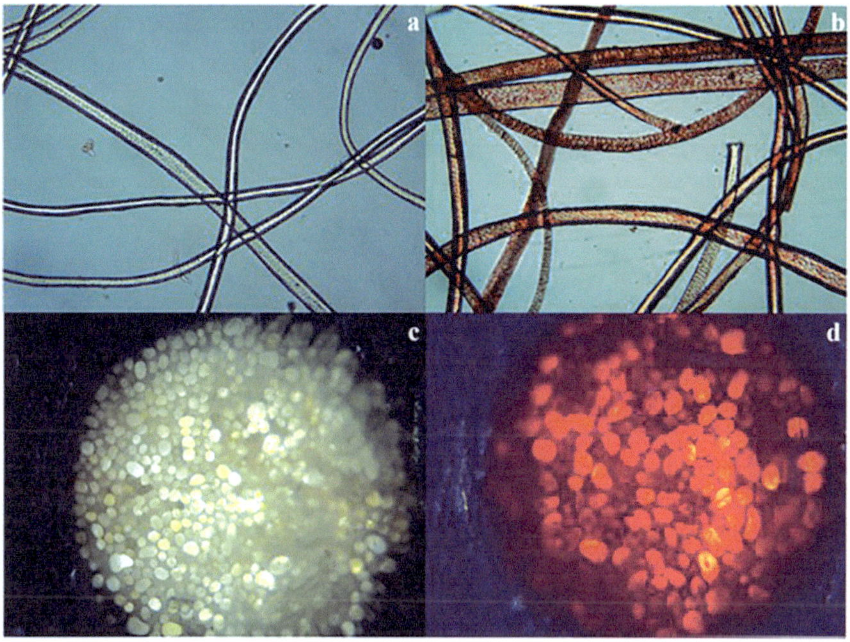

Fig. 4.7 Optical microscope images of wool samples in longitudinal view: **a** undyed, **b** dyed; cross-section: **c** undyed, **d** dyed (Chajii et al. 2025)

The microscopic study of dyed and undyed wool applying *Rubia tinctorum L.* has displayed a prominent colour change in dyed fibre. It is because of the absorption of the dye on the wool fabric. Figure 4.7 illustrates the constant and long-lasting colour along with enhanced homogeneity across the whole fabric. The images verify profound dispersion and homogeneous permeation of the dye into the fibre (Chajii et al. 2025). Also, microscopic properties describe functional characteristics like antimicrobial activity and mechanical strength of the dyed substrates (Luo and Zhang 2021). Therefore, it acts as a major factor in outlining the structural quality along with the interaction of dye and substrate. This facilitates modernising the potent utilisation of bio-dyes and pigments.

4.2.5 Thermal Techniques

Various characterisation techniques like thermogravimetric analysis (TGA), differential scanning calorimetry (DSC), and dynamic mechanical analysis (DMA) are used to study the thermal nature of dyes. Among these, TGA is broadly utilised to measure thermal stability. It works under a controlled atmosphere, either in the presence of nitrogen or oxygen. It observes the weight change in a sample as a function of temperature. Samples sensitive to oxidative decay are generally examined in the presence of inert nitrogen to prevent ignition (Nguyen and Bechtold 2021; Raji et al. 2023). The decayed or evaporated sample gives a negative mass variation during thermal contact. Similarly, a positive value occurs if the sample is oxidised. Definite decomposition steps can be studied using DTG. Such special methods are used in the presence of a sensitive thermobalance. It allows both physical and chemical transformation by evaluating continuous change in mass during heating. The physical transformation refers to a change in phase, whereas chemical transformation involves the release of volatile molecules and the decomposition of molecules (Nguyen and Bechtold 2021; Raji et al. 2023; SenthilKumar et al. 2023).

Differential scanning calorimetry (DSC) is also used to investigate thermal transitions to determine the melting, glass transition, and crystallisation of natural dye samples. It observes the temperature and heat flux during phase changes to determine the thermal response of dyes. It operates by passing energy to the sample and reference to balance the temperature in an inert atmosphere. A specified thermal program, such as heating or cooling at 10 °C per minute is applied. During this process, heat is regulated to maintain thermal equilibrium. Specifically, liquid nitrogen is used as a cooling agent. It functions in an oxygen-free-environment. A DSC thermogram shows the peak of the thermal process, consisting of enthalpic variation, the peak direction, and area, which determine the heat absorption or release. The technique results in significant details on molecular reconfiguration, phase transformation, and structural change in dyes in the presence of thermal stress (Nguyen and Bechtold 2021; Chajii et al. 2025).

Various studies have illustrated the thermal properties of natural dyes obtained from plant sources. Figure 4.8 shows the thermal evaluation of the precipitated dye derived from Canadian goldenrod and onion skin in the 30–800 °C range. The TGA curve denoted an initial weight loss at around 120 °C, because of the evaporation of water present in the surface, structure, and crystal lattice. Then the second change was seen at 120–150 °C, about 6–7% weight loss. It is mainly due to the chemical change and partial degradation of flavonoids within the dye material. Consequently, DCS data displayed an endothermic effect between 30 and 240 °C. That means the first and second stages exhibit polished water loss and a molecular rearrangement process. This discovery specifies the definite stability pattern of flavonoid-rich biodye (Nguyen and Bechtold 2021).

The extract acquired from *Acacia* flowers was analysed through the TGA and DSC methods in Fig. 4.9. It shows four separate phases of weight reduction in TGA. The initial 6% decrease illustrates a DTG maximum at 75 °C associated with the liberation of water. A second phase has a 0.7% difference attributed to cyclic structural reconfiguration. Similarly, a 3.94% reduction and apparent thermal activity at 200 °C were noted, indicating the primary degradation of the colourant, and the interference of chromophore fragments. The final reduction was found near 270 °C, resulting in an extensive 19.73% mass loss. An endothermic effect at 110 °C indicated moisture vaporisation, a glass transition between 70 and 100 °C, and crystallisation at 160 °C occurred as corresponding DSC scans. These three specific changes were

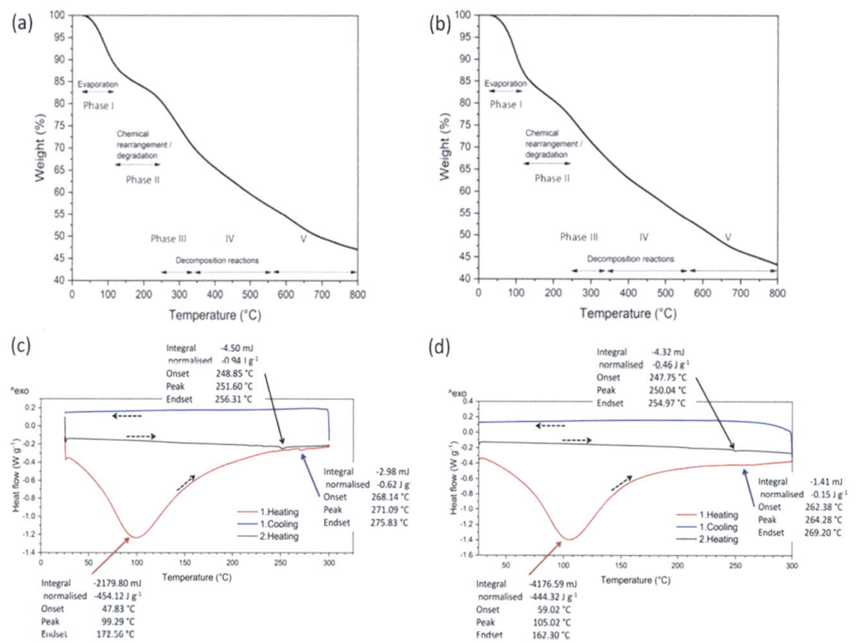

Fig. 4.8 TGA and DSC curves of natural dye lakes prepared from Canadian goldenrod and onion peel: **a** TGA curve of Canadian goldenrod; **b** TGA curve of onion peel; **c** DSC measurement of aluminium dye lake from Canadian goldenrod; and **d** DSC measurement of aluminium dye lake from onion peel. Black curve represents the first heating cycle, red for cooling, and blue for the second heating cycle (Nguyen and Bechtold 2021)

observed between 20 and 200 °C. The extra endothermic peak around 190 °C is referred to as a loss of an essential structural part within the dye complex (Fajraoui et al. 2019). Overall, TGA–DSC, which is an integrated technique, yields refined thermal behaviour of dyes applied for textile and other purposes.

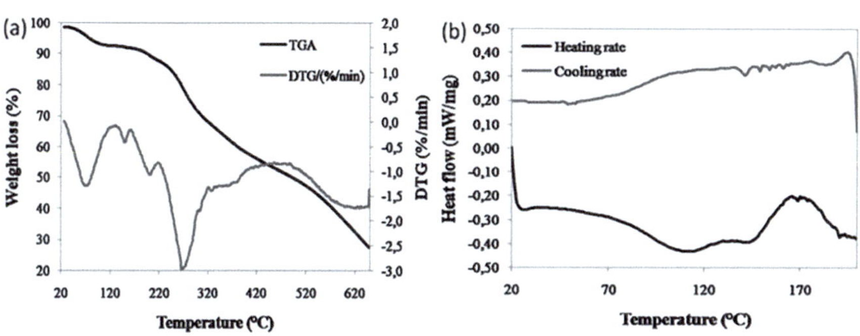

Fig. 4.9 **a** TGA and **b** DSC curves of dye extracted from *Acacia cyanophylla* flowers (Fajraoui et al. 2019)

4.3 Stability of Natural Dyes

Natural dyes and pigments are sensitive to light, heat, and reactive chemicals, which show limited resistance. Their stability can be substantially improved by using strategies including an antioxidant-mediated photostabilisation, metal-chromophore chelation, and hybridisation with inorganic frameworks. However, their hydrophilic nature makes natural dyes highly susceptible to oxidation in ultraviolet light, and the colour fades away very easily (Li et al. 2012; Shin and Yoo 2012; Degani et al. 2017). This intrinsic vulnerability can be alleviated through mechanisms of photostabilisation, and it is known as UV absorption. In this process, excited molecules are quenched, and free radicals are scavenged, allowing UV absorption to act in synergy, additively, or even opposition within a single system. However, given the rapid generation of excited states in natural dyes, small quantities of antioxidants generally fail to provide effective stabilisation.

The intensity of the dye in solution and thin polymer films was enhanced significantly by mixing indigo and antioxidants such as butylated hydroxytoluene or methylene blue with p-benzoquinone or triethylamine. Its effectiveness is due to an intermolecular reaction rather than UV shielding. It decreases the speed of light decomposition (Li et al. 2012). Also, the colour stability and duration are elevated at low pH and temperature by coordinating inorganic metal salts like aluminium, zinc, tin, etc. The process is studied in chlorophyll and anthocyanin matrix complexes (Shin and Yoo 2012). Moreover, incorporating bio dyes into inorganic complexes like zeolites Socony Mobil-5 or zeolite catalysts results in mixed pigment. It shows excellent stability towards chemical attack and degradation. Such durability can be observed in ancient pigments like Maya blue (Dejoie et al. 2010; Ouellet-Plamondon et al. 2015). These combined methods that are fixing, stabilising, and optimising extraction serve as a potent way for intense, durable natural dyes.

4.4 Conclusions

Natural dyes are widely adopted by various industries. Hence, the identification of the molecular structure present in natural dyes is essential. Their characterisationis fundamental to understanding their strength and effectiveness in the textile industry and other matrices. Physical characterisation helps to determine the stability of natural dyes and fastness properties, while spectroscopic, chromatographic, microscopic, and thermal techniques confirm the colouring component and functional group. Analytical techniques help indentify the factors that degrade and destabilise the colour. They facilitate the development of schemes like proper fixing agents and antioxidants for effective dyeing. Moreover, such advanced instruments enable accurate molecular comprehension; they govern enhanced extraction, preparation, and inspection. Therefore, the characterisation of natural dyes ensures enhanced strength and utility to sustain and upgrade them in several industries. It is essential for identifying their molecular structures.

References

Ahn C, Obendorf SK (2006) GC-MS analysis of dyes extracted from turmeric. Fibers Polym 7:158–163. https://doi.org/10.1007/BF02908260

Ahn C, Zeng X, Li L, Obendorf SK (2014) Thermal degradation of natural dyes and their analysis using HPLC-DAD-MS. Fash Text 1:1–13. https://doi.org/10.1186/s40691-014-0022-5

Ainfatin KR, Syahida S, Mohd Azman H et al (2020) Optimization of dye extraction from purple cabbage and cordyline fruticosa in dye-sensitized solar cell. Solid State Phenom 307 SSP:192–200. https://doi.org/10.4028/www.scientific.net/SSP.307.192

Alegbe EO, Uthman TO (2024) A review of history, properties, classification, applications and challenges of natural and synthetic dyes. Heliyon 10:e33646. https://doi.org/10.1016/j.heliyon.2024.e33646

Ali S, Hussain T, Nawaz R (2009) Optimization of alkaline extraction of natural dye from Henna leaves and its dyeing on cotton by exhaust method. J Clean Prod 17:61–66. https://doi.org/10.1016/j.jclepro.2008.03.002

Aluko E (2024) Ethanol-based extraction of Annatto (*Bixa Orellana* L.) and characterization of the Bixin and Norbixin. ACS Omega 9:18273–18277. https://doi.org/10.1021/acsomega.3c10120

Asen S, Stewart RN, Norris KH (1972) Co-pigmentation of anthocyanins in plant tissues and its effect on color. Phytochemistry 11:1139–1144. https://doi.org/10.1016/S0031-9422(00)88467-8

Bekele N, Abewaa M, Mengistu A et al (2024) Optimization of natural dye extraction from the root of Rumex abyssinicus plant using response surface methodology. Results Chem 8:101577. https://doi.org/10.1016/j.rechem.2024.101577

Bhandari NL, Bist K, Ghimire J et al (2020a) Feasibility study of the *Euphorbia pulcherrima* plant extract as natural dye with different mordants for fabric dyeing. J Inst Sci Technol 25:30–36. https://doi.org/10.3126/jist.v25i1.29421

Bhandari NL, Ghimire J, Shrestha S et al (2021a) Green extraction, characterization, applications and antimicrobial analysis of natural dye from phyllanthus emblica. Asian J Chem 33:404–410. https://doi.org/10.14233/ajchem.2021.23008

Bhandari NL, Pokhrel B, Bhandari U et al (2020b) An overview of research on plant based natural dyes in Nepal: scope and challenges. J Agric Nat Resour 3:45–66. https://doi.org/10.3126/janr.v3i2.32328

Bhandari NL, Shrestha S, Bhandari G et al (2021b) Extraction of dye from *Castanopsis indica* for its use in textile dyeing and medicinal purpose with natural mordant. Egypt J Chem 64:6681–6690. https://doi.org/10.21608/EJCHEM.2021.48519.2991

Chajii O, Chemchame Y, Zourif A et al (2025) Microwave-assisted extraction of natural dye from *Rubia tinctorum* L. plant for dyeing wool fabrics with multiple effects. Fibers Polym 26:2427–2448. https://doi.org/10.1007/s12221-025-00956-7

Chang H, Kao MJ, Chen TL et al (2013) Characterization of natural dye extracted from wormwood and purple cabbage for dye-sensitized solar cells. Int J Photoenergy 2013. https://doi.org/10.1155/2013/159502

Colombini MP, Andreotti A, Baraldi C et al (2007) Colour fading in textiles: a model study on the decomposition of natural dyes. Microchem J 85:174–182. https://doi.org/10.1016/j.microc.2006.04.002

Cook CC (1982) Aftertreatments for improving the fastness of dyes on textile fibres. Rev Prog Color Relat Top 12:73–89. https://doi.org/10.1111/j.1478-4408.1982.tb00228.x

Cristea D, Vilarem G (2006) Improving light fastness of natural dyes on cotton yarn. Dye Pigment 70:238–245. https://doi.org/10.1016/j.dyepig.2005.03.006

Datta S, Uddin M, Afreen K et al (2013) Assessment of antimicrobial effectiveness of natural dyed fabrics. Bangladesh J Sci Ind Res 48:179–184. https://doi.org/10.3329/bjsir.v48i3.17327

Degani L, Gulmini M, Piccablotto G et al (2017) Stability of natural dyes under light emitting diode lamps. J Cult Herit 26:12–21. https://doi.org/10.1016/j.culher.2017.02.002

Degano I, Ribechini E, Modugno F, Colombini MP (2009) Analytical methods for the characterization of organic dyes in artworks and in historical textiles. Appl Spectrosc Rev 44:363–410. https://doi.org/10.1080/05704920902937876

Dejoie C, Martinetto P, Dooryhée E et al (2010) Association of Indigo with zeolites for improved color stabilization. Appl Spectrosc 64:1131–1138. https://doi.org/10.1366/000370210792973622

Deveoğlu O, Torgan E, Karadağ R (2012) High-performance liquid chromatography of some natural dyes: analysis of plant extracts and dyed textiles. Color Technol 128:133–138. https://doi.org/10.1111/j.1478-4408.2012.00358.x

Estiningtyas IW, Kusumawati N, Setiarso P et al (2023) Effect of natural dye combination and pH extraction on the performance of dye-sensitized photovoltaics solar cell. Int J Renew Energy Dev 12:1054–1060. https://doi.org/10.14710/ijred.2023.56172

Fajraoui A, Ben Nasr J, Lacoste C et al (2019) Coloration of the polylactic acid with the natural dye extracted from acacia cyanophylla flowers. Polym Test 78:105988. https://doi.org/10.1016/j.polymertesting.2019.105988

Geelani SM, Ara S, Mir NA et al (2017) Dyeing and fastness properties of *Quercus robur* with natural mordants on natural fibre. Text Cloth Sustain 2. https://doi.org/10.1186/s40689-016-0019-0

Harsito C, Prabowo AR, Prasetyo SD, Arifin Z (2021) Enhancement stability and color fastness of natural dye: a review. Open Eng 11:548–555. https://doi.org/10.1515/eng-2021-0055

Hassan MN, Fariha FT, Zayee Z, Hasan M (2024) Analyzing different functional and dyeing performance of natural blended fabric utilizing natural dyes and quality prediction by fuzzy logic. Heliyon 10:e40399. https://doi.org/10.1016/j.heliyon.2024.e40399

Hosseinnezhad M, Gharanjig K, Jafari R et al (2021) Cleaner colorant extraction and environmentally wool dyeing using oak as eco-friendly mordant. Environ Sci Pollut Res 28:7249–7260. https://doi.org/10.1007/s11356-020-11041-2

Jamadar D, Sannapapamma K (2018) GCMS analysis and colour fastness properties of silk yarn dyed with Acacia nilotica pods. 7:474–479

Kaur G (2020) Gas chromatography—a brief review. Int J Inf Comput Sci 5:8

Kemp W (1991) Ultraviolet and visible spectroscopy BT—organic spectroscopy. In: Education M (ed) Kemp W. UK, London, pp 243–283

Kumar Gupta V (2020) Fundamentals of natural dyes and its application on textile substrates. In: Samanta, Awwad N, Algarni HM (eds) Chemistry and technology of natural and synthetic dyes and pigments. IntechOpen, London, United Kingdom, pp 0–32

Li D, Calvert P, Mello C et al (2012) Stabilization of natural dyes by high levels of antioxidants. Adv Mater Res 441:192–199. https://doi.org/10.4028/www.scientific.net/AMR.441.192

Li J, Ji A, Lei M et al (2023) Hypsochromic shift donor-acceptor NIR-II dye for high-efficiency tumor tmaging. J Med Chem 66:7880–7893. https://doi.org/10.1021/acs.jmedchem.3c00253

Luo Y, Zhang X (2021) Effects of yellow natural dyes on handmade Daqian paper. Herit Sci 9:1–10. https://doi.org/10.1186/s40494-021-00560-x

Mandal S, Venkatramani J (2023) A review of plant-based natural dyes in leather application with a special focus on color fastness characteristics. Environ Sci Pollut Res 30:48769–48777. https://doi.org/10.1007/s11356-023-26281-1

Mongkholrattanasit R, Kryštůfek J, Wiener J (2010) Dyeing and fastness properties of natural dyes extracted from eucalyptus leaves using padding techniques. Fibers Polym 11:346–350. https://doi.org/10.1007/s12221-010-0346-8

Morales Y, Reyes-Trujeque J, Hermosin B, Azamar-Barrios J (2012) Characterization of a natural dye by spectroscopic and chromatographic techniques. MRS Proc 1374:49–59. https://doi.org/10.1557/opl.2012.1377

Nguyen HL, Bechtold T (2021) Thermal stability of natural dye lakes from Canadian Goldenrod and onion peel as sustainable pigments. J Clean Prod 315:128195. https://doi.org/10.1016/j.jclepro.2021.128195

Oda H (2001) Improvement of light fastness of natural dyes. Part 2: effect of functional phenyl esters on the photofading of carthamin in polymeric substrate. Color Technol 117:257–261. https://doi.org/10.1111/j.1478-4408.2001.tb00071.x

Okonkwo S, Ohanuzue C, Onuegbu G et al (2019) Extraction of natural dyes from *Whitfieldia lateritia* plant and its application on cotton fabric. J Text Sci Eng 9:1–4. https://doi.org/10.4172/2165-8064.1000392

Ouellet-Plamondon C, Aranda P, Favier A et al (2015) The maya blue nanostructured material concept applied to colouring geopolymers. RSC Adv 5:98834–98841. https://doi.org/10.1039/c5ra14076e

Padfield T, Landi S (1966) The light-fastness of the natural dyes. Stud Conserv 11:181. https://doi.org/10.2307/1505361

Pal A, Barkale H V, Dey N (2025) Dual-mode ratiometric ESIPT probes for vapor-phase detection of a pulmonary agent: Leveraging machine learning for high-precision quantitative analysis. Chem—An Asian J. https://doi.org/10.1002/asia.202401676

Poulin J (2018) A new methodology for the characterisation of natural dyes on museum objects using gas chromatography–mass spectrometry. Stud Conserv 63:36–61. https://doi.org/10.1080/00393630.2016.1271097

Raji Y, Nadi A, Chemchame Y et al (2023) Eco-friendly extraction of flavonoids dyes from Moroccan (*Reseda luteola* L.), wool dyeing, and antibacterial effectiveness. Fibers Polym 24:1051–1065. https://doi.org/10.1007/s12221-023-00019-9

Samanta AK, Agarwal P (2009) Application of natural dyes on textiles. Indian J Fibre Text Res 34:384–399

Samanta AK, Agarwal P, Singhee D, Datta S (2009) Application of single and mixtures of red sandalwood and other natural dyes for dyeing of jute fabric: studies on colour parameters/colour fastness and compatibility. J Text Inst 100:565–587. https://doi.org/10.1080/00405000802125246

Semwal RB, Semwal DK, Kapoor P (2012) Dyeing properties of *Berberis aristata* DC with natural and synthetic mordants. Trends Appl Sci Res 7:392–399. https://doi.org/10.3923/tasr.2012.392.399

Senthil Kumar B, Sundar M, Ramasubbu R et al (2023) Thermal stability and theoretical analysis of Madder dye absorption pattern on cotton fabric. Heliyon 9:e22188. https://doi.org/10.1016/j.heliyon.2023.e22188

Sharmila S, Kowsalya E, Kamalambigeswari R et al (2017) Production of natural dye by solid state fermentation. Int J ChemTech Res 10:219–225

Shin Y-S, Yoo D-I (2012) Storage stability and color reproducibility of yellow and red dyes extracted from *Carthamus tinctorius* L. Text Color Finish 24:165–172. https://doi.org/10.5764/tcf.2012.24.3.165

Skoog DA, Holler FJ, Crouch SR (2017) Principles of instrumental analysis. Cengage Learning, Cengage Learning, Boston, Massachusetts, USA, p 992

Umbreen S, Ali S, Hussain T, Nawaz R (2008) Dyeing properties of natural dyes extracted from turmeric and their comparison with reactive dyeing. Res J Text Appar 12:1–11. https://doi.org/10.1108/rjta-12-04-2008-b001

Yadav S, Tiwari KS, Gupta C et al (2023) A brief review on natural dyes, pigments: recent advances and future perspectives. Results Chem 5:100733. https://doi.org/10.1016/j.rechem.2022.100733

Zarkogianni M, Mikropoulou E, Varella E, Tsatsaroni E (2011) Colour and fastness of natural dyes: revival of traditional dyeing techniques. Color Technol 127:18–27. https://doi.org/10.1111/j.1478-4408.2010.00273.x

Chapter 5
Mordants and Mordanting in Natural Dyeing

Abstract Mordants are fixing agents that play a significant role in the natural dyeing process of fibres and textiles. The present chapter provides a comprehensive study of mordants and mordanting in eco-friendly dyeing processes. Mordants reinforce their role in improving dye intensity, light, and wash fastness on textile fabrics. Dyes obtained from natural sources show minimal interaction behaviour with substrates. Therefore, this results in poor colour fastness, high leachability, and low stability. These core factors necessitate the integration of mordants. They act as a chemical bridge to enable the coordination between dye molecules and fibre functional groups. The chapter classifies mordants into natural (tannin-based and oil-based), metallic (synthetic), and nano-mordants. Natural mordants like tannic acid serve as an environmentally friendly substitute to enhance dye intensity. They operate through hydrogen bonding, ionic interaction, or chelation. Oil-based mordants form complexes with metallic salts to improve dye stability. Metallic mordants such as alum, ferrous sulphate, and copper sulphate function by forming a coordination bond with the fibre and dye functional groups. Metallic mordants provide a variation of colour shades, but they possess diverse environmental and human health-related issues because of the heavy metal residues. Nano-mordants, prepared from metallic or bio-synthesised nanoparticles, are found to be an emerging, recent research field showing excellent surface reactivity with multifunctional properties like UV protection, enhanced colour intensity, and promising alternatives for sustainable and clean dyeing technologies. The chapter also focuses on the method of mordanting, such as pre-, meta-, and post-mordanting, along with the working mechanism. It also addresses the role of mordants in protein- and cellulose-based fibres with a mechanistic principle. Collectively, this study shows the potential of next-generation natural, bio-, and nano-modifiers to advance eco-friendly, high-performance fabric colouration approaches, which promote sustainable and multifunctional natural dyeing applications. Nowadays, several scientific collaborative research investigations are increasingly focused on finding environmentally friendly dyes and mordants. They aim to reduce the toxicity and other factors associated with bio- and nano-mordants with optimized concentration to develop the desired colour shades suitable for various purposes, such as in the food industry, cosmetics, pharmaceuticals, and medicine.

N. L. Bhandari et al., *Natural Dyes and Pigments of Nepal Himalaya*,
https://doi.org/10.1007/978-3-032-23229-8_5

5.1 Background

Natural dyes are obtained from plants, minerals, and insects, becoming an important part of human civilisation. They are applicable in various sectors such as traditional remedies, fabrics, cuisines, cosmetics, and artwork (Bishal et al. 2023). However, natural dyes have limited reactive sites and are not able to bind with substrates easily. Because of this, the dyes show poor colour fastness, low stability, and high leachability (Patel 2011). Thus, mordants are used in dyeing. A mordant is a chemical substance that is utilised in a colourant to fix the dye to fabric. It forms a coordination linkage with the dye along the fabric to obtain intense colour (Bhandari et al. 2020c). Nowadays, nanomodifiers have also been used for multi-applications like ultraviolet protection, antibacterial activity, and effective thermal stability (Assefi Pour et al. 2020). The majority of mordants applied in natural dyeing industries are synthetic ones. Synthetic mordants result in heavy metal residues that harm the environment and human health (Gupta 2019; Bhandari et al. 2020d). Most of the natural dyeing industries still use synthetic mordants, even in natural colouring, which is not a greener technique. To solve this problem, alternatives are required such as natural mordants on natural dye to bind easily. Therefore, comprehensive research in this area is essential (Taheri et al. 2014).

5.2 Chemistry of Mordants and Their Significance in Dyeing

Natural dyes show a low tendency for textile and fabric affinity. To increase their bonding with dye, mordants are required. Mordants are chemical compounds that interact to form a chemical bond between the fabric and the dye (Gupta 2019; Bhandari et al. 2020d). This is a general phenomenon that has been applied for over 2000 years. The term mordant originated from the Latin word *mordere*. It signifies either ‘to bite’ or ‘to grasp hold of’. This mentions that a mordant helps a dye attach to fabric to maintain its tone while washing (Singh and Bharati 2014a, b). The frequently used mordants are ferrous sulphate ($FeSO_4.7H_2O$), potassium dichromate ($K_2Cr_2O_7$), aluminium (Al^{3+}), iron (Fe^{2+}), tin (Sn^{2+}), chromium (Cr^{3+}), and copper (Cu^+) ions in their salts (Gill and Singh 2005; Prabhu and Bhute 2012). In earlier times, potash alum [$KAl(SO_4)_2.12H_2O$] was a commonly used mordant, also called aluminium potassium sulphate. With respect to this, ferrous sulphate ($FeSO_4.7H_2O$) and tin chloride ($SnCl_2$) were also applied as mordants. (Mongkholrattanasit et al. 2016). Essentially, mordants have many benefits: they escalate the colour bonding to fibres, strengthen colour tone and make them durable (Ding and Freeman 2017).

Several steps in mordanting are pre-treatment, meta-mordanting, and post mordanting (Repon et al. 2024) and they are explained as:

- **Pre-mordanting**: In this process, first the material is treated with the mordant, followed by the dye. The coordination complex forms on the fibre. In the pre-mordanting process, the fabric is coloured with a natural dye in an aqueous medium. It is a two-bath method in which the fabric is mordanted in the first bath and dyed in the second bath. Usually, mordanting and dyeing are carried out at the same temperature (~60–70 °C). The mordants are complexing agents that, if used together in the same bath, may cause dye precipitation. The quality of coloured textiles and their fastness are deteriorated as a result of this process (Gupta 2019).
- **Meta-mordanting**: The chemicals used for mordanting are mixed with a natural dye in the same bath during the meta-mordanting process. Dyeing and mordanting occur at the same time. The temperature of 80–90 °C is maintained for this process. While treating the substrate in one step is simpler, meta-mordanting only works with a few dyes (Gupta 2019).
- **Post-mordanting**: In this process, the fabric is dyed first and then dipped with a mordant in the same bath. As with pre-mordanting, the coordination complex forms on the fibre. The chroming temperature is normally between 80 and 90 °C. The temperature is decreased to 60 °C after chroming, and the mixture is run for about 15 min before the liquor is drained. Pad-dry laundry and pad-dry-steaming-laundry are the two methods for applying natural colours to cellulosic textiles. Meanwhile, high-temperature curing is not recommended in this regard as dye molecules are prone to thermal decomposition. Natural dyes can be used to colour fibres and yarns in the same way as synthetic dyes (Gupta 2019).

There are three types of mordants used for the formation of a complex with cotton fabrics, i.e., metallic mordants, tannic mordants, and oil mordants (Manian et al. 2016; Repon et al. 2024). Synthetic mordants are toxic, hazardous, and carcinogenic to human beings and the environment (Gill and Singh 2005). Researchers have reported non-toxic and eco-friendly natural mordanting agents such as *Aloe vera*, lemon juice, mango (*Mangifera indica*) bark extract, and peepal *(Ficus religiosa)* bark extract (Wangatia et al. 2015). Research on the use of natural mordants in natural dyeing has been increasing in recent years as a new scope in fabric colouring, representing a fascinating natural and green dyeing technique.

5.3 Natural Mordants

Mordants obtained from sources other than metals (natural sources) are known as natural mordants. These are either tannin-based (tannin or tannic acid; $C_{76}H_{52}O_{46}$) or oil-based, and help to bind natural dyes onto fibres by hydrogen bonding or the formation of coordination complexes (Benli 2024).

5.3.1 Tannin-Based Mordants

Tannins are mainly used as fixing agents, binders, and cleaning catalysts to preserve leather and related products. They belong to a category of complex polyphenols that may be hydrolysable or condensed. They bind proteins and precipitate them. Gallo-ellagic tannins and tea tannins are typical hydrolysable tannins (Fig. 5.1). Tannins are naturally occurring mordants that are widely released by plant tissue in both gymnosperms and angiosperms, and derive their name from the French word tan, which means 'bark of oak'. Tannins exist as large molecular mass polyphenols with an ortho-dihydroxy benzene, known as catechol. They contain a bidentate ligand formed by deprotonation and are responsible for imparting various colours when used with different metals in textile fibre dyeing (Shabbir et al. 2019). Most tannins obtained from plants, particularly through bark and other parts like leaves, fruits, and galls, exhibit acidic and constricting chemicals. The tannin ingredients are acquired by extracting directly or in a more potent form for use in tanning purposes. They are generally used in the textile industry as mordants. To dye textile fibres, several tannin-rich compounds are applied as mordants. Due to their reduced price, they are seldom employed as a tannic acid alternative during mordanting. Myrabolans and sumach are the most significant tannins used so far (Srivastava 1912; Gill and Singh 2005).

Gill and Singh (2005) reviewed the effect of mordants on dyeing and reported the effect of Harda (*Terminalia chebula*) and synthetic mordants on dyeing wool with henna (*Lawsonia inermis*). Alum, Harda, and tannin increased the red-coloured shade of henna-dyed wool, while other mordants decreased it. They reported the eco-friendly nature of alum and Harda mordants. Therefore, the use of alum or Harda alone or in combination as a substitute is a viable approach for producing different shades of colours (Gill and Singh 2005). Wangatia et al. (2015) investigated the

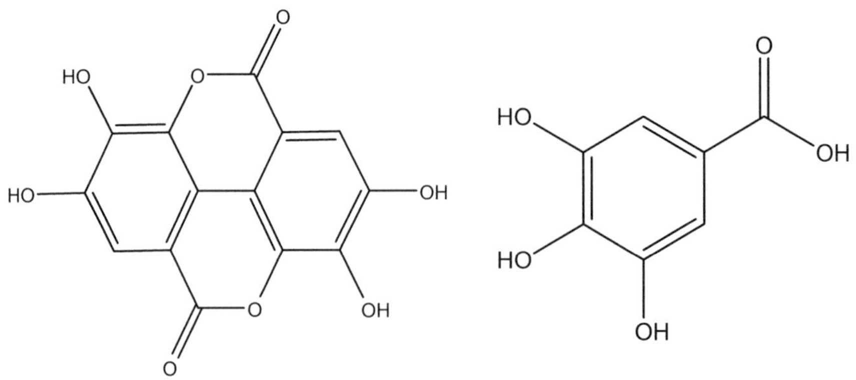

a. b.

Fig. 5.1 Molecular structure of a hydrolysable **a** gallo-ellagic tannin (Banc et al. 2023) and **b** tea tannin (Zhang et al. 2022) possibility of dyeing cotton fabrics using mango bark juice as a mordant. They reported that the mango mordant increased the colouring effect on the fabrics. The colour values were affected by the addition of mordants, resulting in different hues being obtained from the same amount of dye extracts using different mordants (Ding and Freeman 2017).

5.3.1.1 Myrobalan

Myrobalan is a Harda fruit, the scientific name of which is *Terminalia chebula*. The dried fruits are approximately an inch long, extremely stiff, with a harsh, strong taste like a small shrivelled plum. The skin of the fruit contains the majority of the tannins. Ellagitannic acid is the tannin in myrabolan. When the yellowish brown colouring matter in Myrabolan is eliminated, the substance becomes an acceptable alternative for gallotannin. It is utilised for tanning, cotton, and silk black colouration (Srivastava 1912; Manian et al. 2016).

5.3.1.2 Sumach

Sumach, rich in gallotannic acid, is derived from the foliage and stems of various *Rhus* species. *Rhus coriaria* contains the most refined sumach. Commercially, it is available as a whole leaf or pulverised material. The best quality product contains 15–20% tannin and a moss green colour. Sumach includes reddish colouring pigment, which makes it unsuitable for dyeing bright and dazzling hues (Patel 2011).

5.3.1.3 Tannic Acid

Tannic acid is found in a variety of tannin-containing compounds, including nut galls and sumach. It is reported that 60–77% of tannic acid is found in nut galls. It is an amorphous light yellow powder with a strong astringent flavour, which turns yellow on exposure to light and can be dissolved in six parts of water. Because of the prolonged duration of fermentation, the hydrous solution progressively degrades. The addition of boric acid prevents this fermentation-induced decomposition. Atmospheric oxygen is easily absorbed by the tannate solution, which then decomposes into a beige tint (Gill and Singh 2005; Prabhu and Bhute 2012). The chemical structures of commonly used tannin-based and other bio-mordants are shown in Fig. 5.2.

5.3.1.4 Applications of Tannin-Based Mordants

The majority of natural dyes possess a low affinity for cotton. Cotton fabrics are extensively dyed by tannin-based mordants, especially tannic acid. They were commonly

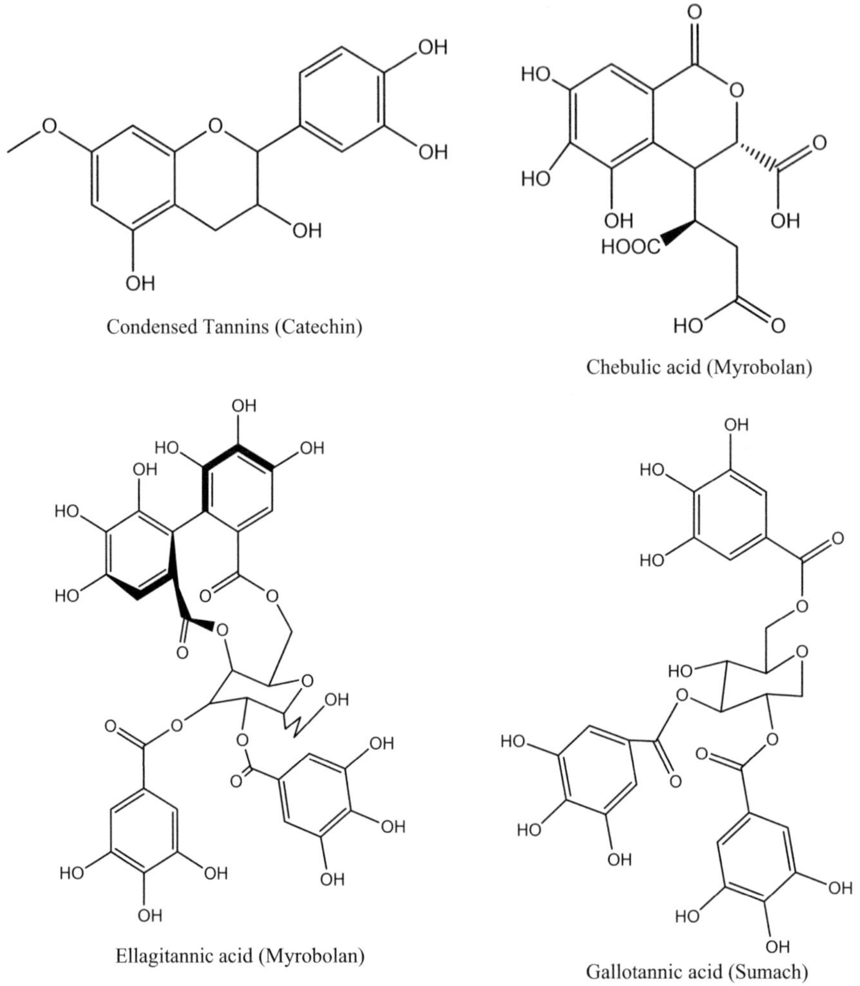

Fig. 5.2 Chemical structures of tannin-based bio-mordants used in natural dyeing (Banc et al. 2023; Lee et al. 2010; Zhang et al. 2022; Uemori et al. 2025)

employed to prepare cotton, allowing it to retain its colour for a long time. The goal of tanning the vegetable fibres to bind specific inorganic salts of metals, especially those of chromium, aluminium, potassium, etc., as tannin complexes, rather than to fix the colouring matter. In the process of dyeing, natural dyes and metal tannates on the fabric surface acquire non-soluble coordination adducts (Adeel et al. 2023; Ghosh et al. 2024).

Tannic acid is the finest among all tannins for mordanting the cotton fabric. It does not include useless or harmful inherent impurities during fixing and colouring. It is commonly applicable for both soft and bright shade dyeing. Gallnut, sumach, and

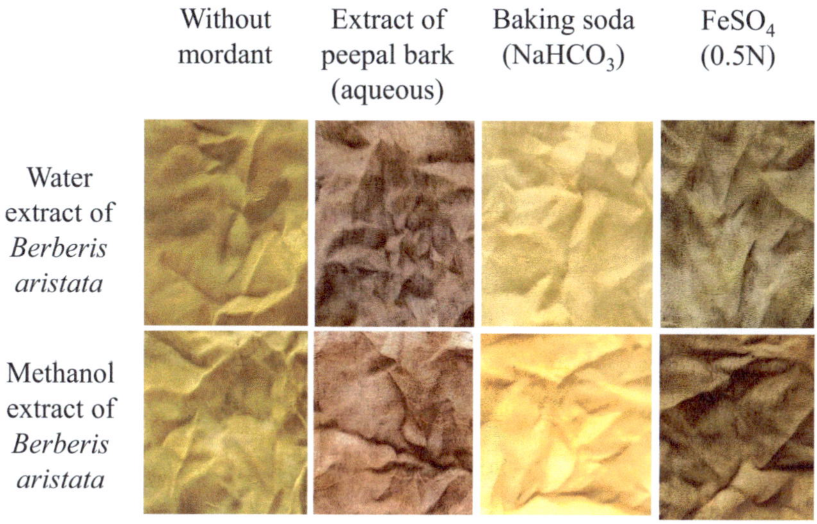

Fig. 5.3 Effect of mordants on cotton fabrics with different colour shades developed by dyeing with colour extracted from *Berberis aristata* in water (above) and methanol (below)

myrabolan extracts are commonly used for dark colours. Tannic acid in the range of 2–5% is commonly used for light hues, whereas 5–10% is required for dark colours. The substance is placed in a bath briefly before being soaked overnight (Singh and Bharati 2014a, b). This effect of tannin-based mordants on cotton is further illustrated in Fig. 5.3, which shows the colour shade obtained after dyeing with marigold flower extracts. The upper panel represents fabrics treated with water extracts, while the lower panel shows those dyed with methanol extracts, demonstrating the impact of mordants and solvent on the resulting colour.

5.3.2 Oil Mordants

Alum, used as the main mordant, forms a combination with oil-based mordants. This is because oil is insoluble in water and has no affinity for cotton, making it difficult to wash out the treated fabrics and fibres. Commonly used oil mordants are castor oil, coconut oil, and sesame oil. Their most common application is in madder dyeing to obtain the Turkey red shade. Oil alone has a low binding capacity with cotton fabric; thus, it faded quickly during dyeing. To improve dyeing, oil is combined with alum, which gives better results by forming a chemical complex. During this process, oil itself easily removes from the fabric.

Natural oils contain palmitic, stearic, oleic, ricinolic, and other fatty acids, along with their glycerides. The (–COOH) carboxylic acid of fatty acids and metallic salts combines to get –COOM. The M signifies the metal in –COOM. On treating the

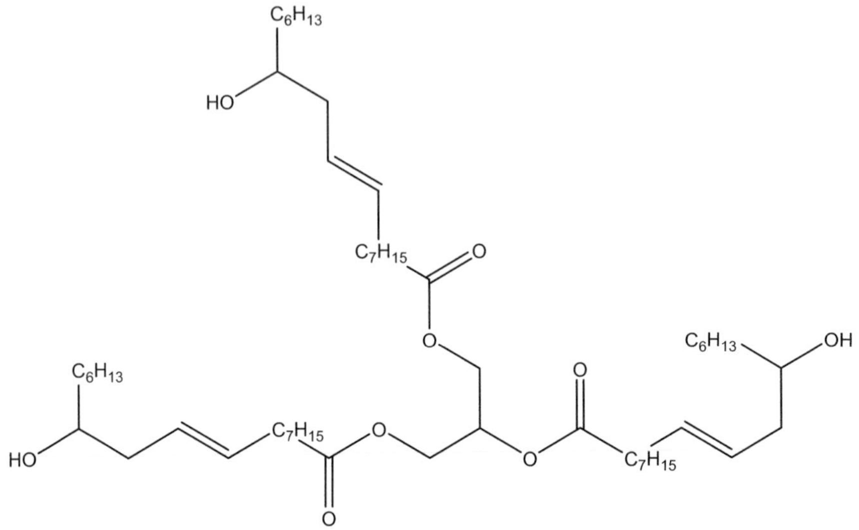

Fig. 5.4 Castor oil as a mordant (El-Azab et al. 2010)

sulphonic group (–SO_3H) of Conc.H_2SO_4, it forms sulphonic oils with enhanced metal bonding affinity than natural oils. Such an acid becomes –SO_3H when it reacts with metal ions (Patel 2011; Singh and Bharati 2014a, b; Haddar et al. 2014). Figure 5.4 represents the structure of castor oil, commonly used as a mordant in natural colouring, mainly in cosmetic products.

5.3.2.1 Applications

The bonded metal forms a chelate with a mordant dye, producing a fast and vibrant turkey-red colour produced from madder. Turkey-Red Oil (TRO), known as sulphonated oil, is a word applied extensively to refer to the sulphonic castor oil. In 1876, a French–English company introduced it in the market. It is regarded as a predecessor of manmade surfactants used today (Vankar and Dixit 2011; Singh and Bharati 2015).

5.3.3 Advantages of Natural Mordants

Natural colours and natural or biomordants have significant environmental potential as they are sustainable, eco-friendly, and health-compatible compared with synthetic dyes and colours. They work well with a variety of natural fibres, including jute, cotton, silk, and wool (Ding and Freeman 2017; Gupta 2019). Natural dyes are mild and result in non-corrosive equipment. They can be replenished and decrease the

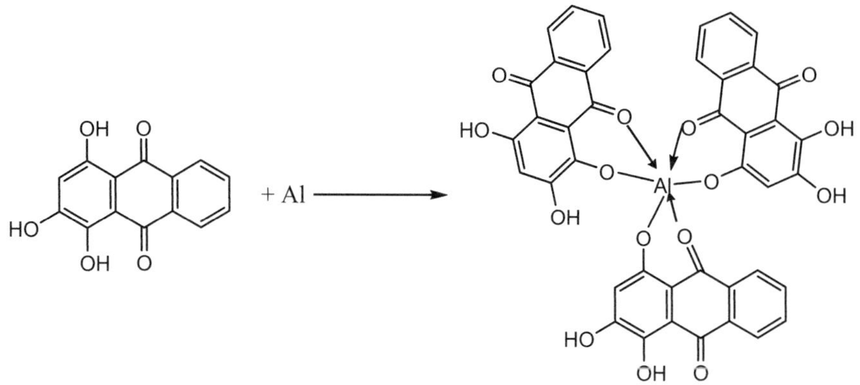

Fig. 5.5 Mechanism of fixation of natural dyes through aluminium mordant (Gedik et al. 2014)

dependence on nonrenewable sources, which are in restricted amounts. Compared to synthetic mordants, natural mordants break down in nature more easily and rapidly. The shades that occur are milder, brighter, and mix uniformly with natural dyes. Because of their softer nature with fibres, they help to maintain intensity and consistency. One of the major advantages of using natural mordants is their ability to impart superior colour strength and richness compared with synthetic mordants, as demonstrated in the following Fig. 5.5 (Bhandari et al. 2020a, b, c). The figure illustrates the process of colour shades differing with the interaction of mordannt variation on dye extracted from the *Euphorbia pulcherima* flower. It also shows that the natural mordants (such as lemon juice, yeast used in fermentation, and aloe vera juice) act as effective mordants compared to the synthetic mordants (such as $FeSO_4$, $CuSO_4$, and $K_2Cr_2O_7$).

5.4 Metallic (Synthetic) Mordants

Metallic mordants have historically been essential in textile dyeing, referred to as artificial mordants. They are prepared from a variety of metallic salts (Manian et al. 2016). Synthetic mordants have the potential to directly impact fibres and the environment. They can generate a wide range of hues with various natural dyes and enhance durability under washing, rubbing, and light exposure. Iron and tin-based mordants can have detrimental effects, such as fibre breakage, making the yarns brittle, as well as affecting the tensile properties of woollen yarn (Manian et al. 2016; İşmal and Yıldırım 2019). Naturally available inorganic salts of metals (such as potassium, aluminium, copper, iron, etc.) were utilised as mordants in the early days. Aluminium, chromium, iron, copper, and tin are often used as mordants. Alum ($Al_2K_2(SO_4)_4 \cdot 24H_2O$), potassium dichromate ($K_2Cr_2O_7$), ferrous sulphate ($FeSO_4.7H_2O$), copper sulphate ($CuSO_4.5H_2O$), stannous chloride ($SnCl_2$), and stannic chloride ($SnCl_4$), etc., are commonly used salt-based mordants (Prabhu and Bhute 2012). Some of the major metallic mordants are described below.

5.4.1 Alum

Alums are the hydrated sulphate salts of aluminium. The word 'alum' was first used to name a bi-salt of sulphates of aluminium and potassium ($Al_2K_2(SO_4)_4$) that crystallised in the form of an octahedron with 24 water molecules. It is known as potash alum, comprising around 36% aluminium sulphate. The term 'alum'eventually expanded to a broad category of amorphous dual cation sulphates with comparable constituents. As a result, several alums that are aluminiun sulphate salts like ammonium alum $Al_2(NH_4)_2(SO_4)_4.24H_2O$, chrome alum $Cr_2K_2(SO_4).24H_2O$, ferric alum $Fe_2(NH_4)_2(SO_4).24H_2O$, soda alum $Al_2Na_2(SO_4)_4.24H_2O$, etc., are reported (Singh and Bharati 2015; Evitasari et al. 2019).

All of these alums crystallise into octahedral crystals and cannot be separated by the process of fractional crystallisation. Around 22 g of aluminium potassium sulphate could be dissolved in 100 mL of water at 30 °C, whereas 357 g of it could be dissolved in the same amount of boiling water. Boiling or diluting alum with water does not dissociate it into individual ions. The term 'neutral alum' refers to an alum solution that is prepared as a basic and feasible solution without precipitation. Hence, the so-called neutral alum is prepared by adding sodium hydroxide or carbonate to an alum solution. Temporary precipitates form by the gradual addition of the alkali, which then dissolves upon stirring. However, it is a basic salt, prepared by combining 12 parts of alum and 1 part of calcium hydroxide in a boiling solution. A precipitate is formed when the solution is heated above 40 °C (Malla 1994; Patel 2011; Singh and Bharati 2015; Pranta and Rahaman 2024).

Chemically, aluminium sulphate or other salts of aluminium used as mordants attach to any fibres, combine with particular mordantable functional groups contained in natural dyes, and are bound by coordinate/covalent bonds, hydrogen bonds, and other forces, as illustrated in Fig. 5.5 (Morshed et al. 2016).

5.4.1.1 Applications

Alum generates around 13% alumina (Al_2O_3) on cotton. When base is added to the aluminium potassium sulphate solution, an alkaline alum solution is formed, which is employed as a mordant. To prepare an alkaline alum of the needed strength, 10 kg of cotton and 1 kg of alum are taken. Then it is dissolved in 200 L of water along with 0.5 kg NaOH. There are two techniques for fixing cotton with basic aluminium sulphate (Malla 1993; Singh and Bharati 2015). The following points signify the use of mordants in colouring.

- A mordant is directly infused into the material fabric, and the latter is fixed with additional chemicals or by ageing.
- At first, a sulphate solution is impregnated into a material, which interests the alkaline sulphate to obtain an insoluble combination with alumina, as in oil mordants and tannic acid.

The direct mordanting process can be carried out with a basic aluminium sulphate solution. The indicated that basic sulphate solution is absorbed by a piece of cotton for a few minutes before being rotated and kept in the bath for about 3–5 h before drying. Non-homogeneous drying causes the basic salt to decompose unevenly, resulting in nonuniform colouration. As a result, it is recommended that the alumina be fixed on the fibre, allowing it to precipitate freely. The mordanting of alumina is accomplished by immersing the substance for half an hour in a lukewarm fixing bath. Each litre of fixing liquor must contain 5–10 g of each of the salts that are sodium carbonate, sodium phosphate, sodium silicate (100 °C), ammonium carbonate, or ammonia liquor (50 g). A solution containing 1% neutral soap can be applied as well (Singh and Bharati 2014a, b).

5.4.2 *Ferrous Sulphate*

Green vitriol ($FeSO_4.7H_2O$) is another name for ferrous sulphate. Iron pyrites are used to prepare ferrous sulphate on a large scale. Commercially, it is a bluish-green compound with crystals of the monoclinic system that crystallise in the atmosphere and oxidise into a yellowish coloured ferric sulphate despite its stability in pure form. Generally, at room temperature, 115 parts of crystalline ferrous sulphate ($FeSO_4.7H_2O$) are dissolved by 100 parts of water. However, ferrous sulphate solution does not ionise when heated or cooled. It is regarded as among the most essential and earliest mordants known, and it is widely used today as well (Malla 1993; Singh and Bharati 2015; Arora et al. 2025). Fixation of ferrous sulphate with natural dyes and wool is presented in Fig. 5.6 (Bukhari et al. 2017).

5.4.2.1 Applications

Iron salts are widely used in colouring and printing. The use of inorganic salts of metal in colouring, fixing, and processing silk results in dyed fibres, but black is particularly crucial. These mordants are used on cotton and wool fabrics that are treated with tannin. The most popular method of using copper or ferrous sulphate in cotton dyeing (particularly for blacks) is to first treat the material with tannin (see Fig. 5.7). Then, the cloth is passed through lemon water and finally immersed in the copper as solution. Ferrous tannate is fixed onto the fibre by double decomposition of calcium tannate and copper, and calcium sulphate is produced, which is washed off. Ferrous tannate is quickly oxidised into ferric tannate by exposure to air, which is

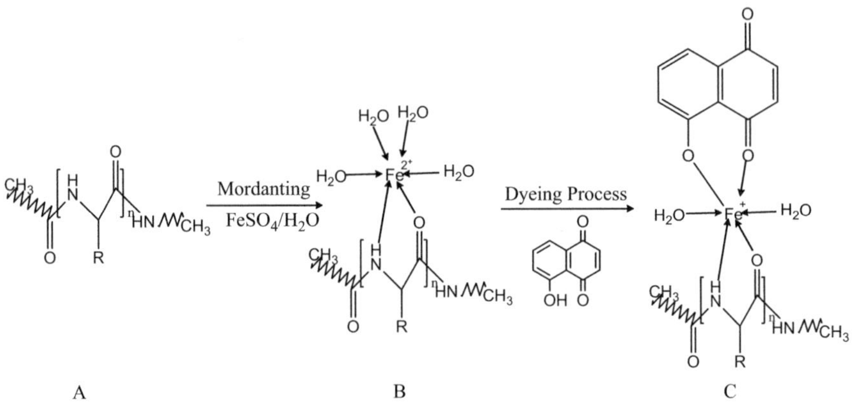

Fig. 5.6 Schematic representation of wool–Fe mordant–dye interaction (Bukhari et al. 2017)

the active mordant. If ferrous sulphate is used alone, it leaves iron spots on the cloth. However, the sample is initially treated with a boiling solution of a bio-dyestuff or a combination of dyestuffs, like logwood, cutch, etc., and then transferred to a cold ferrous sulphate mixture (Singh and Bharati 2015; Bukhari et al. 2017; Nivedha and Kalaiarasi 2023). The resulting variations in colour are shown in Fig. 5.7, which presents wool fabrics dyed with *Berberis aristata* bark extracts in both aqueous and alcoholic media. The figure highlights the effect of iron-based mordants, including the natural mordant derived from *Ficus religiosa* bark juice.

5.4.3 *Potassium Dichromate*

Red chromate, potash dichromate ($K_2Cr_2O_7$), or chrome are the different names of potassium dichromate. These are orange-coloured triclinic prisms, their large size resulting from the growth of potassium dichromate crystallites. It possesses toxic properties similar to those of chromic acid as well as its salts. Its dust causes a skin allergy by forming severe blisters. Dichromate solutions are light-sensitive (Malla 1993; Patel 2011; Singh and Bharati 2015; Muthu 2023). The chemical interaction between the dichromate mordant and natural dyes with nylon is given in Fig. 5.8.

5.4.3.1 Applications

The conversion of dichromate into chromic acid prior to dyeing the mordanted substance for effective application of the chrome mordant is crucial. If the fibres

Fig. 5.7 Photographs showing colouring effect with dye extracted from *Berberis aristata* bark in aqueous and alcoholic media. The figure shows the differences in colour shades with variation in mordants, including the natural mordant from *Ficus religiosa* bark juice

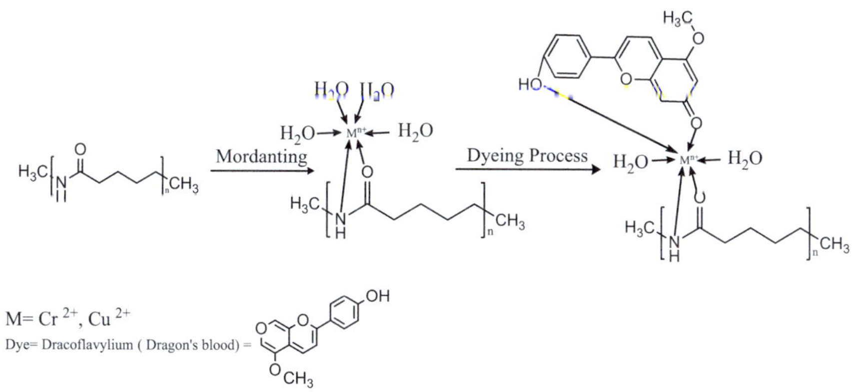

Fig. 5.8 Chemical interaction between dichromate mordant and natural dyes on nylon (Islam et al. 2025)

are dyed with dichromate thereafter, the chromic acid obtained acts as an oxidising agent, whereas the chromic oxide formed functions as a mordant. The chromate mordant is applied onto cotton, for which 800 g of potassium dichromate, 600 mL of ammonia are dissolved in 200 mL of water. To it, 800 g of sodium thiosulphate 500 mL of magnesium acetate along with 4000 mL of gum thickening are added to form the chromate mordant during dyeing (Muthu 2023).

Figure 5.9 shows a comparison of cotton dyeing with natural and synthetic mordants. It provides evidence that the natural mordants are equally effective in producing the specific colour shades as metallic mordants.

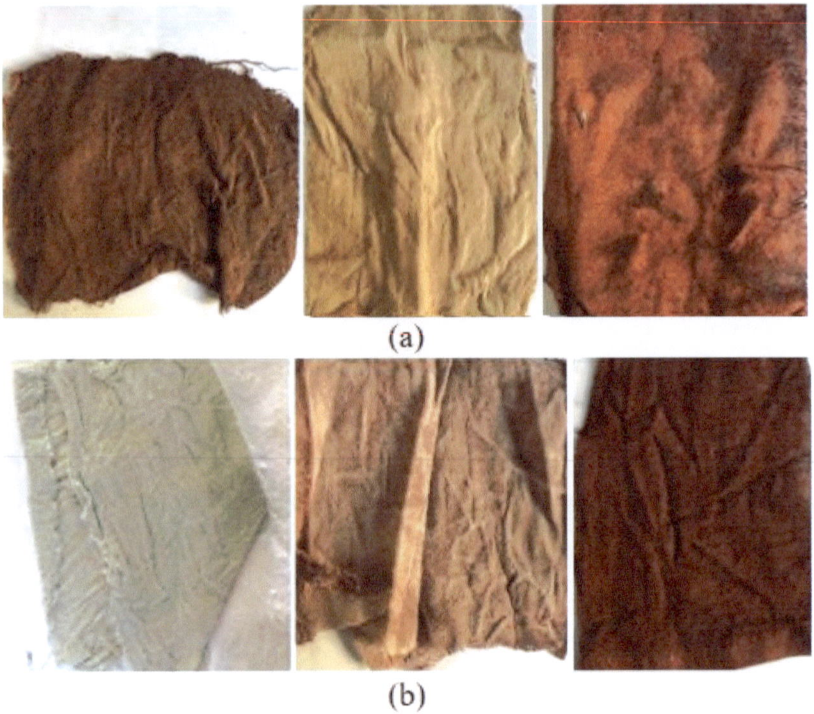

Fig. 5.9 Photographs showing cotton dyeing with **a** natural mordants: concentrated black lemon juice (Chuk), yeast used for fermentation (Mercha), and *Aloe vera*; and **b** synthetic mordants: $CuSO_4$, $FeSO_4$, and $K_2Cr_2O_7$ on crude dye of *Euphorbia pulcherrima* flower, respectively (Bhandari et al. 2020b)

Using the resultant mixture thus prepared, the cloth is padded, dried, and steamed for 30–120 min. The cloth is coated with chromium oxide and magnesium oxide as mordants. This is a time-consuming operation due to the extended heating of the impregnated material, and one may end up with weakened fabric. As a result, it is recommended that plant fibres, such as cotton, be first fixed with tannin and then reacted with a boiling dichromate mixture. It is shown that 20% tannic acid is needed for complete colouring. Three parts of dichromate along with one part of sulphuric acid are mixed to prepare the chroming bath (in 100 mL) (Basavaiah and Swamy 2001; Pandey et al. 2020). Care should be taken that the use of excess acid may damage the quality of cotton fabrics.

5.4.4 *Cupric Sulphate*

Cupric sulphate pentahydrate ($CuSO_4.5H_2O$), also called copper sulphate or blue vitriol, is prepared by burning copper ores and soaking them in sulphuric acid. Crystals of copper sulphate with varying levels of purity are produced by this route. It crystallises as translucent blue triclinic crystals containing five water molecules for crystallisation, which dissolve easily mixed in water at 30 °C. Under general conditions, 49 parts of cupric sulphate can be dissolved by 100 parts of water (Malla 1993; Muthu 2023; Shahid et al. 2023). A stable dye-mordant fibre combination is created when these metal ions attach to hydroxyl or carbonyl groups found in the natural dye molecules. Figure 5.10 shows a schematic illustration of this chemical interaction.

5.4.4.1 Applications

Treatment of fabrics and textiles with copper sulphate improves the fastness properties of several hues, such as logwood. Copper salts are used as oxidants in the manufacture of *Acacia* wood brown and in cotton dyeing due to their low cost. They are commonly used to colour black tones on cotton that has been fixed with tannin (Bhandari et al. 2021).

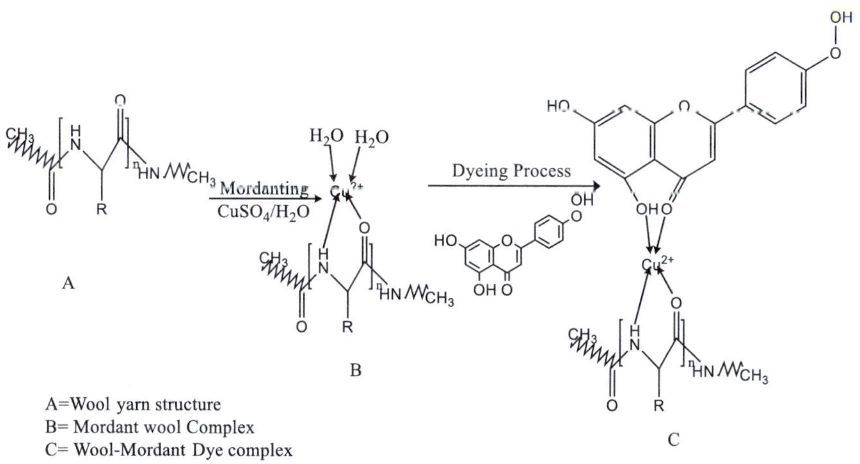

Fig. 5.10 Illustration of coordination bonding between wool peptide groups, natural dye, and copper mordant (Khan et al. 2015)

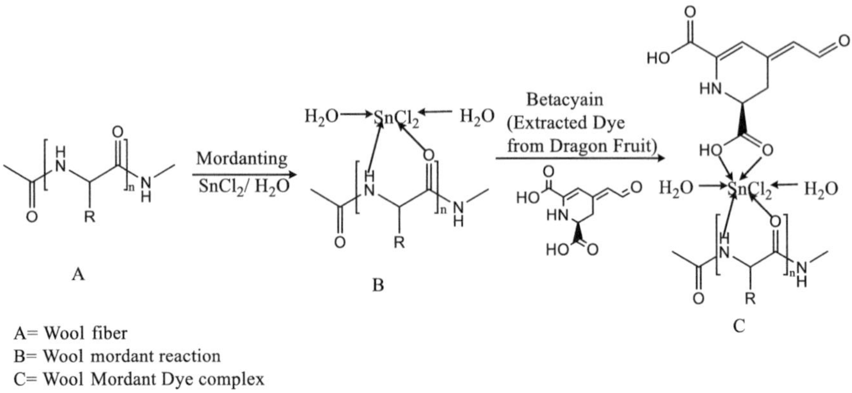

Fig. 5.11 Complex reaction mechanism of wool–stannous chloride mordant for natural dye extracted from dragon fruit (Sadannavar et al. 2024)

5.4.5 *Stannous Chloride*

Tin is dissolved in concentrated hydrochloric acid (HCl) to prepare stannous chloride ($SnCl_2$). Monoclinic crystals, including two molecules of water of crystallisation, commercially known as tin crystals, are formed from such an aqueous solution. The solubility of $SnCl_2$ in water is quite high. On diluting the clear solution with water, a basic stannous chloride is produced (Mongkholrattanasit et al. 2016). It undergoes a hydrolysis reaction with water. When tin crystals are exposed to air, they oxidise in the same way as stannous chloride dihydrate ($SnCl_2 \cdot 2H_2O$), which is the common solid form, is dissolved in water.

$$SnCl_2 + 5H_2O \rightarrow SnCl(OH) + HCl$$

It is applied as a fixing agent to enhance the shade of natural dyes. This effect is demonstrated in Fig. 5.11, which shows the chemical interaction between the natural dye obtained from dragon fruit and wool fibres in the presence of stannous chloride as a metallic mordant (Sadannavar et al. 2024).

5.4.6 *Stannic Chloride*

Stannic chloride ($SnCl_4.5H_2O$) can be produced by oxidizing stannous chloride ($SnCl_4$) with chlorine or potassium chlorate. Monoclinic crystals of $SnCl_4$ are its most common forms. It is water-soluble and used as a fixer in dyeing cotton and silk (Benli 2024).

5.4.6.1 Applications of Tin Mordants

The colours imparted by the tin mordants are comparatively brighter. Discharge printing uses stannous salts, which are strong reducing agents, and hence, they cannot be used for copper sulphate, like oxidising mordants. These mordants shouldn't be applied to the bio dyes, which are prone to reduction as they can wash out quickly. They aren't commonly used to mordant cotton fabric. *Persian berries* produce a pleasant yellow to orange shade on the substances already mordanted with tannin and stannous chloride (Vankar and Dixit 2011; Manian et al. 2016).

Cotton is generally mordanted using stannic salts. Many bio-dyes, such as logwood, fustic, quercitron, and weld, are mordanted and set onto cotton using stannic oxide. The material is usually mordanted with a natural mordant, treated in the presence of a weak solution of stannic chloride, and then rinsed prior to dying. Tannic acid present in the tannin works as a mordant for the stannic hydroxide, which is the real fixer under these conditions (Mongkholrattanasit et al. 2016). The effect of tin mordants, along with other metallic mordants, is illustrated in Fig. 5.12, which shows the colour variations obtained on wool after being treated with various mordants, in comparison to the untreated control sample. Each metal mordant, such as tin, chromium, and copper, produces its characteristic shade due to differences in dye mordant interaction. In the absence of mordant, a pale colour was observed in the controlled sample, representing poor dye fixation (Sadannavar et al. 2024).

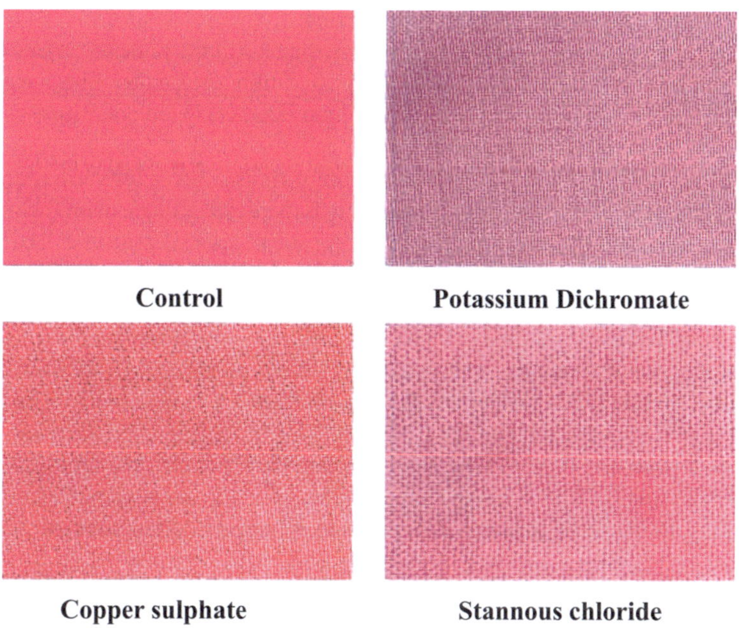

Fig. 5.12 Colour shades obtained by wool samples after mordanting, dyeing, and control treatment (Sadannavar et al. 2024)

5.4.7 *Environmental and Health Concerns*

Despite the efficient dye affinity, metallic mordants pose a significant difficulty in the sustainable dyeing process due to their environmental resistance. Synthetic mordants, specifically heavy metals, present an ecological threat due to weak compostability. The waste residue often obtained from textile dyeing contains heavy metals, which can hinder environmental degradation. Therefore, the replacement of metallic mordants with natural ones is necessary, which results in fewer environmental impacts (İşmal and Yıldırım 2019).

5.5 Nano-Mordants

Nano-mordants operate as a mordant in the nano range and can be metallic or bio-based. They have a lower ecological threat compared to conventional metal salts. They show long-lasting colour strength, homogeneity, and dye binding affinity. Stronger interaction with dye molecules and fibre functional groups is made possible by their high surface area to volume ratio, which results in improved colour fastness and lowers the need for mordant concentration (Pour et al. 2020). The use of nano-mordants in natural dyeing, whether prepared by green methods like plant extracts, polysaccharides, or biopolymers, or by chemical methods through gold, copper, silver, iron, zinc, and carbon black, is increasing.

The resulting nanoparticles (typically less than 100 nm) possess surface-active sites capable of coordinating bonds with dye and fibre molecules. These nanoparticles' crystalline structure and functional group interaction are confirmed by characterisation methods that are spectroscopic and analytical techniques. The use of metallic and non-metallic nano-mordants from various sources for various applications has been in use since ancient times, such as *thanka* paintings, monument paintings, temples, statues, ornaments, gold painting and plating (Pereira et al. 2025), but they have not been explored fully to date. Nowadays, they are a hot topic of research to study the interaction phenomenon of nanoparticles and their concentrations to form coordination complexes with dye molecules in developing different colour shades.

The remarkable example is the study by Taheri et al. (2014), who examined zirconium dioxide (ZrO_2) nanoparticles as a mordant for dyeing wool using thyme extract (Fig. 5.13). The integration of ZrO_2 nanoparticles improved the wool fibre's ability to absorb and fix dye. The nano-mordant demonstrated the multifunctional potential of nanoparticles in natural dyeing applications by influencing flammability and adding antibacterial qualities in addition to enhancing colour strength.

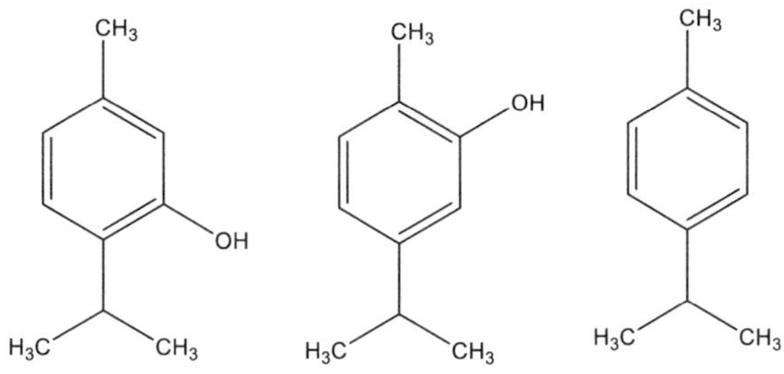

Fig. 5.13 Chemical structure of the main compounds of *Thymus* species (Taheri et al. 2014)

5.5.1 *Advantages of Nano-Mordants*

Nano-mordants require a lower concentration, about one-tenth of traditional mordants, to provide similar or higher colour strength because of their enormous surface area and reactivity. Green-synthesised nano-modifiers help reduce chemical waste and metal discharge. In addition to better colour fastness, fabrics treated with nano-mordants show antimicrobial, UV-blocking, and self-cleaning qualities. Natural dyes (such as anthocyanin, curcumin, and tannin) are made more stable by nano-mordants without causing colour fading (Pour et al. 2020; Du et al. 2022; Pereira et al. 2025).

5.5.2 *Challenges and Future Prospects*

Regardless of the benefits of nano-mordants, several challenges exist with scalability, cost, nanoparticle aggregation, and long-term environmental consequences. To guarantee uniformity and safety in the dyeing industries, synthesis and application procedures must be standardised and optimised. Future studies will concentrate on bio–nano hybrid mordants, which combine nanoparticles and plant-based polyphenols for optimal performance and low toxicity (Pereira et al. 2025). In this context, nanoparticles obtained from various plant extracts through eco-friendly methods show a favourable environment for more research to support the clean dyeing industry and technology.

5.6 Mechanism of Mordanting

The mordanting mechanism is the interaction of chemical and physical properties between the colourant, mordant, and fibre. Mordant forms either stable coordination or chelation complexes with dye molecules and the fibre by functioning as a bridge. The mechanism of mordanting can differ for cellulosic like cotton fabrics, fibres, and for protein materials, such as wool, hair etc. In many textile fibres, the stability, colour fixing, and fastness properties are enhanced by this procedure (Geelani et al. 2016).

5.6.1 *General Principle of Mechanism*

Natural dyes most often have hydroxyl (-OH), carbonyl (>C = O), and amino (–NH_2) functional groups that create a coordination bond between metal ions or biomordants. The mordant interacts with the active sites, like the hydroxyl group in cellulose and the amino or carbonyl groups in protein materials (Ashrafi et al. 2018). Therefore, mordant functions as a bridge to attach the dye to the fibre surface through chemical coordination. The general mechanism is derived as follows:

$$\text{Fibre}-\text{M} + \text{Dye} \rightarrow \text{Fibre}-\text{M}-\text{Dye},$$

where M is the metal or bio-mordant ion that connects the fibre and dye (Shahmoradi Ghaheh et al. 2021).

5.6.1.1 Mechanism in Natural Mordants

Ionic attraction, hydrogen bonding, and chelation are the major stabilising mechanisms for biomordants like tannins and polyphenols. Cellulosic -OH group or protein fibres' –NH and –CO groups form a hydrogen bond with polyphenolic hydroxy groups. Similarly, $\pi - \pi$ stacking occurs between aromatic biomordants and aromatic dyes. Chelation is a process of connecting the phenolic –OH group directly with dye molecules that contain carbonyl or hydroxyl groups (İşmal and Yıldırım 2019). Tannin molecules form a more stable channel and improve washing resistance through crosslinking with the fabric and colourant (Adeel et al. 2023).

Singh et al. (2019) studied tannin-based natural mordant by taking the extract of the tamarind (*Tamarindus indica*) seed coat for wool fibre and anthocyanin as a natural dye derived from Kapok (Shwet semal; *Ceiba pentandra*) flower extract (KFE). Wool fibre is composed of protein keratin that has amino (–NH_2), carboxyl (–COOH), and hydroxyl (–OH) groups. Initially, it reacts with the polyphenolic components of tamarind seed coat extract. The fibre surface with a tier of bio-mordant is covered by functional groups of wool, the phenolic groups (–OH) and aromatic ring of tannin due to ionic interaction and formation of a hydrogen bond. The tannin molecule adhesion enhances the binding tendency of wool's reactivity and dye. The natural colouring compounds, such as anthocyanins and flavonoids, interact with the phenolic sites of the tannins through hydrogen bonding, $\pi - \pi$ stacking, and occasionally weak coordinate interactions between carbonyl and hydroxyl groups when the tannin-treated wool is exposed to the KFE extract. As a result, a stable wool-tannin-dye complex is created when the tannin serves as a molecular bridge

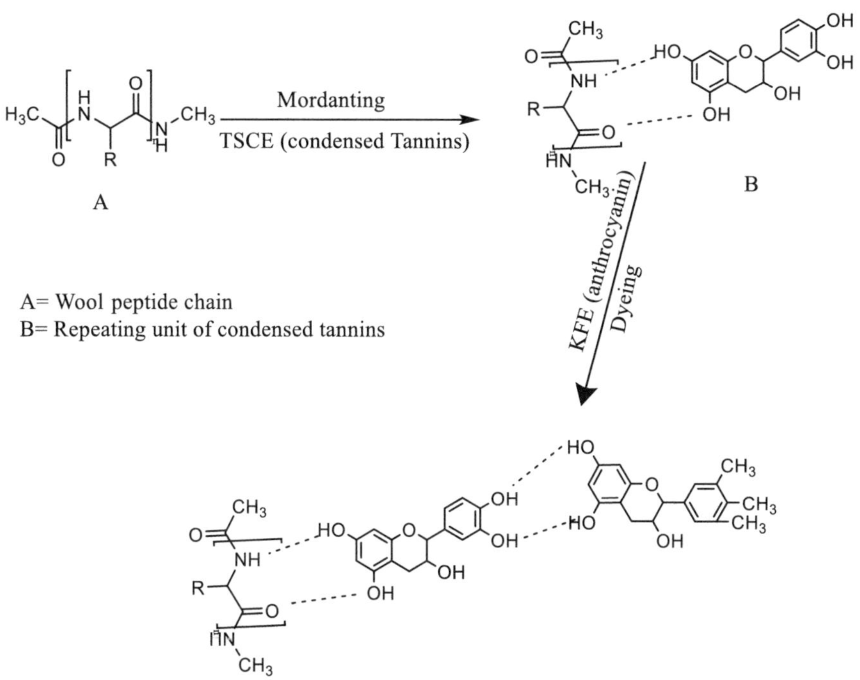

Fig. 5.14 The interaction of anthocyanin from Kapok flower extract (KFE), condensed tannin from tamarind seed coat extract (TSCE), and the wool peptide chain is illustrated (Singh et al. 2019)

connecting the dye molecules to the wool fibre. Phenolic groups partially oxidise and condense during drying and heating, strengthening these interactions and fixing the dye to the fibre more securely. The working mechanism of the bio-based mordant (tannin) with wool is illustrated in Fig. 5.14 (Singh et al. 2019).

Similar mechanisms are followed by other bio-mordants as well. Moreover, the difference in the fabric's appearance before and after natural mordanting is explained by the corresponding mechanism illustrated in Fig. 5.15. Compared to the unmordanted sample, the post-mordanted one exhibits enhanced dye absorption and greater colour intensity. The increased colour strength resulting from post-mordanting provides clear evidence that the natural mordant layer helps to form a more stable and long-lasting dye–fibre combination (Geelani et al. 2016).

5.6.1.2 Mechanism in Metallic Mordants

Metallic mordants operate mainly through coordination complexes with metal ions, dye molecules, and the fibre. Metal ions such as aluminium (Al^{3+}), ferric ion (Fe^{3+}),

Fig. 5.15 Dyed fabrics in combination with natural mordants (Geelani et al. 2016)

cupric ion (Cu^{2+}), and stannous ions (Sn^{2+}) create covalent bonds with electron-donating groups such as –OH, $–NH_2$, –COOH in dyes and fibre substrate (Khan et al. 2015).

A metal fibre complex (Fibre-M^{n+}) is developed in a pre-mordanting method by treating the fibre with the mordant solution, which allows the metal ions to link to the available reactive sites of the fibre. In the next step, mordanted fibre is placed in a dye bath to connect the dye molecule with the metal ions which are already attached to the fibre to obtain coordination complexes.

The fibre is initially dyed and treated with the mordant in the post-mordanting method, where the interaction between the dye and fibre is more enhanced. In such a situation, the metal ion functions as a link between the colourant that has already been absorbed and the fibre matrix (Manian et al. 2016). Due to differences in ligand field splitting and shifts in the electronic transition in the dye molecules, every metal ion gives a distinct colour. The reaction mechanism of metallic mordants with dyes and fibres can be demonstrated below (Fig. 5.16).

The metallic mordanting mechanism is further supported by the intense tone that occurs because of coordination bonding between the metal ion, dye, and fabric. The metallic mordanting mechanism is additionally illustrated by Fig. 5.17. It illustrates the process of cotton and wool fabrics dyeing using *Targets erecta* extract in the presence of various metallic mordants. Several colours are developed from a single dye material by utilising a varied ratio of fixing agents, like potassium dichromate,

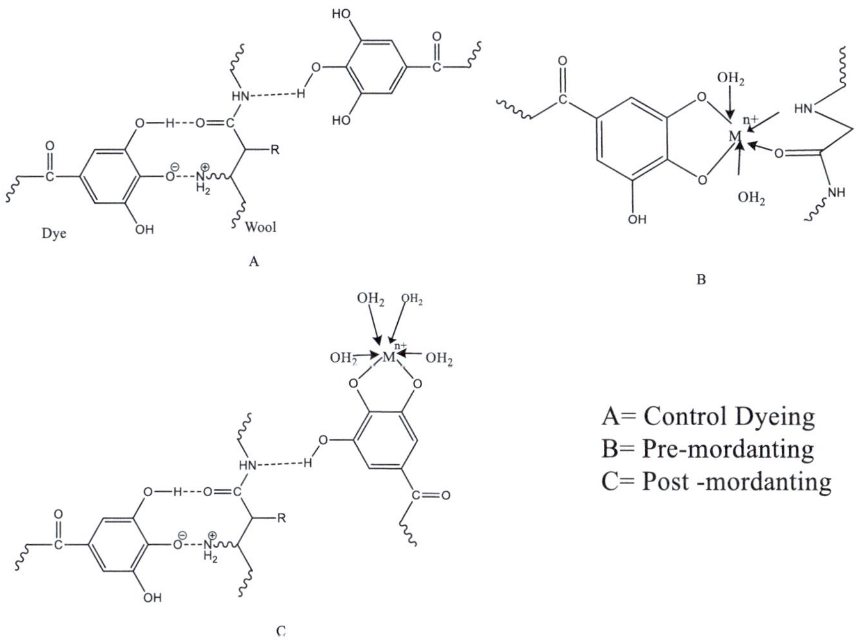

Fig. 5.16 Schematic representation of interaction among dye, metal, and wool: **a** control dyeing, **b** pre-mordanting, **c** post-mordanting (İşmal and Yıldırım 2019)

tin mordant, potash alum, copper mordant, and ferrous sulphate. Figure 5.17 (ii–vi) show the sample premordanted with potassium dichromate, stannous chloride, potash alum, copper sulphate, and ferrous sulphate, respectively. The sample dye in the absence of the mordant is derived from Figure 5.17 (i). The overall figure shows that the colour strength of the mordanted sample is higher than that of the unmordanted fabric, which proves that the metallic mordants enhance dye fixation and colour intensity (Manian et al. 2016).

5.6.1.3 Mechanism in Nano-Mordants

The mechanism of nano-mordanting is similar to the conventional coordination mechanism of metallic mordanting. However, it is distinctly promoted through the improved surface area and the reactivity of nanoparticles (Pereira et al. 2025). Nanoparticles, such as Fe_3O_4, TiO_2, or Al_3O_4, contain various reactive sites for binding the dye, which is absorbed on the surface of the fibre homogeneously. Nanoparticles enhance the coordination of the dye molecules along with the functional groups of the fabric by providing various linking sites (Du et al. 2022). After that, the surface atoms of nanoparticles are coordinated by the dye molecules through

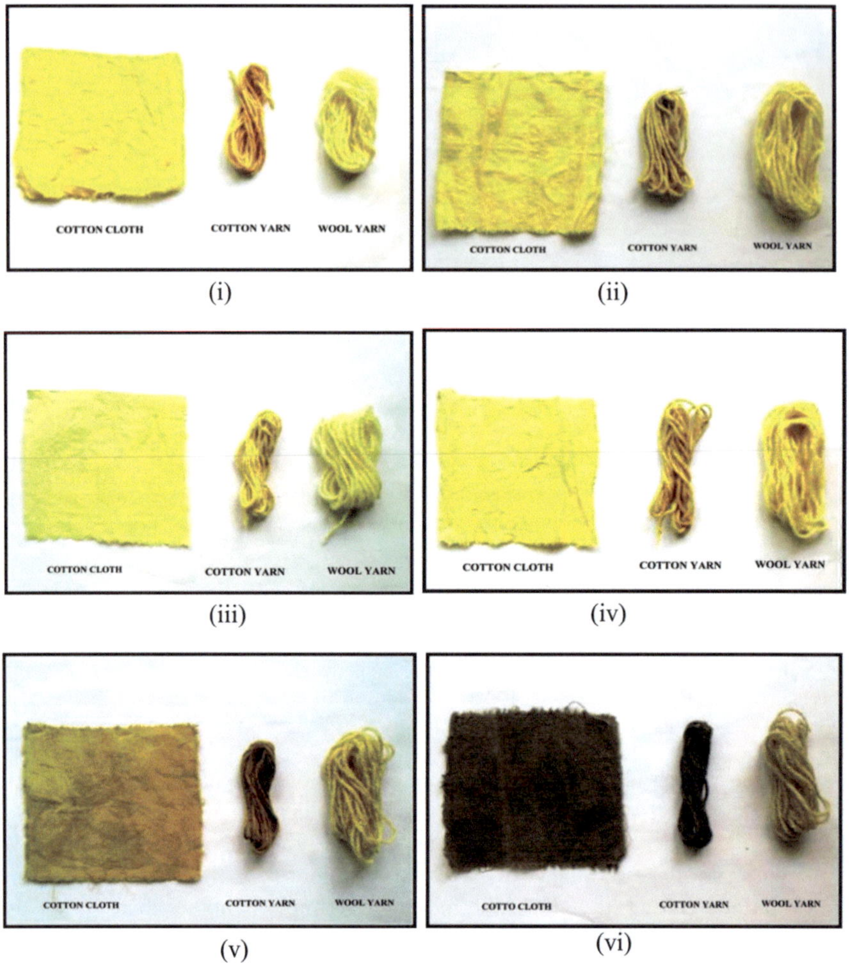

Fig. 5.17 Dyeing of cotton and wool fabrics with *Tagetes erecta* extract using different metallic mordants, where: (i) colouring effect of dye (no mordant), (ii) potassium dichromate as mordant, (iii) stannous chloride as mordant, (iv) potash alum as mordant, (v) copper sulphate as mordant, and (vi) ferrous sulphate as mordant (Jha et al. 2015)

chelation and bonding to form concise nano-dye complexes that are more thermodynamically stable than those obtained by traditional mordants (Shahmoradi Ghaheh et al. 2021).

The mechanism of the nano-mordant is studied by Pour et al. (2020). They applied bentonite nanoparticles and microbial transglutaminase (m-TGAS) to enhance the dyeability of wool fibres with *Rubia tinctorum* (mader). These enzymes increase more reactive sites on the fibre's surface by crosslinking glutamine and lysine residues with wool peptide chains. Because of the larger surface area of bentonite nanoparticles, they stick to those reactive sites through coordination with metal ions in the clay,

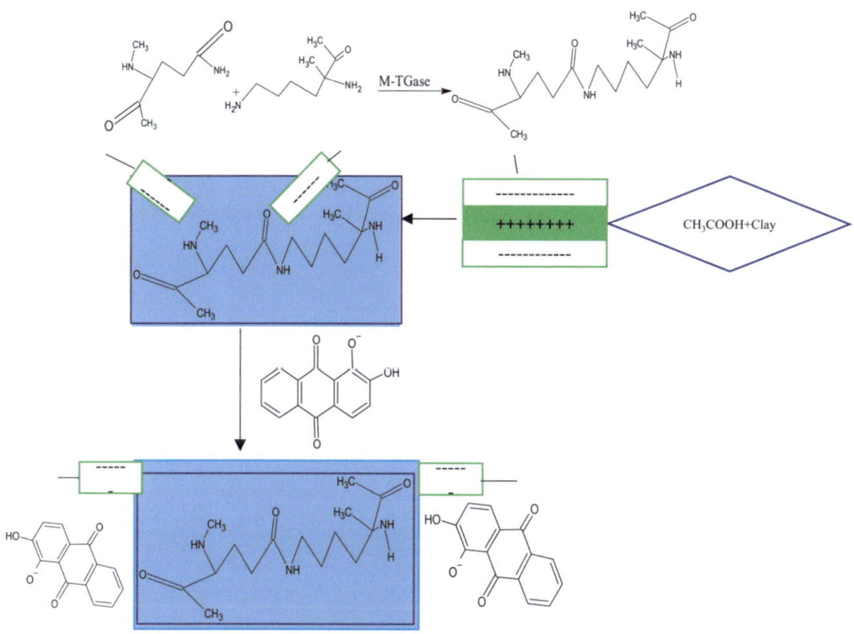

Fig. 5.18 Chemical mechanism of reactions involved during surface functionalisation of wool via M TGase and bentonite to improve its dyeability with madder (Aseefi Pour et al. 2020)

hydrogen bonding, and van der Waals interaction to obtain a bio-nano-mordant tier. Diagrammatic description of the suggested mechanism for bentonite and enzyme-treated wool is illustrated in Fig. 5.18.

The wool sample treated with the combined enzyme and bentonite system showed a visible colour shift, going from pink to deep red, as seen in Fig. 5.19. This suggests that increased colour uptake results from the improved interaction between the dye molecule and the functionalised fibre surface. Furthermore, a lighter red hue was seen when the concentration of m-TGase was increased. This indicates that excessive crosslinking may reduce the number of accessible sites for nanoparticle attachment or change the fibre surface in a way that slightly reduces dye absorption. These findings highlight the bio-nano-mordant's potential for controlled and high-performance natural dyeing (Pour et al. 2020).

5.7 Cellulose and Protein Mordanting

Cellulose-based fibres originate from plants, such as cotton, flax, and bamboo (Benli 2024). They are composed of cellulose, a homopolymer containing β-1,4-glycosidic linkages between D-glucose units. The chain is straight and unbranched because glucose is the repeating unit. It contains three hydroxyl groups, allowing for a compact

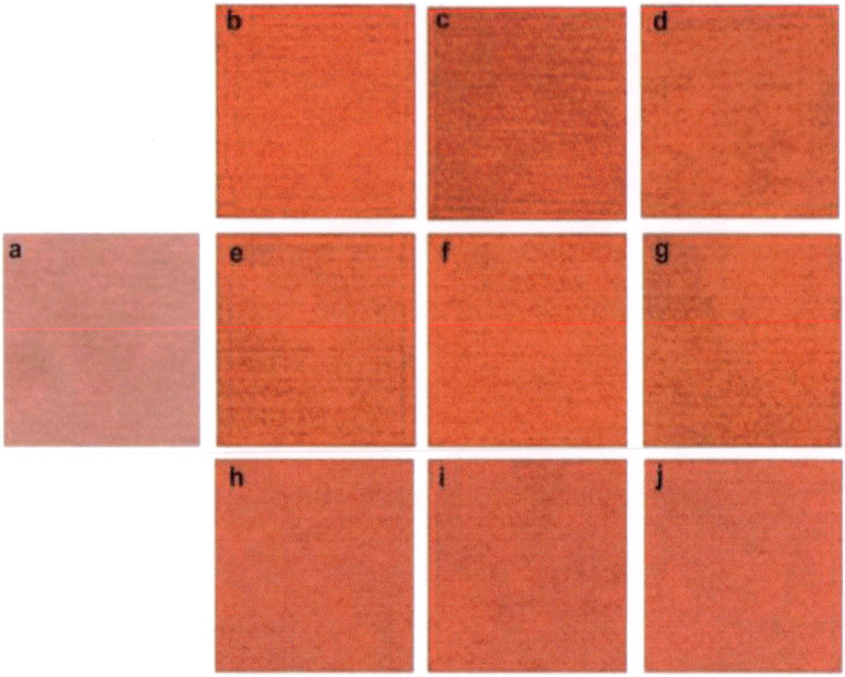

Fig. 5.19 Colour of dyed wool: **a** untreated wool; treated wool with: **b** 5% m-TGase + 3% clay, **c** 5% m-TGase + 5% clay, **d** 5% m-TGase + 10% clay, **e** 10% m-TGase + 3% clay, **f** 10% m-TGase + 5% clay, **g** 10% m-TGase + 10% clay, **h** 20% m-TGase + 3% clay, **i** 20% m-TGase + 5% clay, and **j** 20% m-TGase + 10% clay (Assefi Pour et al. 2020)

network of hydrogen bonds, which gives cellulose its high tensile strength, crystalline microfibrils, and relative chemical inertness (Kumpikaitė et al. 2021).

The surface of cellulose fibres is slightly anionic in aqueous solution because of the deprotonation of the -OH group; as a consequence, it does not intrinsically have strong cationic sites for dye bonding under normal dye bath conditions (see Fig. 5.20). Hence, a particular dye system or conditions such as alkali, reactive groups, pre-treatments, and salts are necessary for long-term fixing (Periyasamy 2022).

Due to the presence of a wide range of dye-molecular structures, diverse fibres behave differently during dyeing. For cellulose fibres, five major dye classes are involved as each class reacts individually through hydrogen bonding, covalent bonding, and insoluble pigment trapping, and requires distinct processes (pH, temperature, auxiliary salts) (Burkinshaw 2021).

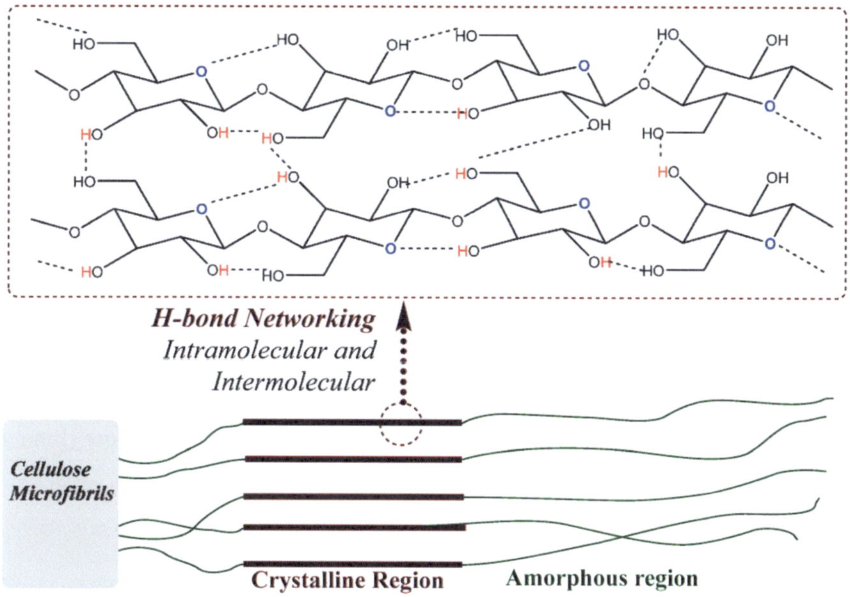

Fig. 5.20 The crystallinity of cellulose materials (fibre) exhibits a high intramolecular as well as intermolecular H-bonding network (Negi 2025)

5.7.1 *Classification of Dyes for Cellulose Fibres*

5.7.1.1 Direct Dyes

These are anionic dyes that dissolve in water and can be used directly on cellulose fibres without the need for a mordant. For binding, they mostly rely on the van der Waals forces and hydrogen bonds. The addition of electrolytes to the dye bath increases their absorption by encouraging aggregation and decreasing dye solubility (Burkinshaw 2021).

5.7.1.2 Reactive Dyes

These dyes contain reactive groups that form covalent bonds with the hydroxyl groups of cellulose under alkaline conditions, leading to excellent wash fastness (Lewis 2014).

5.7.1.3 Vat Dyes

They are insoluble dyes that undergo chemical reduction with cellulose fibres to become their soluble leuco forms. After that, the dyes revert to their insoluble form, resulting in mechanical and physical entrapment and excellent fastness (Božič and Kokol 2008; Morin 2017).

5.7.1.4 Sulphur Dyes

They are similar to vat dyes, but they are more affordable and are frequently employed for cellulosic fibres or blends (Singh Shekhawat et al. 2024).

5.7.1.5 Azoic Dyes

These dyes are created in situ by combining diazonium salts with naphthol infused into the fibre, resulting in permanent, insoluble azo dyes within the cellulose (Bafana et al. 2011).

5.7.2 Role of Inorganic Salts or Inorganic Electrolytes

Many dyes, especially reactive and direct dyes, are anionic, and cellulose fibres naturally have a slightly anionic charge in water because of the ionisation of –OH groups, creating electrostatic repulsion that reduces dye adsorption. The use of inorganic salts (NaCl, Na_2SO_4) improves exhaustion for direct and reactive dyes on cellulose fibres. The major role of inorganic electrolytes is to neutralise negative charges at the fibre–aqueous interface, thereby increasing dye uptake by facilitating the approach of anionic dye molecules (Burkinshaw 2021).

5.7.3 Mordant Interaction

The majority of –OH (hydroxyl) groups on the surface of cellulose fibres cannot align with many natural dyes due to their chemical inertness. This causes repulsion, which lowers dye absorption and effectiveness (Thapliyal et al. 2023). Mordants are therefore introduced to modify the fibre surface, create new binding sites, and promote the formation of stable dye–fibre complexes. Traditionally, metallic mordants (alum, iron salts) coordinate with dye molecules to obtain stable metal–dye complexes, which then interact with cellulose hydroxyl groups by coordinative complex bonding (Kumpikaitė et al. 2021). However, metallic mordants are avoided due to ecological threats. On the other hand, natural mordants such as tannin from plants react with cellulose and dye through hydrogen bonding, leading to a more sustainable replacement (Kovačević et al. 2021). For instance, the shade of a chlrophyll derivative is enhanced by reacting cotton fabric with naturally occurring polysaccharide, namely chitosan, as a natural mordant. The positively charged amino groups of chitosan react with the dye molecules containing anionic groups through hydrogen bonds with cellulose hydroxyls to achieve a stable cotton–bio-mordant–dye complex (İşmal and Yıldırım 2019).

Furthermore, Periyasamy (2022) studied an eco-friendly method by applying soy protein as a bio-mordant. It provides functional groups like amine and carbonyl, which react with fibre, along with colourant molecules extracted from *Syzgium cumini* fruits. Through ultrasonic-assisted treatment, the soy protein attaches to a fibre surface, increasing the intensity of the colour by enhancing dye binding and facilitating fixation.

Protein fibres have distinct behaviours such as stretchable, hard, and water absorption. They are composed of proteins like keratin or fibroin (Feughelman 2002). These are primarily derived from animal sources. For example, silk from silkworms and wool from the fleece of sheep are widely used in the textile industry due to their comfort and durability. They are generally classified as keratin-based animal hair fibres and fibroin-based silk fibres (see Fig. 5.21) (Thapliyal et al. 2023).

Protein fibres are polymers of amino acids linked by peptide bonds, making them chemically and structurally distinct from cellulose fibres (See Fig. 5.22). Their properties generally depend upon the type of protein (Dulo et al. 2021). Wool has a complex hierarchical structure with cuticle scales and a cortex. It exhibits elasticity and chemical resistance due to the presence of cystine disulphide bonds. In contrast, silk fibroin contains a β-sheetcrystalline structure formed by a repetitive glycine-alanine sequence, resulting in high strength and lustre (Feughelman 2002). Thus, protein fibres are amphoteric, acting as a cationic or anionic substrate depending on pH (Bukhari et al. 2017).

They contain functional groups such as amino, carboxyl, and hydroxyl groups, as shown in Fig. 5.23, which inhibit high affinity towards many classes of dyestuff. Dyeing protein fibres requires precise control of pH, temperature, and dye class (reactive dyes, vat dyes, acid dyes, chrome dyes) because they are sensitive to alkaline conditions, which can cause peptide-bond hydrolysis and fibre damage. Protein fibres show a significant affinity for acid dyes because the fibres become positively charged in acidic environments, promoting electrostatic attraction for anionic dye ions (Nadeem et al. 2024).

Protein fibres interact strongly with mordants because the fibres contain multiple functional groups that can chelate metal ions. Traditionally, after dyeing, chromium is applied to fix the dye, a process known as chrome-dyeing. These metal-fibre complexes bind with natural dye molecules, enhancing both wash and light fastness (Khan et al. 2015). These interactions between protein fibres, metallic mordants, and natural dye are shown in Fig. 5.24, depicting the protein–metallic mordant-quinone dyeing mechanism and the formation of metal–fibre complexes that enhance the binding of natural dye molecules, improving both wash and light fastness (Dulo et al. 2021).

However, metals such as hexavalent chromium (Cr^{6+}) possess environmental and human health risks. Hence, there is a shift towards biomordants (Park and Park 2010). Natural mordants are rich in tannins and chitosan are composed of large macromolecules and protein fibres possess a hydrophobic surface layer that

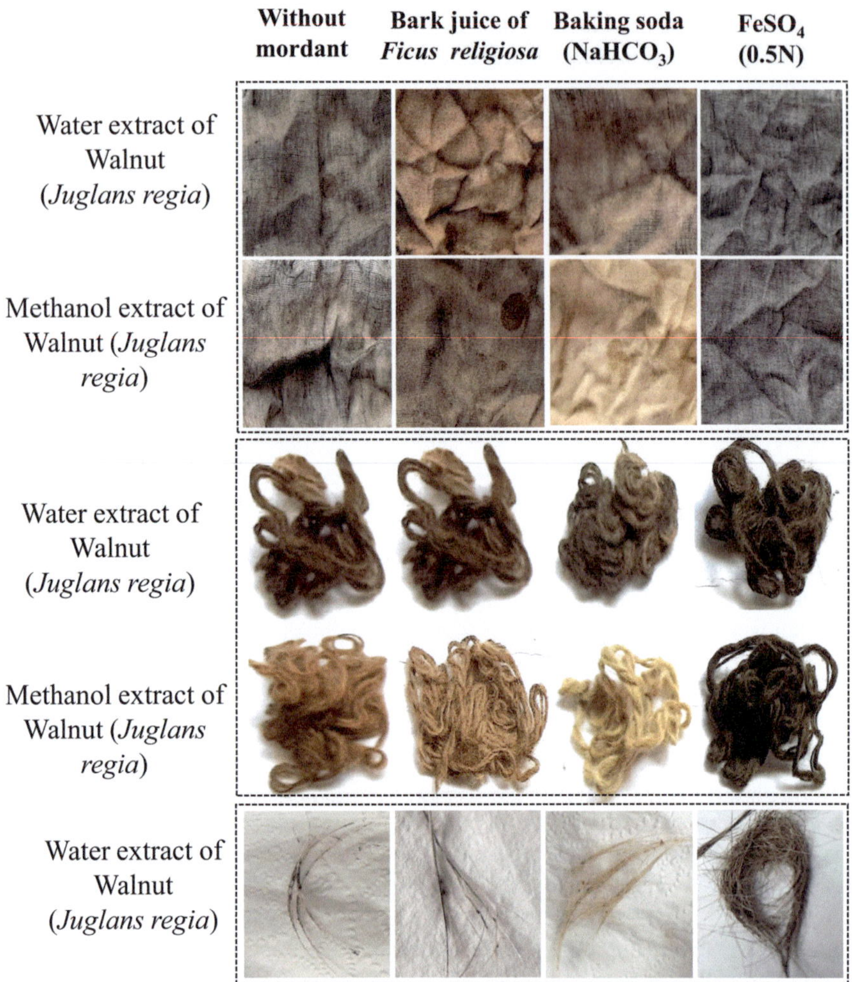

Fig. 5.21 Effect of mordants with water and methanol extract of *Juglans regia*: **a** cotton cloth (cellulose) dyeing, **b** wool colouring, and **c** human hair dyeing

physically blocks the adsorption of large biomordants, preventing them from interacting with the fibre structure (Fig. 5.25). They require careful pre-treatment, such as soaking, acidic conditions, and heat, to increase fibre accessibility and improve fixation (Teklemedhin 2018). Figure 5.25 shows the interaction of bio-mordants with protein fibres, illustrating the effect of fibre hydrophobicity and the role of pretreatment.

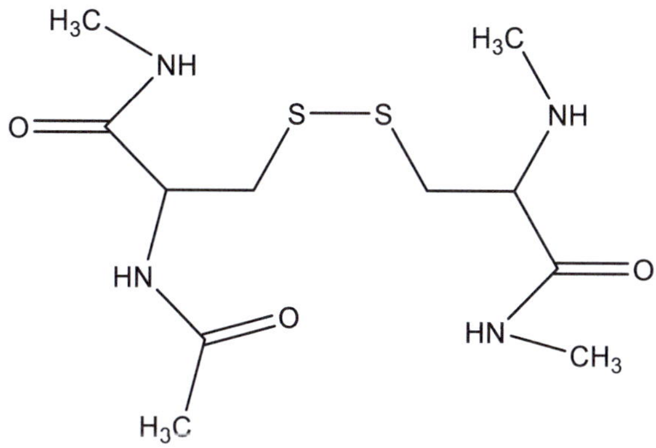

Fig. 5.22 Molecular structure of a polypeptide (Dulo et al. 2021)

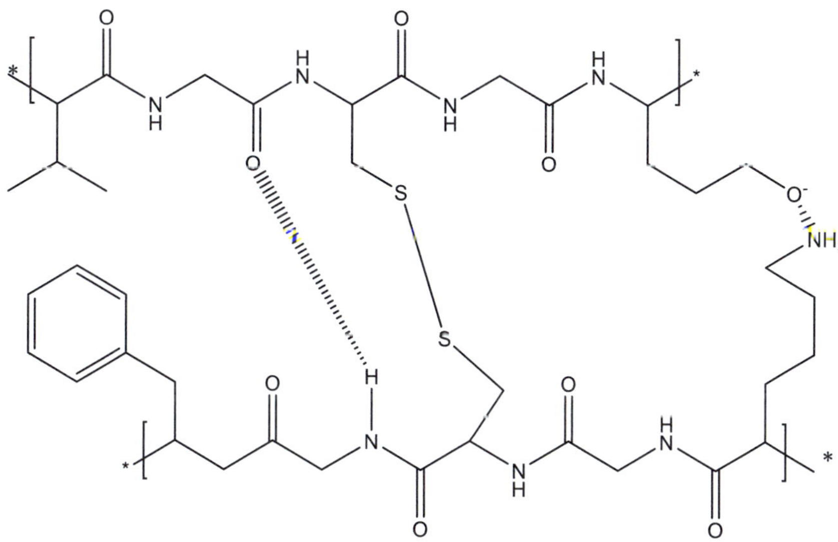

Fig. 5.23 Protein-based fibre structure (Dulo et al. 2021)

5.8 Conclusions

Mordants have a significant impact on natural dyeing because they act as chemical bridges that connect dye molecules to fibre substrates. This improves dye uptake, colour depth, and fastness (Geelani et al. 2017). The chemical nature, mode of interaction, and the functional groups in the fibre and dye play a major role in determining their success. Most mordants are polyvalent or polydentate metal ions that react with the functional groups of fibres and dyes or colour stains, forming a colloidal coordination complex, which is the foundation of mordant chemistry (Benli 2024).

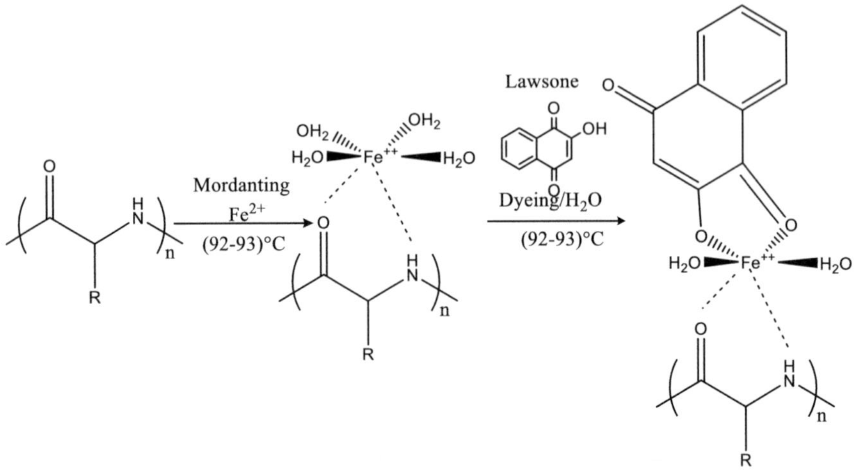

Fig. 5.24 Protein–metallic mordant–quinone dyeing mechanism (Dulo et al. 2021)

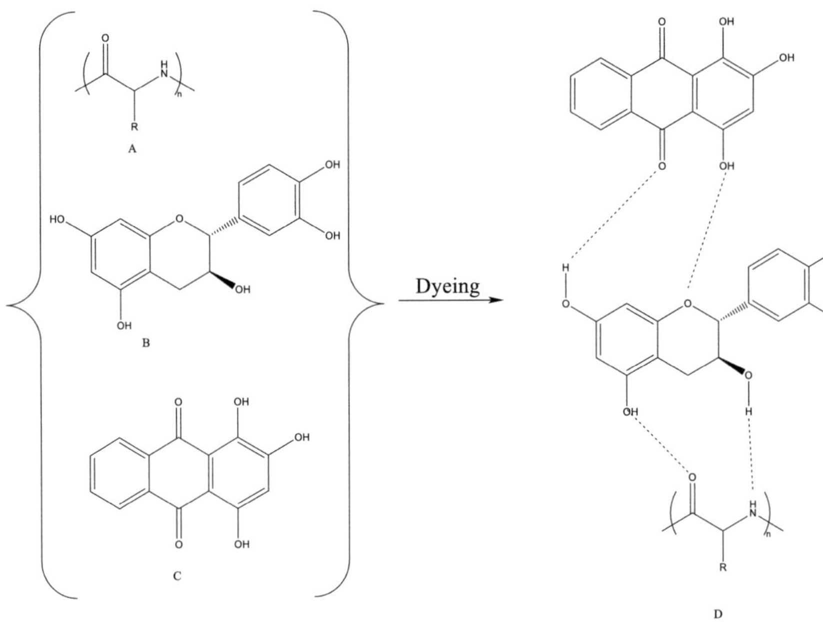

A= Wool
B= *Accacia catechu* mordant
C= Purpurin
D= Wool Biomordant Dye interaction

Fig. 5.25 Interaction of wool, bio-mordant, and dye (Dulo et al. 2021)

As environment friendly substitutes for synthetic metal salts, natural mordants, including tannins, organic acids, and polyphenols derived from natural sources, have drawn more attention and interest in research (Ashrafi et al. 2018). These biodegradable, non-toxic bio-mordants can create a variety of hues while lowering pollution levels in the environment. There are a few disadvantages and difficulties associated with natural colouring with mordants despite their numerous benefits (Manian et al. 2016).

The most difficult aspect of natural colouring with mordants is that the consistency of colour shades and fastness characteristics can differ significantly across batches and suppliers. This problem of inconsistency may be problematic for sectors that depend on product regularity, especially when accepting huge orders. However, because natural colouring requires larger amounts and takes longer to process than synthetic colours, its production costs are significantly higher. Hence, based on the environmental and health-related issues, the bio-mordants are an even better option for sustainability, even better than metallic mordants (Dutta et al. 2021).

Lately, nano-mordants have become a sophisticated alternative because of their increased surface area, potent adsorption capacity, and potential to produce extremely stable fibre-nano-dye complexes. It would be highly beneficial if researchers could develop such effective bio-nanomordants in the near future that can fix the desired colour shades in natural colouring, and the complex degrades easily in nature (Pereira et al. 2025).

The investment in the next-generation bio-mordants and natural dyes is potentially a new direction in sustainable textile production. If the toxicity and other factors associated with bio-mordants with natural colour can be analysed and optimised, then the natural colours also could be utilized for multipurposes, such as in the food industry, cosmetics, pharmaceuticals, and medicine (Taheri et al. 2014). Scientific investigation is increasingly focused on finding more environmentally friendly dyes and mordants these days, investigating a variety of options. Numerous microbial pigments, which are made by bacteria, fungi, and algae, are a novel field of study as well. An eco-friendly alternative to conventional dyes is also being explored as a new area of research, and ultimately a poly-chromatic rainbow with several tones can be produced by bacteria (Dutta et al. 2021).

References

Adeel S, Yameen M, Asghar F et al (2023) Extraction and application of plant-based Tannins as sources of natural colourants. In: Muthu SS (ed) Natural dyes and sustainability. Springer Nature, Switzerland, Cham, pp 103–126

Arora J, Aggrawal P, Gupta G (2025) Rainbow of natural dyes on textiles using plants extracts: sustainable and eco-friendly processes. Green Sustain Chem 7:35–47. https://doi.org/10.4236/gsc.2017.71003

Ashrafi N, Gharanjig K, Hosseinnezhad M et al (2018) Dyeing properties and color fabrics using natural dye and mordant. Progress Color, Colorants Coatings 11:79–83. https://doi.org/10.30509/pccc.2018.75742

Assefi Pour R, Bagheri R, Naveed T et al (2020) Surface functionalization of wool via microbial-transglutaminase and bentonite as bio-nano-mordant to achieve multi objective wool and improve dyeability with madder. Heliyon 6:e04911. https://doi.org/10.1016/j.heliyon.2020.e04911

Bafana A, Devi SS, Chakrabarti T (2011) Azo dyes: past, present and the future. Environ Rev 19:350–371. https://doi.org/10.1139/a11-018

Banc R, Rusu ME, Filip L, Popa D-S (2023) The impact of ellagitannins and their metabolites through gut microbiome on the gut health and brain wellness within the gut–brain axis. Foods 12:270. https://doi.org/10.3390/foods12020270

Basavaiah K, Swamy JM (2001) Application of potassium dichromate and iron–thiocyanate in the spectrophotometric investigations of phenothiazines. Il Farmaco 56:579–585. https://doi.org/10.1016/S0014-827X(01)01096-5

Benli H (2024) Bio-mordants: a review. Environ Sci Pollut Res 31:20714–20771. https://doi.org/10.1007/s11356-024-32174-8

Bhandari N, Shrestha S, Bhandari G et al (2021) Extraction of dye from *Castanopsis indica* for Its use in textile dyeing and medicinal purpose with natural mordant. Egypt J Chem 64:6681–6690. https://doi.org/10.21608/ejchem.2021.48519.2991

Bhandari NL, Bist K, Ghimire J et al (2020a) Feasibility study of the euphorbia pulcherrima plant extract as natural dye with different mordants for fabric dyeing. J Inst Sci Tech 25:30–36. https://doi.org/10.3126/jist.v25i1.29421

Bhandari NL, Bist K, Ghimire J et al (2020b) Feasibility study of the Euphorbia pulcherrima plant extract as datural dye with different mordants for fabric dyeing. J Inst Sci Tech 25:30–36. https://doi.org/10.3126/jist.v25i1.29421

Bhandari NL, Pokhrel B, Bhandari U et al (2020c) An overview of research on plant based natural dyes in Nepal: scope and challenges. J Agric Nat Res 3:45–66. https://doi.org/10.3126/janr.v3i2.32328

Bishal A, Ali KA, Ghosh S et al (2023) Natural dyes: its origin, categories and application on textile fabrics in brief. Natural dyes

Božič M, Kokol V (2008) Ecological alternatives to the reduction and oxidation processes in dyeing with vat and sulphur dyes. Dyes Pigm 76:299–309. https://doi.org/10.1016/j.dyepig.2006.05.041

Bukhari MN, Shahid-ul-Islam SM et al (2017) Dyeing studies and fastness properties of brown naphtoquinone colorant extracted from *Juglans regia* L on natural protein fiber using different metal salt mordants. Text Cloth Sustain 3:3. https://doi.org/10.1186/s40689-016-0025-2

Burkinshaw SM (2021) The role of inorganic electrolyte (salt) in cellulosic fibre dyeing: Part 1 fundamental aspects. Color Technol 137:421–444. https://doi.org/10.1111/cote.12547

Ding Y, Freeman HS (2017) Mordant dye application on cotton: optimisation and combination with natural dyes. Color Technol 133:369–375. https://doi.org/10.1111/cote.12288

Du H, Yue M, Huang X et al (2022) Preparation, application and enhancement dyeing properties of ZnO nanoparticles in silk fabrics dyed with natural dyes. Nanomaterials 12:3953. https://doi.org/10.3390/nano12223953

Dulo B, Phan K, Githaiga J et al (2021) Natural quinone dyes: a review on structure, extraction techniques, analysis and application potential. Waste Biomass Valor 12:6339–6374. https://doi.org/10.1007/s12649-021-01443-9

Dutta P, Mahjebin S, Sufian MA et al (2021) Impacts of natural and synthetic mordants on cotton knit fabric dyed with natural dye from onion skin in perspective of eco-friendly textile process. Mater Today: Proc 47:2633–2640. https://doi.org/10.1016/j.matpr.2021.05.229

El-Azab WIM, Aiad I, Azzam EMS, Gad EAM (2010) Molecular, surface, and thermodynamic properties of nonionic surfactants based on Castor oil. J Dispersion Sci Technol 31:1150–1156. https://doi.org/10.1080/01932690903218021

Evitasari RT, Rahayuningsih E, Mindaryani A (2019) Dyeing of cotton fabric with natural dye from peristrophe bivalvis extract. AIP Conf Proc 2085:020055. https://doi.org/10.1063/1.5095033

Feughelman M (2002) Natural protein fibers. J Appl Polym Sci 83:489–507. https://doi.org/10.1002/app.2255

Gedik G, Avinc O, Yavas A, Khoddami A (2014) A novel eco-friendly colorant and dyeing method for poly(ethylene terephthalate) substrate. Fibers Polym 15:261–272. https://doi.org/10.1007/s12221-014-0261-5

Geelani SM, Ara S, Mir NA et al (2016) Dyeing and fastness properties of *Quercus robur* with natural mordants on natural fibre. Text Cloth Sustain 2:8. https://doi.org/10.1186/s40689-016-0019-0

Geelani SM, Ara S, Mir NA et al (2017) Dyeing and fastness properties of Quercus robur with natural mordants on natural fibre. Text Cloth Sustain 2:8. https://doi.org/10.1186/s40689-016-0019-0

Ghosh J, Khan F, Noor T et al (2024) Union dyeing of cotton-silk blended woven fabric using Tanin bio-mordant and Thyme as a colourant. Textile Leather Rev 7:303–326. https://doi.org/10.31881/TLR.2023.208

Gill P, Singh OP (2005) Dyeing wool with natural dyes and natural mordants. Colourage 52:43–49

Gupta VK (2019) Fundamentals of natural dyes and its application on textile substrates. In: Samanta AK, Nasser A, Algarni HM (eds) Chemistry and technology of natural and synthetic dyes and pigments, 1st edn. IntechOpen, London, UK, p 304

Haddar W, Baaka N, Meksi N et al (2014) Optimization of an ecofriendly dyeing process using the wastewater of the olive oil industry as natural dyes for acrylic fibres. J Clean Prod 66:546–554. https://doi.org/10.1016/j.jclepro.2013.11.017

Islam S, Belowar S, Das S, Rahamatolla M, Datta SC (2025) Chemistry of natural and synthetic dye material with metal mordants in various fabrics for sustainable textile application: a comprehensive review. Environ Sci Pollut Res 32(45):25511–25540. https://doi.org/10.1007/s11356-025-37102-y

İşmal ÖE, Yıldırım L (2019) Metal mordants and biomordants. In: Butola S (ed) The impact and prospects of green chemistry for textile technology, 1st edn. Woodhead Publishing, pp 57–82

Jha CK, Ratan Kumar RK, Kumar SV, Rajeswari VD (2015) Extraction of natural dye from marigold flower (*Tagetes erecta* L.) and dyeing of fabric and yarns: a focus on colorimetric analysis and fastness properties. Der Pharmacia Lettre 7:185–195

Khan SA, Shahid-ul-Islam SM et al (2015) Mixed metal mordant dyeing of wool using root extract of *Rheum emodi* (Indian Rhubarb/Dolu). J Nat Fibers 12:243–255. https://doi.org/10.1080/15440478.2014.919893

Kovačević Z, Sutlović A, Matin A, Bischof S (2021) Natural dyeing of cellulose and protein fibers with the flower extract of *Spartium junceum* L. Plant Mater 14:4091. https://doi.org/10.3390/ma14154091

Kumpikaitė E, Varnaitė-Žuravliova S, Tautkutė-Stankuvienė I, Laureckienė G (2021) Comparison of mechanical and end-use properties of grey and dyed cellulose and cellulose/protein woven fabrics. Materials 14:2860. https://doi.org/10.3390/ma14112860

Lee H-S, Koo Y-C, Suh HJ et al (2010) Preventive effects of chebulic acid isolated from *Terminalia chebula* on advanced glycation endproduct-induced endothelial cell dysfunction. J Ethnopharmacol 131:567–574. https://doi.org/10.1016/j.jep.2010.07.039

Lewis DM (2014) Developments in the chemistry of reactive dyes and their application processes. Color Technol 130:382–412. https://doi.org/10.1111/cote.12114

Malla L (1993) The natural dyes of Nepal. In: Research centre for applied science and technology. Tribhuvan University, Kirtipur, Nepal

Malla SB (1994) Medicinal herbs in the Bagmati zone. In: International centre for integrated mountain development (ICIMOD). Kathmandu, Nepal

Manian AP, Paul R, Bechtold T (2016) Metal mordanting in dyeing with natural colourants. Color Technol 132:107–113. https://doi.org/10.1111/cote.12199

Mongkholrattanasit R, Saiwan C, Rungruangkitkrai N et al (2016) The effect of Alum, Ferrous, Stannous and NaCl on silk fabric dyed with natural dye from Laccifer Lacca Kerr. Appl Mech Mater 848:141–144. https://doi.org/10.4028/www.scientific.net/AMM.848.141

Morin J-F (2017) Recent advances in the chemistry of vat dyes for organic electronics. J Mater Chem C 5:12298–12307. https://doi.org/10.1039/C7TC03926C

Morshed MN, Deb H, Azad SA et al (2016) Aqueous and solvent extraction of natural colorants from *Tagetes erecta* L., *Lawsonia inermis*, *Rosa* L for coloration of cellulosic substrates. Am J Polym Sci Technol 2:34–39

Muthu SS (ed) (2023) Natural dyes and sustainability. Springer, Cham, Switzerland

Nadeem T, Javed K, Anwar F et al (2024) Sustainable dyeing of wool and Silk with *Conocarpus erectus* L. leaf extract for the development of functional textiles. Sustainability 16:811. https://doi.org/10.3390/su16020811

Negi A (2025) Environmental impact of textile materials: Challenges in riber–dye chemistry and implication of microbial biodegradation. Polymers 17:871. https://doi.org/10.3390/polym17070871

Nivedha K, Kalaiarasi K (2023) Sustainable colorants from natural resources. In: Muthu SS (ed) Novel sustainable raw material alternatives for the textiles and fashion industry. Springer Nature, Switzerland, Cham, pp 23–34

Pandey DP, Parajuli N, Devkota A et al (2020) Eco-friendly methods of natural dye extraction from bark of *Ficus religiosa* and their impacts on dyeing technology in Nepal. NUTA Jnl 7:42–51. https://doi.org/10.3126/nutaj.v7i1-2.39930

Park SJ, Park YM (2010) Eco-dyeing and antimicrobial properties of chlorophyllin copper complex extracted from *Sasa veitchii*. Fibers Polym 11:357–362. https://doi.org/10.1007/s12221-010-0357-5

Patel BH (2011) Natural dyes. In: Clark M (ed) Handbook of textile and industrial dyeing: principles, processes and types of dyes, 1st edn. Woodhead Publishing, Cambridge, UK, pp 395–424

Pereira PHF, Costa SA, Costa SM, Arantes V (2025) Efficient production of lignin nanoparticle colloids and their feasibility for eco-friendly dyeing of natural and synthetic textile fabrics. Ind Crops Prod 225:120390. https://doi.org/10.1016/j.indcrop.2024.120390

Periyasamy AP (2022) Natural dyeing of cellulose fibers using *syzygium cumini* fruit extracts and a bio-mordant: a step toward sustainable dyeing. Sustain Mater Technol 33:e00472. https://doi.org/10.1016/j.susmat.2022.e00472

Pour RA, Bagheri R, Naveed T et al (2020) Surface functionalization of wool via microbial-transglutaminase and bentonite as bio-nano-mordant to achieve multi objective wool and improve dyeability with madder. Heliyon 6. https://doi.org/10.1016/j.heliyon.2020.e04911

Prabhu KH, Bhute AS (2012) Plant based natural dyes and mordnats: a review. 2

Pranta AD, Rahaman MdT (2024) Extraction of eco-friendly natural dyes and biomordants for textile coloration: a critical review. Nano-Struct Nano-Objects 39:101243. https://doi.org/10.1016/j.nanoso.2024.101243

Repon MdR, Dev B, Rahman MA et al (2024) Textile dyeing using natural mordants and dyes: a review. Environ Chem Lett 22:1473–1520. https://doi.org/10.1007/s10311-024-01716-4

Sadannavar MK, Periysamy A, Islam SR et al (2024) Natural dyeing and antimicrobial functionalization of wool fabrics dyed with Chinese dragon fruit extract to enhance sustainable textiles. Sustainability 16:6832. https://doi.org/10.3390/su16166832

Shabbir M, Rather LJ, Mohammad F (2019) Exploring the potential of tannin based colorants towards functional value addition of wool textiles. Fibers Polym 20:1812–1819. https://doi.org/10.1007/s12221-019-9059-9

Shahid M, Ali A, Zahid N et al (2023) Copper-treated environmentally friendly antipathogenic cotton fabric with modified reactive blue 4 dye to improve its antibacterial and aesthetic properties. Coatings 13:133. https://doi.org/10.3390/coatings13010133

Shahmoradi Ghaheh F, Moghaddam MK, Tehrani M (2021) Comparison of the effect of metal mordants and bio-mordants on the colorimetric and antibacterial properties of natural dyes on cotton fabric. Color Technol 137:689–698. https://doi.org/10.1111/cote.12569

Singh G, Mathur P, Singh N, Sheikh J (2019) Functionalization of wool fabric using kapok flower and bio-mordant. Sustain Chem Pharm 14:100184. https://doi.org/10.1016/j.scp.2019.100184

Singh HB, Bharati KA (2014) Mordants and their applications. In: Handbook of natural dyes and pigments. Elsevier, pp 18–28

Singh HB, Bharati AK (2014) Mordants and their applications. In: Singh HB, Bharati, Avinash Kumar (eds) Handbook of natural dyes and pigments, 1st edn. Woodhead Publishing India, Delhi, India, pp 18–28
Singh HB, Bharati AK (eds) (2015) Handbook of natural dyes and pigments. Woodhead Publishing India, Delhi, India, First
Singh Shekhawat N, Kumar Patra S, Kumar Patra A, Bag B (2024) A novel environmental friendly and sustainable process for textile dyeing with sulphur dyes for cleaner production. Chem Eng J 479:147329. https://doi.org/10.1016/j.cej.2023.147329
Srivastava JP (1912) On the relative affinities for cotton of the tannins of nut galls, divi-divi, Myrabolans, and Sumach. J Soc Dyers Colour 28:286–290. https://doi.org/10.1111/j.1478-4408.1912.tb00676.x
Taheri M, Moghaddam MRA, Arami M (2014) A comparative study on removal of four types of acid azo dyes using electrocoagulation process. Environ Eng Manag J 13:557–564
Teklemedhin TB (2018) Dyeing of wool fabric using natural dye and natural mordant extracts. TTEFT 4. https://doi.org/10.31031/TTEFT.2018.04.000593
Thapliyal D, Verma S, Sen P et al (2023) Natural fibers composites: origin, importance, consumption pattern, and challenges. J Compos Sci 7:506. https://doi.org/10.3390/jcs7120506
Uemori H, Inoue A, Suzuki S et al (2025) Ellagitannin oligomers from *Eucalyptus camaldulensis* leaves and their role in the detoxification of aluminum. Molecules 30:2216. https://doi.org/10.3390/molecules30102216
Vankar PS, Dixit S (2011) Natural dyeing of cotton, wool and silk with the stem and leaves extract of *Illicium griffithii*. Res J Text Appar 15:77–83. https://doi.org/10.1108/RJTA-15-02-2011-B009
Wangatia LM, Tadesse K, Moyo S (2015) Mango Bark mordant for dyeing cotton with Natural dye: fully eco-friendly natural dyeing. Int J Textile Sci 4:36–41
Zhang W, Wang X, Weng J et al (2022) Eco-dyeing and functional finishing of wool fabric based on *Portulaca oleracea* L. as colorant and *Musa basjoo* as natural mordant. Arab J Chem 15:103624. https://doi.org/10.1016/j.arabjc.2021.103624

Chapter 6
Applications of Natural Dyes in Our Daily Life

Abstract This chapter deals with the advantages of using natural dyes. These dyes are non-toxic, biodegradable and do not negatively affect the environment. It also details their application on textile fibres, cosmetics, food, pharmaceuticals, leather, polymers, etc. In addition, applications of natural dyes are found in dye-sensitised solar cells and as corrosion inhibitors, emphasising their multi-functional uses in modern technology. However, there are challenges in standardisation, reproducibility, and proper production of natural dyes. Thus, further studies are needed to overcome these problems.

6.1 Advantages of Natural Colouring

Natural dyes and pigments are preferred over synthetic alternatives due to their greater eco-friendliness (Dave et al. 2022). These natural dyes and pigments are obtained from plants, animals, and minerals and can be extracted using water or organic solvents as a medium. Synthetic dyes and pigments are obtained from petrochemicals, and they pose risks to human and animal health as they are toxic and cause allergies. This makes natural dyes more desirable. Because of this, natural dyes are used in food, children's products, cosmetics, and many other products that we use in our daily lives (Gürses et al. 2016; Bhandari et al. 2020; Alegbe and Uthman 2024).

6.2 Cellulose and Protein Dyeing

Textile fibres are generally classified into two categories on the basis of their chemical composition. They are cellulose- and protein-based textile fibres. Cellulose-based textile fibres are obtained from plant species, such as cotton, linen (from flax), and hemp. They have long chains of glucose units and are strong, as well as chemically inactive and negatively charged in water. Protein fibres are obtained from animals, such as wool, silk, and cashmere (Chinga-Carrasco 2011). They contain polypeptide

N. L. Bhandari et al., *Natural Dyes and Pigments of Nepal Himalaya*,
https://doi.org/10.1007/978-3-032-23229-8_6

chains made up of amino acids. Amino acids are amphoteric in nature and can react both with acids and bases (Brooks 2009).

The dyeing processes for both cellulose and protein-based textile fibres differ from one another, as they have contrasting chemical properties. Cellulose lacks innate ionic attraction to dyes, and this is the primary challenge in achieving colourfastness. It is possible to overcome this problem by using fibre-reactive dyes. This process involves the creation of a permanent covalent bond with the cellulose's hydroxyl groups in an alkaline medium. This gives exceptional resistance to washing and bright colours (Lewis 2014). The next method involves the use of vat dyes. These dyes are insoluble in nature, and they are chemically reduced to a water-soluble, colourless 'leuco' compound. The 'leuco' compound diffuses into the fibre and subsequently oxidises to its original insoluble form. This makes the colour permanent (Maulik et al. 2022). In contrast to the above-mentioned processes, the third method is the direct dye method, which is simple and involves physical adsorption via hydrogen bonding and van der Waals forces. This method is cheap and economical, but it results in an inferior wash-fastness of coloured textile fabrics (Keraani et al. 2022).

Dyeing of protein fibres is done by creating ionic bonding. The most common method used in the dyeing of protein fibres is the use of acid dyes. These dyes form ionic bonds within an acidic medium. During the dyeing process, when a protein fibre is submerged in a hot acidic dye bath, the functional amino groups on the fibre develop a positive charge, and this attracts and bonds with sulfonic acid groups of the acidic dyes. The optimisation of the pH value and temperature is necessary to achieve a better quality of colour after finishing dyeing (Rippon and Evans 2020).

If mordants are used, then metal salts (alum or tin) are used to form a stable and insoluble complex, and they give fine and desired colour on the fibre (Manian 2023). In this process, both natural dyes, like logwood, or synthetic dyes are used for obtaining a range of colours after washing and exposing them to light. The selection of proper dye and medium chemicals, together with the precise optimisation of pH and temperature, is necessary to achieve durable, vibrant, and beautiful colour.

6.2.1 *Dyeing Fabrics/Textiles*

There is a wide preference for synthetic dyes in the textile industries, but they are considered to be the primary source of water pollution because synthetic dye-based pollutants are non-biodegradable and produce coloured effluents (Walpola et al. 2011). Thus, there is a huge environmental cost, and debate about the use of natural dyes is resurging in the modern world, as natural dyes are ecologically sustainable (Rungruangkitkrai and Mongkholrattanasit 2012). Generally, natural dyes are beneficial to use as they absorb ultraviolet (UV) rays, blocking harmful radiation and providing protection. It is estimated that fabrics or textiles that use natural dyes are capable of blocking up to 80% of UV light and can protect the skin from damage and ageing (Bhandari et al. 2020).

The dyeing process comprises different steps, such as pre-mordanting, dyeing, and after-treatment. In all these processes, it is important to maintain the optimum temperature, specific duration of the dyeing period, and pH value (Eyupoglu et al. 2025). Cotton textiles are dyed at or near boiling temperatures because high temperatures swell the fibre and facilitate the penetration of dye molecules into it. Protein fibres are generally delicate and are dyed at low temperatures to prevent damage. Thus, dyeing of protein fibres begins at normal room temperature and ends as the temperature is gradually increased to the required levels. To obtain a uniform colour, the total duration of the dyeing process can last from an hour to several hours. During the dyeing process, temperature and pH value are important, and so are mordants that help to bind the dye to the fibre.

Semwal et al. (2012) investigated the dyeing properties of *Berberis aristata* (Indian barberry) roots using natural and artificial mordants to dye cotton and thread. In the study, they used the bark of *Myrcia esculenta* as a natural mordant, and tin dichloride ($SnCl_2$) and iron trichloride ($FeCl_3$) as artificial mordants, showing a high absorption percentage by the natural mordant (*Myrcia esculenta,* 47.05%) compared to that by artificial mordants (46.02% for $SnCl_2$ and 36.73% for $FeCl_3$). The colour shades present in cotton fabrics and woollen cloth dyed with *Berberis aristata* are illustrated in Fig. 6.1.

Extracts obtained from *Clitoria ternatea* (butterfly pea flower), *Tagetes erecta* (marigold), and *Punica granatum* (pomegranate peel) were tested on fabrics in the presence of different mordants. Mordants used were lemon juice, ferrous sulphate ($FeSO_4{\cdot}7H_2O$), and copper sulphate ($CuSO_4{\cdot}5H_2O$) and it was found that colour fastness was achieved even with washing, rubbing, and light (Kanchana et al. 2013). Figure 6.2 illustrates the range of colour shades achieved on fabric dyed with *Acacia catechu* extract when using mordants. The effect of mordants such as *chūk* (concentrated lemon juice), *fitkīrī* (alum), *chunā* (lime), and ferrous sulphate in cotton fabrics with different dye extracts at 80 °C for 60 min is presented in Fig. 6.3. For mordanting coloured cotton garments, they used 3% concentrations of natural mordants (bark of *Mangifera indica,* and leaf extracts of *Aloe vera*) and commercial blue vitriol ($CuSO_4{\cdot}5H_2O$). They used a 1:20 material-to-liquor ratio at 80 °C for 60 min, and the colour shades obtained with these different mordants are presented in Fig. 6.4 (Bhandari et al. 2021).

The use of natural dyes has significant challenges because there is difficulty in colour reproducibility, limited shade ranges, low colour yield, and inferior fastness to light and washing. The properties of natural dyes are very much dependent on the extraction methods. For example, conditions involving aqueous, acidic, alcoholic, or alkaline liquids for the extraction of natural dyes from turmeric (*Curcuma longa*) change their phytochemical profile and result in ranges of colour performance (Selvam et al. 2015). Dyes extracted from turmeric using an aqueous medium and from fenugreek (*Trigonella foenum-graecum*) using an alkaline medium showed good antibacterial activity, and these dyes are good for fabric dyeing.

Fig. 6.1 Dyeing performance of *Berberis aristata*: **a** Pure dye; **b** Pre-mordanting—dye + *M. esculenta*; **c** Simultaneous mordanting—dye + *M. esculenta*; **d** Dye + tin dichloride ($SnCl_2$); **e** Dye + ferric chloride ($FeCl_3$); **f** Post-mordanting—dye + *M. esculenta*; and **g** *M. esculenta* (Semwal et al. 2012)

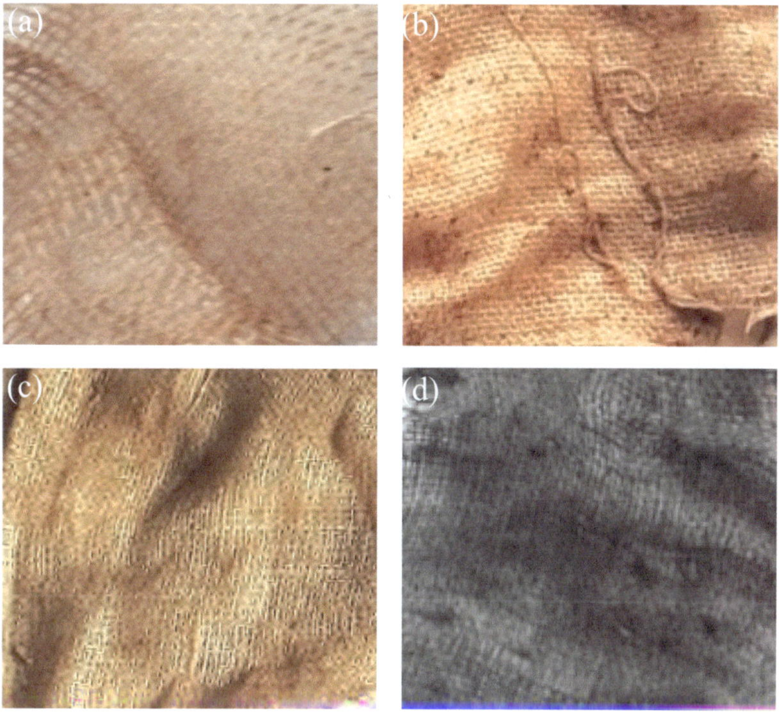

Fig. 6.2 Fabric dyeing with *Acacia catechu* flower extract: **a** Without mordant; **b** Lemon mordant + extract; **c** *Fitkīrī* mordant + extract; and **d** $FeSO_4$ mordant + extract

6.2.2 Dyeing Cotton

Cotton is a cellulose-based fibre with a crystalline structure, and it does not allow large dye molecules to penetrate the surface. Thus, it is essential to swell the fibres and use specific dye classes that can form strong bonds with cellulose. Without the use of appropriate dyes, cotton cannot be properly dyed, and colour can easily fade away during washing and upon exposure to light (Siddiqua et al. 2017).

There are three main methods for dyeing cotton. Reactive dyes in an alkaline bath at temperatures between 60 and 80 °C are used for wash fastness. Examples of reactive dyes are Procion MX for cold dyeing or Remazol for hot dyeing, which give a wide range of bright shades (e.g., Congo Red and Direct Black 22). This process generally uses a hot aqueous bath with added electrolytes to reduce electrostatic repulsion, allowing the dye molecule to attach through hydrogen bonding. However, this process yields lower wet fastness, often necessitating a post-dyeing treatment that needs some fixing agents (Hande et al. 2022). To achieve the highest level of durability, vat dyes are used. Vat dyes are insoluble compounds known for their exceptional fastness to washing and light. The classic examples of vat dyes are the use of indigo for denim textile and synthetic vat dyes like Ciba Blue textile.

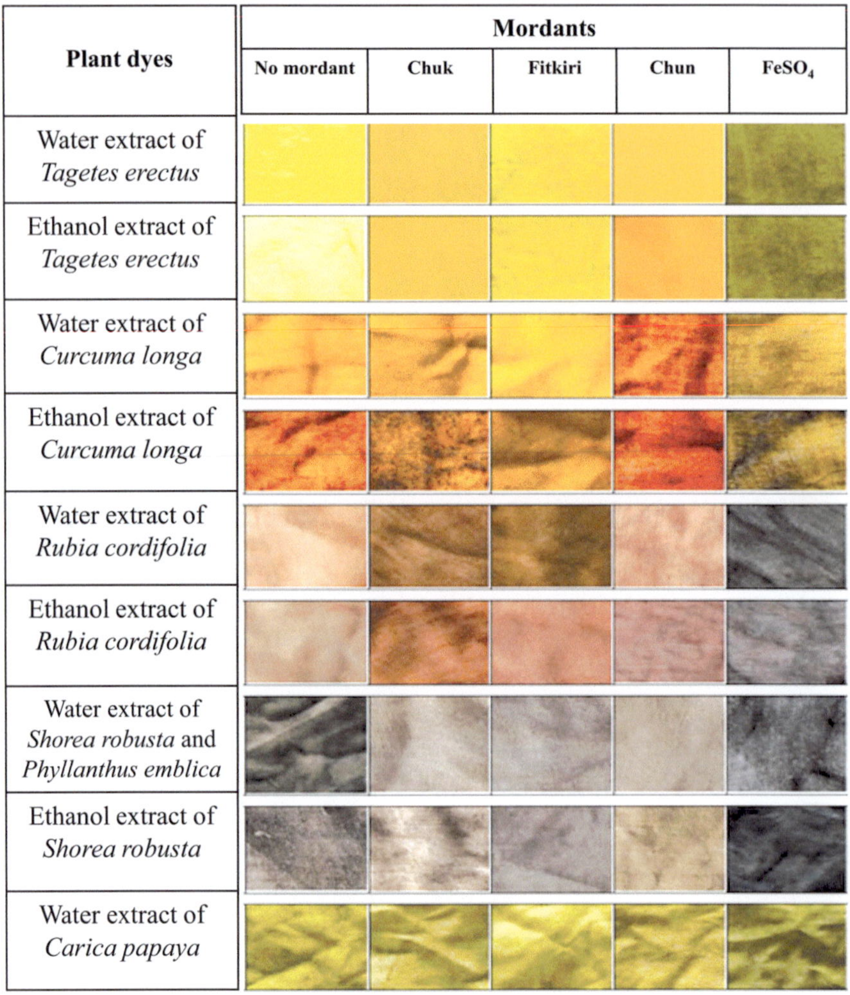

Fig. 6.3 Effect of mordants on cotton fabrics dyed with different plant extracts

Natural dyes obtained from plants such as turmeric or madder are also used for cotton, but they have no inherent affinity for the fibre. Thus, a mordant is necessary, and metal salt mordants are used to fix the dye to the fibre (Rehman et al. 2022).

6.2.3 *Dyeing Wool*

Wool is a protein fibre, and the protein is known as keratin. Keratin has a complex structure, and it contains both acidic (–COOH) and basic ($–NH_2$) groups in the

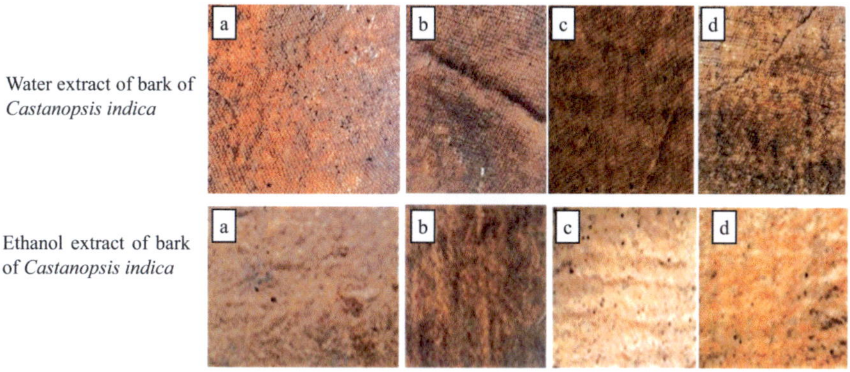

Fig. 6.4 Fabric dyed in various shades using water and ethanol extracts from *Castanopsis indica*: **a** Without mordant; **b** Mango extract mordant; **c** Aloe vera extract mordant; and **d** Copper sulphate ($CuSO_4 \cdot 5H_2O$) mordant (Bhandari et al. 2021)

polymer chains. The scaly surface of wool generally hinders the penetration of dye molecules effectively. This results in an inferior quality finish. Hot and acidic baths are not used because wool fibres mat together. This matting causes felting, yellowing and loss of strength of the wool fibre (Baseri 2022).

The dyeing of wool is done in the presence of an acidic bath. The acid (e.g., acetic acid, sulphuric acid) protonates the amino groups on the wool, giving the fibre a positive cation group ($-NH_3^+$), which then attracts and bonds with the negatively charged dye anion ($-SO_3^-$). Wool dyeing typically starts at around 40–50 °C, and the temperature is then gradually raised to a boiling or near-boiling point (80–98 °C) to allow dye molecules to diffuse into wool. The level of acidity determines the dye's migration and final fastness (Luo et al. 2023). For example, levelling acid dyes such as Acid Red 1 with moderate molecules moderate wet fastness for pale shades. In contrast, milling acid dyes such as acid blue 80 with larger molecules are applied in weakly acidic baths and offer better wet fastness. Thus, Acid Blue 80 is suitable for machine-washable woollen clothes.

Metal-complex dyes that are mixed with different metal atoms are used to produce dull but very deep and exceptionally fast shades (Repon et al. 2024). These are similar to acid dyes, but they require a stronger acid bath, such as Irgalan or Neolan dyes. Wool has a much higher affinity for many natural dyes than cotton, and dyeing is carried out at a lower temperature than that used for cotton to prevent damage. However, different mordants are still used to improve colour fastness and achieve optimum colour shades (Allafi et al. 2022).

6.2.4 Dyeing Silk

Silk is a protein fibre, and the protein present is known as fibroin. As silk is smooth and lustrous, it does not have a scaly surface. It has polymer chains that are densely packed together. It is quite fragile and also more susceptible to damage from chemicals (strong acids and bases) and high temperatures. Thus, care is needed for the dyeing of silk. The boiling of silk for a longer time also damages it by dissolving the natural gum (sericin) present in silk, which holds the threads together. This loss of sericin gum is known as degumming, and this causes loss of fibre weight (Teli 2015; Babu 2018).

Most acid dyes are used for colouring silk, and the method used for dyeing is similar to the method used for dyeing wool. However, colouring of silk is carried out under mild conditions to avoid damage. Temperature is maintained between 80 and 85 °C (never at a rolling boil) in the presence of a weakly acidic to neutral bath (using acetic acid) to avoid silk damage. Examples of acidic dyeing include the use of acid yellow 36 and acid red 18 for colouring silk.

Reactive dyes that form a strong, stable covalent bond with silk are also used because they give bright shades with excellent wash fastness and are cost-effective, too. For this, neutral or weakly acidic baths are used so that amino groups in silk proteins remain positively charged and do not react with the dye molecule. This prevents damage to the silk fibre. However, pH and temperature are important factors for silk dyeing.

Direct dyes that are water-soluble and do not need a mordant as a fixative are also applied to silk. They give bright and attractive colours, but their wet fastness is generally poor in quality. Silk usually has an affinity for several natural dyes, as the protein structure of silk allows dye molecules to penetrate into the fibres, which results in rich and vibrant colours. Mordants like alum are widely used during silk dyeing, and the most widely used natural mordants are tannin-rich dyes (e.g., from pomegranate rind or myrobalan) that give a wide spectrum of colours (Younas et al. 2025).

6.2.5 Dyeing Cosmetics

Cosmetics need a wide range of colours, and there is a high demand for natural dyes that can be used in cosmetics (Guerra et al. 2018). Natural dyes are mainly used for skin applications and have antioxidant and antibacterial properties (Alamgir 2017). An example of a natural dye used in cosmetics is annatto, which is extracted from the seeds of the achiote tree (*Bixa orellana*), and it contains carotenoids like bixin and norbixin. These compounds give an orange-red colour and have a unique dual solubility. Thus, annatto-coloured products are **popular** in different cosmetic items , such as lipstick (Devi et al. 2013). Similarly, henna (*Lawsonia inermis*) is another

Fig. 6.5 Design and colour of *Lawsonia inermis* leaf paste on hands

natural cosmetic dye that has been used for centuries to colour hair, skin, and nails across many cultures (Adeel et al. 2018), as shown in Fig. 6.5. Lawsone, present in henna, gives an orange-red hue and has therapeutic value.

6.3 Leather Colouring

Leather products are often coloured. The leather industries are more reliant on synthetic dyes (of acid, direct, and metal-complex origin). These synthetic dyes give a broad spectrum of bright, consistent colours and excellent fastness properties, but there is increasing use of natural dyes to meet strict environmental regulations and consumer demand for sustainable green products (Covington and Wise 2019). An example of the use of natural dye is the extract from beetroot used in leather, as shown in Fig. 6.6 (Sivakumar et al. 2009). Ferrous sulphate ($FeSO_4$), found in dye derived from the *Cocus lacca* bug, is used as a mordant, and it gives superior quality products compared to other metal mordants, which also produce excellent colour fastness and colour brightness, as seen in Fig. 6.7 (Uddin et al. 2017).

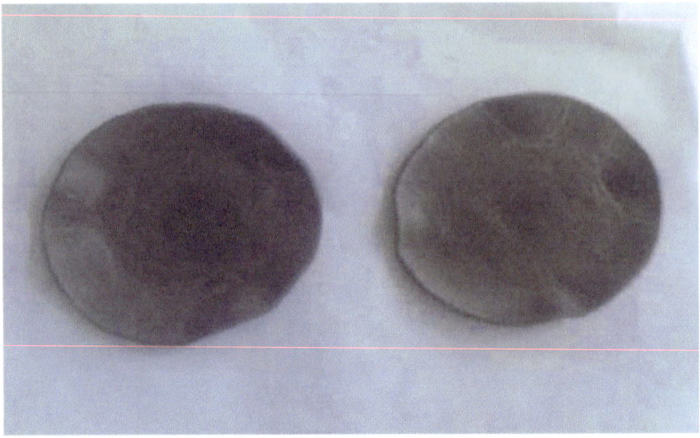

Fig. 6.6 Colour shades in dyed leather with beetroot colourant (Sivakumar et al. 2009)

6.4 Food Colouring

The dye is widely used in food, and dyes increase nutritional value, improve food colour, taste, flavour, and texture (Spence 2015; Bhandari et al. 2021). Their use is mainly to maintain consistent colour and quality in mass-produced food (Silva et al. 2022). Thus, there is a high demand for reliable food additives and chemical processing agents in the world (Spence 2015). In colouring food, both natural and synthetic dyes are used. Colour in food also stimulates appetite and can make a less desirable dish appear more appealing (Chen et al. 2019). For example, peas are coloured green, and strawberry jam is coloured red to make them attractive and more marketable (Duszka et al. 2020).

Numerous studies have been carried out on using natural dyes in food colouration. The use of curcumin from turmeric for yellow colour in dairy products, beverages, cereals, and cosmetics (Saeed et al. 2021) or betanins from beetroot for reddish colour in yoghurts and ice creams is widespread worldwide (Popescu et al. 2023). The colour shades of spaghetti and chicken meat are shown in Fig. 6.8, (a) before dyeing and (b) after dyeing by *Curcuma longa* extract. Likewise, marigold is used as a colouring agent in different baked items, beverages, and dairy products (Maharana et al. 2024). Natural dyes from other plant species, such as annatto (*Bixa orellana*), aidan tree (*Tetrapleura tetraptera*), and roselle (*Hibiscus sabdariffa*) are important food additives (Brudzyńska et al. 2021), and they provide an appealing and consistent appearance.

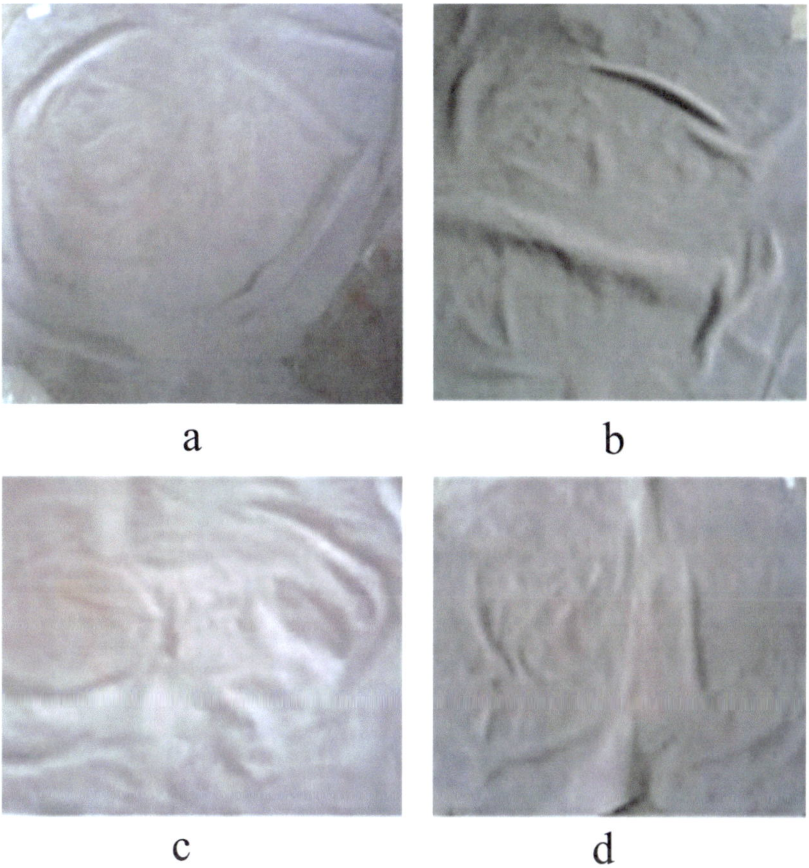

Fig. 6.7 Colour shades of lac dye on leather: **a** Without mordant; **b** With $FeSO_4$ as a mordant; **c** With $Al_2(SO_4)_3$ as a mordant; and **d** With $MgSO_4$ as a mordant (Uddin et al. 2017)

6.5 Drugs or Medicine Colouring

Dyes are used in the pharmaceutical industry as colouring agents for medicines so that used colours can be used for product identification, brand differentiation, and patient compliance. The use of natural dyes in drugs not only provides colour, but also enhances therapeutic properties because a large number of natural dyes used in drugs are obtained from medicinal plant species with inherent bioactive properties (Bhandari et al. 2020; Al Fahad et al. 2024). Dyes obtained from pomegranate (*Punica granatum*), aleppo oak (*Quercus infectoria*), and marigold (*Tagetes erecta*) have demonstrated antimicrobial properties against different bacteria, such as *E. coli*, *Klebsiella* sp., and *Bacillus* sp., and are shown in Table 6.1 (Kanchana et al. 2013; Kamboj et al. 2022).

(a)

(b)

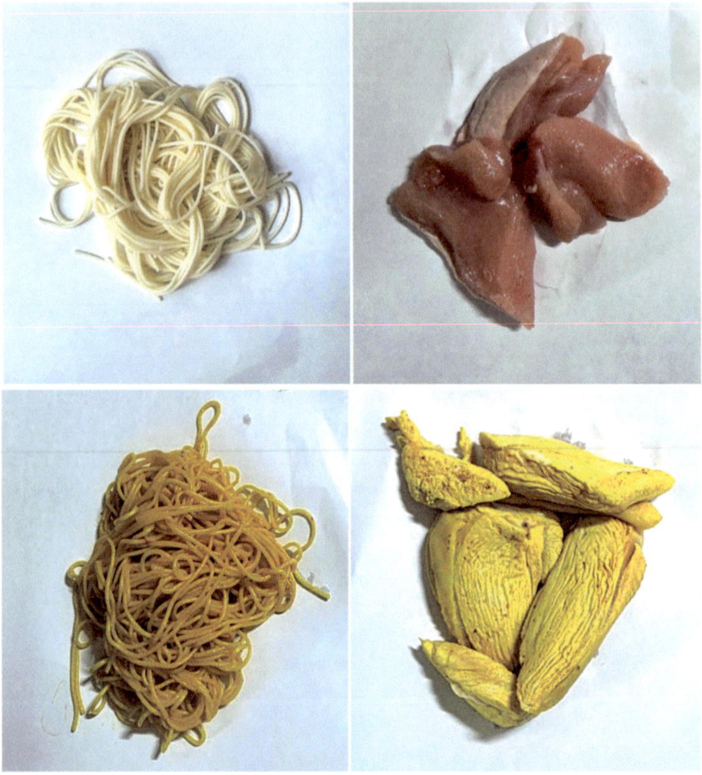

Fig. 6.8 Colour shades produced by *Curcuma longa* extract are shown in **a** before dyeing and **b** after dyeing

Table 6.1 Study focusing on the antimicrobial activity of natural dyes against test bacteria

Diameter of zone of inhibition (mm)					
Bacterial culture	Without mordant	Sample mordanted with fitkiri with e turmeric dye extract	Sample mordanted with lemon with turmeric dye extract	Sample mordanted with ferrous sulphate and dyed with turmeric dye extract	Standard antibiotic (Kanamycin)
S. aureus	3	20	19	24	22
E. coli	3	13	24	17	18
B. subtills	–	10	15	22	16
C. albicans	–	–	5	2	–

The use of mordant in dye-based colouring is quite common and is shown in Fig. 6.9. Fitkiri, lemon, and ferrous sulphate ($FeSO_4$) produced a larger zone of inhibition against *S. aureus, E. coli, B. subtilis*, and *C. albicans* compared to dye without mordant, as illustrated in Table 6.2. Some dyes have antioxidant properties, making drugs more beneficial for ageing and chronic diseases. As presented in Table 6.3, dyes from beetroot and henna demonstrate antioxidant activity (Khan et al. 2018). Carotenoid and anthocyanin pigments are not only used for colouring but also as vitamin supplements for a healthy life (Sass-Kiss et al. 2005; Kapoor et al. 2022). Therefore, the use of colours in pharmaceuticals is widespread, and dyes in medicine have multiple functions. They help medicine to stay stable, perform effectively, and enhance therapeutic properties (Al Fahad et al. 2024).

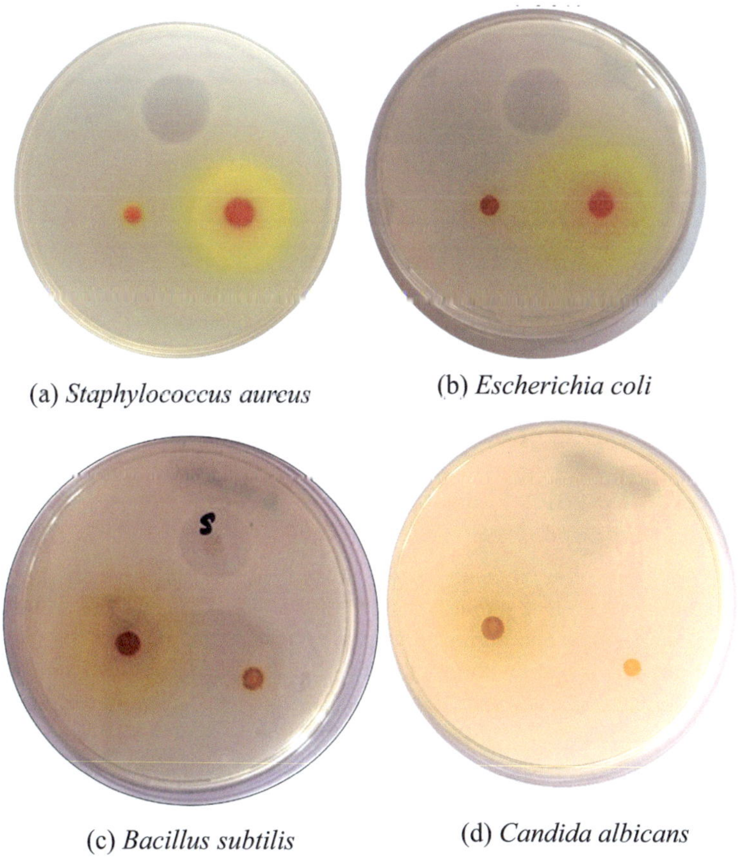

Fig. 6.9 Antibacterial screening of standard kanamycin and dyed cotton fabrics using the disc diffusion method with turmeric dye extract and mordants: **a** Fitkiri; **b** Lemon; **c** and **d** $FeSO_4$

Table 6.2 Study focusing on the antimicrobial zone of inhibition of turmeric dye against test bacteria

Diameter of zone of inhibition (mm)					
Bacterial culture	Without mordant	Sample mordanted with fitkiri with e turmeric dye extract	Sample mordanted with lemon with turmeric dye extract	Sample mordanted with ferrous sulphate and dyed with turmeric dye extract	Standard antibiotic (Kanamycin)
Staphylococcus aureus	3	20	19	24	22
Escherichia coli	3	13	24	17	18
Bacillus subtilis	–	10	15	22	16
Candida albicans	–	–	5	2	–

Table 6.3 Ferric reducing antioxidant power (FRAP) values for different natural dyes, showing antioxidant activity

Dyes	Colour	FRAP value
Beta vulgaris (Beetroot)	Red	2.00
Lawsonia inermis (Henna)	Orange	1.99
Rosa (Rose)	Orange pink	1.98
Daucus Carota (Carrot)	Black/Blue	1.97
Punica granatum (Pomegranate peel)	Pale yellow	1.94

6.6 Dyes in Polymer Colouring

Polymers are also coloured using a wide range of dyes. Polymers are commonly known as plastics and resins that are used for making different items used in our daily lives. These dyes and pigments help polymer products appear transparent or bright with uniform colours without significantly affecting the polymer's mechanical properties. Polymers generally use dyes during various stages of polymer processing, such as mass colouration or post-production dyeing of synthetic fibres. However, proper colours are obtained when there is compatibility between dye and polymer molecules (Fleischmann et al. 2015). Synthetically made disperse dyes, which are non-ionic dyes with very low water solubility, are generally used for hydrophobic polymers like polyester and nylon. They easily sublimate at high temperatures and diffuse into the polymer chains. In contrast, basic (cationic or positively charged) dyes are used for acrylic fibres because they have ionic interactions with the polymer's anionic (negatively charged) molecules.

Polyethene and polypropylene are coloured by using solvent dyes (Gawish et al. 2019). These dyes are mainly used in products like disposable containers, toys, and automotive parts. Polyester and plastic bottles mainly use disperse dyes.

Dyeing of polymers is carried out under high temperature and pressure, which allows dye molecules to penetrate into the crystalline polymer structure. Azo-based disperse dyes, though classified as 'restricted amines', are used to obtain a wide range of vibrant colours in polymers. Titanium dioxide (TiO_2), zinc oxide (ZnO), and calcium carbonate ($CaCO_3$) are used for manufacturing clear white plastic products (Thenepalli et al. 2015; Pinto et al. 2015; Girigoswami et al. 2015). However, white plastic products, such as buckets, bottles, and other appliances, mainly use titanium dioxide pigment. Acid dyes and disperse dyes are used for nylons for their excellent wash fastness. However, there remains a significant challenge in polymer dyeing to ensure the dye's thermal stability during high-temperature processing and its resistance to migration (known as bleeding) and fading during the lifespan of the product.

6.7 Hair Colouring

Hair colouring is done by using both natural and synthetic dyes worldwide. Natural dyes are preferred, as they have no negative side effects (Brown 1997). Hair dyes used for colouring hair are either temporary, semi-permanent, or permanent. Temporary colour molecules deposit on the outer cuticle of the hair, as shown in the cross-section in Fig. 6.10, and they can be removed easily after a few washes. Semi-permanent dyes have molecules with low molecular weight, and they can penetrate slightly into the cortex, and they last up to six washes. Permanent dyes consist of an alkaline medium with an oxidant like hydrogen peroxide (H_2O_2). During colouring, the hair cuticle opens, allowing dyes to diffuse deeply into the cortex, and they last for a long time (Da França et al. 2015). Mordants are also used in hair dyeing. Bhandari et al. (2021) demonstrated that dyes from madder (*Rubia cordifolia*), Saaj (*Terminalia elliptica*), Sal (*Shorea robusta*), and Indian gooseberry (*Phyllanthus emblica*) produced only pale shades without any mordant, but achieved dark, durable colours when used with egg white or ferrous sulphate ($FeSO_4$) as mordants, as illustrated in Fig. 6.11.

6.8 Dye-Sensitised Solar Cells (DSSC)

Dye-sensitised solar cells (DSSCs) that generate electricity also use natural dyes. Thus, DSSCs are cheap and work similar to the photosynthesis process (Khir et al. 2022). Nanocrystalline mesoporous oxide semiconductors are coated onto a transparent conducting glass plate (e.g., fluorine-doped tin oxide, FTO) to form a scaffold for dye adsorption, creating a photoelectrode (photoanode). The different oxides used are titanium dioxide (TiO_2) (anatase), zinc oxide (ZnO), or niobium pentoxide

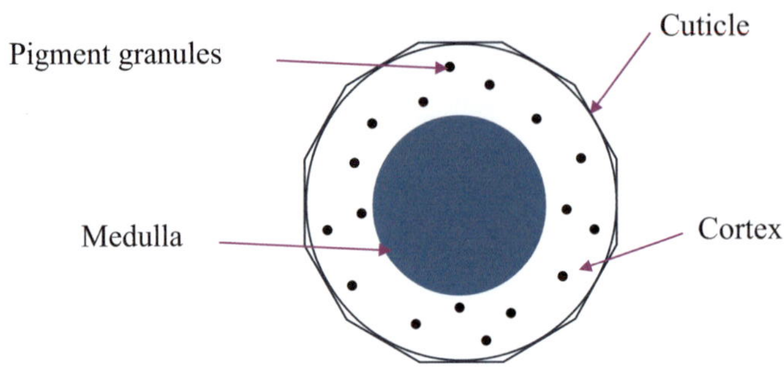

Fig. 6.10 Cross-section of the shaft of a hair strand

(Nb_2O_5). A separate counter electrode, coated with catalyst platinum or carbon, is used to complete the cell (Zulkifili et al. 2015; Ghann et al. 2017). Due to the combination of anion and cation in DSSCs, there is a redox electrolyte, such as the iodide/triiodide redox pair (I^-/I^{3-}), between these two electrodes. These electrolytes regenerate dye molecules for continuous electricity generation.

The basic process used in electricity generation is due to the use of dye molecules (sensitisers) that are attached to semiconductors for capturing light photons. This procedure excites an electron from the Highest Occupied Molecular Orbital (HOMO) and puts it in the Lowest Unoccupied Molecular Orbital (LUMO). The excited electron then goes into the conduction band of the semiconductor, and the dye molecule is oxidised. The redox electrolyte then donates an electron to the oxidised dye molecule, regenerating it. Electrons that move from the external load to the counter electrode complete the circuit, thus generating and conducting electricity. Figure 6.12 shows how this charge separation process works at the interface between a semiconductor and an electrolyte.

As dye molecules, which do not conduct electricity very well on their own, they are chemically bonded to oxide film through functional groups like carboxylic acids, which facilitate charge transfer across the three-dimensional interface (Gong et al. 2016, 2017; Tian et al. 2019).

6.9 Natural Dyes as Corrosion Inhibitors

The emerging application of dyes is in the corrosion prevention of different metals. Corrosion is the degradation of metals, and it causes significant economic loss. Dye molecules that are used as corrosion inhibitors function due to the presence of heteroatoms such as nitrogen (N), oxygen (O), and sulphur (S), π-electrons, and conjugated double bonds found in their molecular structures. These bonds allow them to adsorb onto the metal surface and form a protective film through physical (electrostatic) or chemical (coordinate bond formation) interactions. The protective

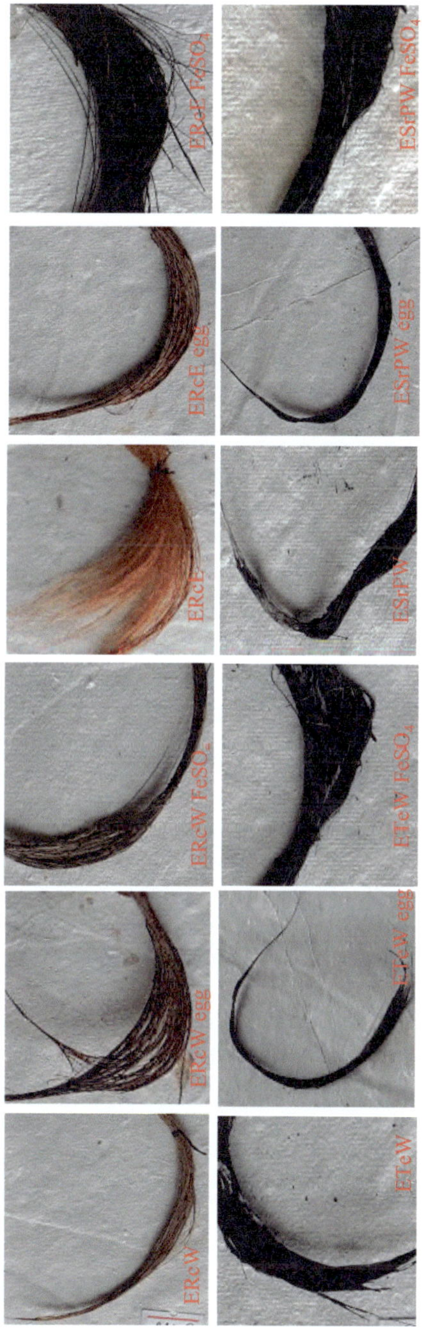

Fig. 6.11 Hair samples dyed with water and ethanol extracts of natural dyes using egg and $FeSO_4$ mordants (ERcW: water extract of *Rubia cordifolia*; ERcE: ethanol extract of *Rubia cordifolia*; ETeW: water extract of *Terminalia elliptica*; ESrPW: water extract of *Shorea robusta* and *Phyllanthus emblica*)

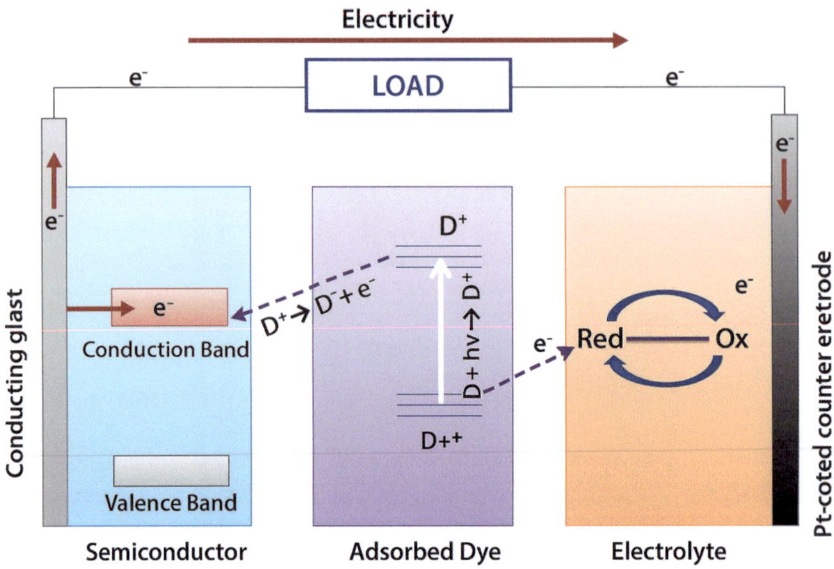

Fig. 6.12 Diagram representing the working principle of dye-sensitised solar cells

film on the metal surface blocks active sites from oxygen or hydrogen ions (H^+) that cause corrosion (Rani and Basu 2012; Nasab et al. 2022; Vashishth et al. 2023).

Lawsone (2-hydroxy-1,4-naphthoquinone) obtained from henna is highly effective as a corrosion inhibitor for mild steel in acidic media. Hydroxyl (–OH) and carbonyl (C = O) groups of different dyes form strong chelating sites and facilitate strong chemisorption onto the iron surface. The yellow pigment in turmeric protects iron and aluminium from corrosion. Its β-diketone moiety and aromatic rings provide multiple active centres for absorption, forming a stable, protective layer. Extracts from safflower flower (*Carthamus tinctorius*) contain the carthamin compound and are effective as a corrosion inhibitor for copper. Similarly, indigo can inhibit corrosion in zinc, and berberine (*Berberis aristata*) is effective in copper and its alloys.

In conclusion, dyes are used in various industrial products. Despite their uses in commonly used products, dyes are used in advanced technological products, such as modern photocells to generate electricity or to prevent corrosion of metals.

6.10 Conclusions

Natural dyes are increasingly preferred for their environmental safety and valuable properties. They provide a sustainable alternative to synthetic dyes, offering ecological and health benefits, and are used in textiles, cosmetics, leather, food, pharmaceuticals, polymers, and advanced technologies. Their effectiveness depends on careful selection of dye types, mordants, pH, and temperature, particularly when colouring cellulose and protein fibres. Although natural dyes can have limited shade ranges and variable fastness, they offer important benefits, such as antimicrobial and antioxidant activity, ultraviolet protection, compatibility with innovative applications, and corrosion inhibition. In conclusion, natural dyes are important resources utilised in both traditional practices and modern industrial applications.

References

Adeel S, Abrar S, Kiran S et al (2018) Sustainable application of natural dyes in cosmetic industry. In: Handbook of renewable materials for coloration and finishing. Wiley, pp 189–211

Al Fahad MdA, Ahamed R, Ahmed T et al (2024) Pharmaceutical applications of natural dyes and pigments. In: Ul Islam S (ed) Renewable dyes and pigments. Elsevier, pp 165–175

Alamgir ANM (2017) Medicinal, non-medicinal, biopesticides, color- and dye-yielding plants; secondary metabolites and drug principles; significance of medicinal plants; use of medicinal plants in the systems of traditional and complementary and alternative medicines (CAMs). In: Alamgir ANM (ed) Therapeutic use of medicinal plants and their extracts, vol 1. Pharmacognosy. Springer International Publishing, Cham, pp 61–104

Alegbe EO, Uthman TO (2024) A review of history, properties, classification, applications and challenges of natural and synthetic dyes. Heliyon 10:e33646. https://doi.org/10.1016/j.heliyon.2024.e33646

Allafi F, Hossain MS, Lalung J et al (2022) Advancements in applications of natural wool fiber: review. J Nat Fibers 19:497–512. https://doi.org/10.1080/15440478.2020.1745128

Babu KM (2018) Silk: processing, properties and applications. Woodhead Publishing

Baseri S (2022) Sustainable dyeing of wool yarns with renewable sources. Environ Sci Pollut Res 29:53238–53248. https://doi.org/10.1007/s11356-022-19629-6

Bhandari N, Shrestha S, Bhandari G et al (2021) Extraction of dye from *Castanopsis indica* for Its use in textile dyeing and medicinal purpose with natural mordant. Egypt J Chem 64:6681–6690. https://doi.org/10.21608/ejchem.2021.48519.2991

Bhandari NL, Pokhrel B, Bhandari U et al (2020) An overview of research on plant based natural dyes in Nepal: scope and challenges. J Agric Nat Resour 3:45–66. https://doi.org/10.3126/janr.v3i2.32328

Brooks MM (2009) Regenerated protein fibres: apreliminary review. In: Eichhorn SJ, Hearle JWS, Jaffe M, Kikutani T (eds) Handbook of textile fibre structure. Woodhead Publishing, pp 234–265

Brown KC (1997) Hair coloring. In: Johnson DH (ed) Hair and hair care, 1st edn. Routledge, New York, USA, p 384

Brudzyńska P, Sionkowska A, Grisel M (2021) Plant-derived colorants for food, cosmetic and textile industries: a review. Materials 14:3484. https://doi.org/10.3390/ma14133484

Chen Y-C, Tsui P, Lee C-S, Chen G (2019) Can plate colour promote appetite and joy while dining? an investigative study in Chinese fine dining restaurants. Asia Pac J Mark Logist 32:105–116. https://doi.org/10.1108/APJML-07-2018-0247

Chinga-Carrasco G (2011) Cellulose fibres, nanofibrils and microfibrils: the morphological sequence of MFC components from a plant physiology and fibre technology point of view. Nanoscale Res Lett 6:417. https://doi.org/10.1186/1556-276X-6-417

Covington AD, Wise WR (2019) Tanning chemistry: the science of leather. R Soc Chem

Da França SA, Dario MF, Esteves VB et al (2015) Types of hair dye and their mechanisms of action. Cosmetics 2:110–126. https://doi.org/10.3390/cosmetics2020110

Dave S, Das J, Varshney B, Sharma VP (2022) Dyes and pigments: interventions and how safe and sustainable are colors of life!!! In: Dave S, Das J (eds) Trends and contemporary technologies for photocatalytic degradation of dyes. Springer International Publishing, Cham, pp 1–20

Devi MVN, Ariharan VN, Prasad NP (2013) Annato: eco-friendly and potential source for natural dye. Int Res J Pharm 4:106–108. https://doi.org/10.7897/2230-8407.04623

Duszka K, Gregor A, Reichel MW et al (2020) Visual stimulation with food pictures in the regulation of hunger hormones and nutrient deposition, a potential contributor to the obesity crisis. PLoS ONE 15:e0232099. https://doi.org/10.1371/journal.pone.0232099

Eyupoglu S, Eyupoglu C, Merdan N (2025) Investigation of dyeing behaviors of bio-mordanted wool fabric dyed with natural dye. Fibers Polym 26:707–721. https://doi.org/10.1007/s12221-025-00858-8

Fleischmann C, Lievenbrück M, Ritter H (2015) Polymers and dyes: developments and applications. Polymers 7:717–746. https://doi.org/10.3390/polym7040717

Gawish SM, Ramadan A, Mosleh SE-S (2019) Improvement of polypropylene (PP) dyeing by modification methods. Egypt J Chem 62:49–62. https://doi.org/10.21608/ejchem.2018.4943.1448

Ghann W, Kang H, Sheikh T et al (2017) Fabrication, optimization and characterization of natural dye sensitized solar cell. Sci Rep 7:41470. https://doi.org/10.1038/srep41470

Girigoswami K, Viswanathan M, Murugesan R, Girigoswami A (2015) Studies on polymer-coated zinc oxide nanoparticles: UV-blocking efficacy and in vivo toxicity. Mater Sci Eng C 56:501–510. https://doi.org/10.1016/j.msec.2015.07.017

Gong J, Sumathy K, Liang J (2016) A simplified electrical model of the dye-sensitised photo-electrochemical cell. Int J Sustain Energy 35:75–87. https://doi.org/10.1080/14786451.2013.870176

Gong J, Sumathy K, Zhou Z, Qiao Q (2017) Modeling of interfacial and bulk charge transfer in dye-sensitized solar cells. Cogent Eng 4:1287231. https://doi.org/10.1080/23311916.2017.1287231

Guerra E, Llompart M, Garcia-Jares C (2018) Analysis of dyes in cosmetics: challenges and recent developments. Cosmetics 5:47. https://doi.org/10.3390/cosmetics5030047

Gürses A, Açıkyıldız M, Güneş K, Gürses MS (2016) Classification of dye and pigments. In: Gürses A, Açıkyıldız M, Güneş K, Gürses MS (eds) Dyes and pigments. Springer International Publishing, Cham, pp 31–45

Hande P, Kulkarni KS, Adivarekar RV et al (2022) A process for dyeing cotton with direct dyes possessing primary aromatic amino groups furnishing wash fastness exhibited by reactive dyes. Color Technol 138:248–254. https://doi.org/10.1111/cote.12586

Kamboj A, Jose S, Singh A (2022) Antimicrobial activity of natural dyes—a comprehensive review. J Nat Fibers 19:5380–5394. https://doi.org/10.1080/15440478.2021.1875378

Kanchana R, Fernandes A, Bhat B et al (2013) Dyeing of textiles with natural dyes—an eco-friendly approach. Int J ChemTech Res 5:2102–2109

Kapoor L, Simkin AJ, George Priya Doss C, Siva R (2022) Fruit ripening: dynamics and integrated analysis of carotenoids and anthocyanins. BMC Plant Biol 22:27. https://doi.org/10.1186/s12870-021-03411-w

Keraani B, Jabli M, Hamdaoui M, Fayala F (2022) An insight into the diffusion of direct dyes into cellulosic materials: effect of fabric characteristics, dye structure, and bath composition. J Nat Fibers 19:5271–5281. https://doi.org/10.1080/15440478.2021.1875369

Khan B, Sindhyan R, Divan A, Rathod S (2018) Extraction, characterization & applications of natural dyes. Ann Plant Sci 7:2463–2467. https://doi.org/10.21746/aps.2018.7.11.4

Khir H, Pandey AK, Saidur R et al (2022) Recent advancements and challenges in flexible low temperature dye sensitised solar cells. Sustain Energy Technol Assess 53:102745. https://doi.org/10.1016/j.seta.2022.102745

Lewis DM (2014) Developments in the chemistry of reactive dyes and their application processes. Color Technol 130:382–412. https://doi.org/10.1111/cote.12114

Luo Y, Cai Z, Pei L, Wang J (2023) Low temperature non-aqueous dyeing of wool fiber with reactive dyes: a sustainable and eco-friendly method. J Clean Prod 432:139803. https://doi.org/10.1016/j.jclepro.2023.139803

Maharana PJ, Maiti S, Kulkarni KS et al (2024) Chemical modification of marigold (*Tagetes erecta*) dye via diazo-coupling reaction for enhanced dyeing on polyester: a sustainable approach. Dyes Pigments 231:112418. https://doi.org/10.1016/j.dyepig.2024.112418

Manian AP (2023) The role of mordants in fixation of natural dyes. In: Handbook of natural colorants. Wiley, pp 507–515

Maulik SR, Chakraborty D, Mohanty M, Debnath C (2022) Vat dye and its evolution in dyeing. In: Textile dyes and pigments. Wiley, pp 87–105

Nasab SG, Yazd MJ, Semnani A et al (2022) Natural corrosion inhibitors. Springer Nature

Pinto D, Bernardo L, Amaro A, Lopes S (2015) Mechanical properties of epoxy nanocomposites using titanium dioxide as reinforcement—a review. Constr Build Mater 95:506–524. https://doi.org/10.1016/j.conbuildmat.2015.07.124

Popescu V, Blaga AC, Cașcaval D, Popescu A (2023) *Beta vulgaris* L.—a source with a great potential in the extraction of natural dyes ontended for the sustainable dyeing of wool. Plants 12:1933. https://doi.org/10.3390/plants12101933

Rani BEA, Basu BBJ (2012) Green inhibitors for corrosion protection of metals and alloys: an overview. Int J Corros 2012:380217. https://doi.org/10.1155/2012/380217

Rehman A, Irfan M, Hameed A et al (2022) Chemical-free dyeing of cotton with functional natural dye: a pollution-free and cleaner production approach. Front Environ Sci 10. https://doi.org/10.3389/fenvs.2022.848245

Repon MdR, Dev B, Rahman MA et al (2024) Textile dyeing using natural mordants and dyes: a review. Environ Chem Lett 22:1473–1520. https://doi.org/10.1007/s10311-024-01716-4

Rippon JA, Evans DJ (2020) Improving the properties of natural fibres by chemical treatments. In: Kozłowski RM, Mackiewicz-Talarczyk M (eds) Handbook of natural fibres, 2nd edn. Woodhead Publishing, pp 245–321

Rungruangkitkrai N, Mongkholrattanasit R (2012) Eco-friendly of textiles dyeing and printing with natural dyes. In: International conference: textiles and fashion. Bangkok, Thailand

Saeed MK, Nisa A, Ahmad I et al (2021) Physico-chemical analysis, total polyphenolic content and antioxidant capacity of yellow dye extracted from *Curcuma longa*: Antioxidant capacity of yellow dye. Pak J Sci Ind Ser B-Biol Sci 64:25–29. https://doi.org/10.52763/PJSIR.BIOL.SCI.64.1.2021.25.29

Sass-Kiss A, Kiss J, Milotay P et al (2005) Differences in anthocyanin and carotenoid content of fruits and vegetables. Food Res Int 38:1023–1029. https://doi.org/10.1016/j.foodres.2005.03.014

Selvam RM, Athinarayanan G, Nanthini AUR et al (2015) Extraction of natural dyes from Curcuma longa, Trigonella foenum graecum and Nerium oleander, plants and their application in antimicrobial fabric. Ind Crops Prod 70:84–90. https://doi.org/10.1016/j.indcrop.2015.03.008

Semwal RB, Semlal DK, Kapoor P (2012) Dyeing properties of *Berberis aristata* DC with natural and synthetic mordants. Trends Appl Sci Res 7:392–399. https://doi.org/10.3923/tasr.2012.392.399

Siddiqua UH, Ali S, Iqbal M, Hussain T (2017) Relationship between structure and dyeing properties of reactive dyes for cotton dyeing. J Mol Liq 241:839–844. https://doi.org/10.1016/j.molliq.2017.04.057

Silva MM, Reboredo FH, Lidon FC (2022) Food colour additives: a synoptical overview on their chemical properties, applications in food products, and health side effects. Foods 11:379. https://doi.org/10.3390/foods11030379

Sivakumar V, Anna JL, Vijayeeswarri J, Swaminathan G (2009) Ultrasound assisted enhancement in natural dye extraction from beetroot for industrial applications and natural dyeing of leather. Ultrason Sonochem 16:782–789. https://doi.org/10.1016/j.ultsonch.2009.03.009

Spence C (2015) On the psychological impact of food colour. Flavour 4:21. https://doi.org/10.1186/s13411-015-0031-3

Teli MD (2015) Advances in the dyeing and printing of silk. In: Basu A (ed) Advances in silk science and technology. Woodhead Publishing, pp 55–79

Thenepalli T, Jun AY, Han C et al (2015) A strategy of precipitated calcium carbonate ($CaCO_3$) fillers for enhancing the mechanical properties of polypropylene polymers. Korean J Chem Eng 32:1009–1022. https://doi.org/10.1007/s11814-015-0057-3

Tian H, Gardner J, Edvinsson T et al (2019) Dye-sensitised. Solar Cells. https://doi.org/10.1039/9781788013512-00089

Uddin T, Razzaq A, Quadery AH et al (2017) Extraction of dye from natural source (lac) and its application on leather. 34

Vashishth P, Mangla B, Bairagi H et al (2023) Natural dyes as corrosion inhibitors. In: Muthu SS (ed) Natural dyes and sustainability. Springer Nature, Switzerland, Cham, pp 265–286

Walpola BC, Subasinghe S, Yoon M-H, Arunakumara KKIU (2011) *Pterocarpus santalinus* Linn. f. (Rath handun): a review of its botany, uses, phytochemistry and pharmacology. J Korean Soc Appl Biol Chem 54:495–500. https://doi.org/10.3839/jksabc.2011.076

Younas U, Safdar F, Maqsood HS et al (2025) A sustainable approach for dyeing silk fabric using dye and mordant extractions from natural resources. Pigment Resin Technol. https://doi.org/10.1108/PRT-02-2025-0018

Zulkifili ANB, Kento T, Daiki M, Fujiki A (2015) The basic research on the dye-sensitized solar cells (DSSC). J Clean Energy Technol 3:382–387. https://doi.org/10.7763/JOCET.2015.V3.228

Chapter 7
Scope and Feasibility Study of Natural Dyeing in Nepal Himalaya

Abstract The use of natural dyes is widespread in the world, and it is a deeply rooted tradition. Using dyes in our daily life often plays a socio-economic role. The global demand for sustainable products has led to the use of natural dyes that often sells at high prices. However, the use of natural dyes is limited, as they are expensive, there is prevalence of fragmented supply chain, and there are serious technological limitations. Thus, synthetic dyes are widely used as alternatives to natural dyes, but they cause pollution. The use of natural dyes can be increased by having scientific collaboration and sustainable supply chains. Some strategic recommendations are given in the end of this chapter so that natural dyes can be wisely used, knowledge is transferred to the next generation, and a competitive industry that supports rural communities is created.

7.1 Traditional Practices of Using Natural Dyeing in Nepal

The history of using natural dyes is interconnected with cultural and societal values in Nepal (Gleba et al. 2016). The use of dyes shows a close relationship between humans and biodiversity. One of the distinctly preserved traditions of using dyes can be seen in religious *thangkas*, temple tapestries, and traditional garments (Malla 1993). Historically, the use of dyes for colouring clothes holds symbolic meanings in religious ceremonies and cultural practices. The practice of using such craft is limited to a few local communities in Nepal, and one of the last remaining communities is the *Ranjitkars* and *Chippas* in the Kathmandu Valley, Nepal, whose history is documented since the early fourteenth century (AsiaInCH 2025). They use very sophisticated techniques passed down through generations of practical experience rather than written manuals. This continuous tradition of using natural dyes needs proper plant identification, seasonal harvesting, preparation, and use of different mordants as fixative agents (Modigh 2018).

Indigenous cultural practice of using dyes in Nepal is supported by the existence of a remarkable population with about 123 castes and ethnicities who speak around 121 languages (CBS 2023) and by the presence of ecological diversity and

N. L. Bhandari et al., *Natural Dyes and Pigments of Nepal Himalaya*,
https://doi.org/10.1007/978-3-032-23229-8_7

bioclimatic zones that support the growth of different plant species (Press et al. 2000; Miehe et al. 2015). More than 182 documented dye-yielding plant species were documented previously in Nepal (Malla 1993), but Chap. 1 has documented more than 200 species found in Nepal and 31 species imported from other countries mainly from India. The plant species provide many dyes and pigments, such as indigo, lac dye, walnut hulls, madder, lichens, and several root dyes, leading to the development of regional specialities used by different communities. Local organisations often support their community with raw material collection, processing services, and knowledge transmission. The *Dhaka*, a traditional handmade fabric of the indigenous people, weaving communities in the Tehrathum district in eastern Nepal, is based on an organised cooperatives, where they sustain the rural economy by sourcing cotton and dyes, providing skilled manpower, and ensuring the intergenerational transmission of their unique craftsmanship (Mabuhang and Khanal 2025). In the Janakpur area, east Nepal, cooperatives and women's self-help groups have transformed traditional *Maithili* art, which is also commonly known as *Madhubani* painting, and have commercialized it from a domestic ritual into a vital rural industry because of support from local cooperatives that support the locals by purchasing locally sourced materials like handmade paper, bamboo brushes, and natural dyes or pigments (Kachhipati 2022). Thus, locals involved in art are able to generate income as well as encourage others to follow them so that the knowledge is transmitted from one generation to the next. Although traditionally made products have faced significant challenges from synthetic alternatives in recent decades, there is growing interest in their preservation and revitalisation, driven by growing global demand for sustainable products, cultural preservation initiatives, and the popularisation of indigenous knowledge.

7.2 Scope and Importance of Dyeing

As a developing nation, the enormous natural wealth in Nepal represents an important opportunity for sustainable economic development that can be achieved by developing natural dye and pigment industries. The country's rich resources are still underutilised and can still be transferred into a significant source of economic growth. However, the viability of these natural resources depends on the growth season, age of the plants, altitudinal range, soil conditions, and water availability of the plant habitat. Neutral pH (6.0–7.5) of water in many areas provides a suitable environment for the growth of a diverse range of dye-yielding trees (Bhandari 2021), herbs, or shrubs (Malla 1993). However, the introduction of synthetic dyes to meet the rising market demand and the limited availability of natural dyes and pigments have seriously threatened indigenous traditions related to colouring. Despite this, natural dyes and pigments remain irreplaceable for some specific applications, such as for lipsticks, which are extracted from *Bixa orellana*, and *Lithospermum erythrorhizon*, and eye shadows obtained from *Indigofera tinctoria* (Sunkyoung 2014; Yameen et al. 2024; Harriman 2025).

Tragically, the valuable indigenous knowledge is now limited to a few surviving elderly practitioners or certain rural communities in Nepal (Rajbhandari 2001; Manandhar 2002; Ghimire et al. 2008). A lack of a well-planned conservation strategy and haphazard collection of plant species from the wild have led to the rapid depletion of both traditional knowledge and plant resources. Therefore, it is important to plan for adequate identification, documentation, and characterisation of dye-producing plant species in Nepal, and further studies are needed to overcome the technical limitations associated with the use of natural dyes (Modigh 2018).

7.3 Feasibility Study of Natural Dyeing Industries in Nepal

As previously mentioned, the botanical wealth of Nepal provides the raw materials needed for the natural dye industry. However, its commercial viability is restricted by high costs, a knowledge-commercial gap, and an informal, fragmented trade chain. The high cost of natural dyes is due to labour-intensive collection, low extraction yields, and complex processing (Panda et al. 2022; Alegbe and Uthman 2024). The indigenous knowledge is not well documented and systematically translated into standardised, scalable industrial protocols. Thus, there is the production of items with inconsistent quality and a lack of trust among big manufacturers who demand reliable products in large volumes (Zahari and Sabran 2025). There is no organised cultivation of dye-yielding plant species, efficient processing centres, or robust quality control mechanisms in the dye industry in Nepal. In addition, the identification of plant resources and the development of value-added strategies need significant improvement in the natural dye industry. There must be a provision for certification of standard products and promotion for '*Nepal Natural Dye*' to establish a new and valuable brand by integrating local harvesters and dyers with stable market outlets. Thus, the future of the natural dye industry depends on natural resources, economic capacity, and technical know-how.

Despite many challenges, there are entrepreneurial efforts towards a viable trade pathway in the natural dye industries. One such industry is Kakani Himalaya Natural Dye Pvt. Ltd., established in 2008 by Shanti Shrestha in Nuwakot, near the Kathmandu valley in Nepal. She was nominated for the prestigious *Surya Nepal Asha Social Entrepreneurship Award* in 2012 for her outstanding work related to the use of natural dyes or pigments in her products. This award specifically honours Nepali social entrepreneurs who apply the principle of *People, Planet, and Profit*, which is the triple bottom line of sustainability. Her nomination highlights a successful model that contributes to societal development through rural empowerment and job creation, providing a reliable market for natural dyes that can be cultivated along with other regular agricultural crops (Bhandari et al. 2020). The industry's production of naturally dyed fabrics is shown in Fig. 7.1.

Fig. 7.1 Kakani Himalaya Natural Dye Pvt. Ltd. is near Kathmandu, Nepal, where textiles are dyed with natural dyes

7.4 International Practices of Natural Dyeing

The use of plant-based dyes is traditional knowledge that is historically integrated into subsistence and has existed as a cultural identity for some communities (Wang et al. 2014; Zahari and Sabran 2025). The indigenous knowledge associated with natural resources is passed down through generations, and such knowledge is mainly based on locally available flora. The examples of using local natural dyes include indigo vats in West Africa and Japan, using cochineal reds in Central America, the weld yellows in Europe, lac resin for scarlet hues in Asia, madder for reds, walnut hulls for brown/black, and *Argeli* (*Edgeworthia gardneri*) for a distinctive yellow in Nepal (Malla 1993; Hunger 2007; Bafana et al. 2011; Waring and Hallas 2013).

There are attempts to have global research funded by big organisations, and studies have systematically investigated and optimised the application of natural dyes. Advanced extraction techniques are applied to improve and standardise the traditional methods, and proper application of dye concentration, temperature, pH, and mordants helps to produce predictable and durable items using both natural and synthetic fibres. However, improvements in the use of specific plant species and mordants are necessary, as natural dyes have significant economic prospects in the global textile and manufacturing industries.

7.5 The Limitations of Natural Dyeing

The development of the natural dyeing industry in Nepal is still at a primitive stage due to different challenges from production to the final processes. For example, the vibrant colours extracted from plant species like *madder* for reds or *walnut hulls* for brown or black are highly susceptible to natural variation. Other natural factors, such as the specific altitude at which a plant species grows, seasonal rainfall, and soil composition, also affect colour quality (Malla 1993; Bhandari et al. 2020). A dyer in eastern Nepal cannot guarantee the same colour even if the same source is brought from central or western Nepal. This variability makes it difficult to fulfil great and consistent demands from the textile industry, which would limit their craft to unique, handicraft pieces rather than large commercial products.

Dye-yielding plant species are often harvested from the wild and are not cultivated, leading to supply uncertainties, and if not managed properly, they could have a devastating effect on the environment due to the loss of plant species and the existing ecosystem (Saxena and Raja 2014; Apau et al. 2025). For example, producing a substantial quantity of dye from rhubarb roots (*Rheum australe*) or the bark of the barberry plant (*Berberis* species) requires harvesting large amounts of raw materials from the wild. It is also a labour-intensive process that makes dyes more expensive. Furthermore, these plant species are available only in specific areas or seasons, making year-round production practically difficult for the producers, hindering the steady workflow of raw materials and complicating the supply chain. Thus, with natural dyes, it is difficult to create a large-scale business for a reliable economy.

Finally, natural dyes often lack optimum performance, mainly with lower colour-fastness, and they fade away easily with repeated washing or exposure to sunlight. Thus, the use of different mordants is required to achieve better, permanently attached colours on the fibre. However, overcoming this difficulty, especially using traditional and safe mordants, adds another layer of technical complexity together with the costs (Salauddin et al. 2021; Repon et al. 2024).

7.6 Socio-Ecological Impacts and Sustainability

Harvesting of plant resources from the wild in Nepal often involves community-based forest management groups and local cooperatives, such as *shahakari* in Nepal. The harvesting of dye-yielding plant species is managed by Community Forestry User Groups (CFUGs), and they are responsible for creating a virtuous cycle of the community's economic incentive to harvest responsibly without destroying the forest to a larger extent (Bhandari et al. 2019). The main challenge associated with natural dyes is that it is difficult to meet the demand without over-exploitation of slow-growing plants from the wild. For sustainability, natural dye-yielding plant species should be cultivated or wisely harvested from forests on a time-rotational basis.

7.7 Colour Pollutions and Their Solutions

Colour pollution is widely associated with dye industries in Nepal and elsewhere in the world. The Bagmati and Dhobi Khola in the Kathmandu Valley historically were not very polluted, but nowadays they are seen with deep blue and black colours due to the direct and untreated discharges from small-scale textile dyeing industries that use synthetic dyes (Green 2003; Gautam et al. 2008). These effluents are not only coloured, but they also contain heavy metals and toxic chemical compounds. To minimise such effluents, it is important to promote *Natural Effluent Treatment Systems* (NEFS) and the *Clean Bagmati* campaign that can help to prevent industrial discharge from being directly discharged into the rivers (Khadka 2020; Giri et al. 2022).

In the natural dyeing sector, discharging their biodegradable wastes into the streams is a cleaner alternative, and this can be a powerful marketing tool. Similar to conditions in Nepal, India faces large-scale colour pollution in textile hubs like Tirupur (Tamil Nadu), Pali (Rajasthan), and Surat (Gujarat) (Bagyaraj 2020). Some rivers in India are biologically dead due to the discharge of industrial effluents into them. To combat this problem, India has implemented stricter environmental regulations known as the *Zero Liquid Discharge* (ZLD) mandate. The main aim behind introducing this regulation is to force dye industries to recycle and reuse almost all their wastewater (Rao and Rao 2022). However, implementation of the ZLD mandate has not been satisfactory as it requires high energy and is expensive.

7.8 Possibilities of Collaboration and Further Opportunities

Scientific studies related to natural dyes in Nepal are not done only for business but also to fulfil academic goals. These studies associated with natural dyes have the potential to drive sustainable development across multiple sectors (Shahid and Shahid-ul-Islam 2013). By moving beyond traditional practices, studies can unlock potential uses far beyond textiles, such as optimising dye-plant cultivation that can create new agricultural income. Analysing phytochemicals of different dyes may reveal medicinal properties that can be used in pharmaceutical industries and also enable the use of pigments for advanced technologies such as dye-sensitised solar cells to generate electricity. Dye-related knowledge can also help in ecotourism and cultural preservation, and dye wastes can be used for different purposes as manure. Thus, by using an integrated approach, local biodiversity can be transformed into an important source of rural economy, technological innovation, and environmental cleanliness (Pizzicato et al. 2023).

To get maximum benefits from natural dyes, a strategic and multi-stakeholder collaboration is necessary. This integrated approach establishes a link between the traditional dyer and entrepreneurs together with scientific institutions that provide knowledge for standardisation and innovation (Casciani and Chen 2024). Key national organisations that must take a leading role in Nepal are research institutions, funding bodies, local government bodies, and non-governmental national and international organisations (NGOs/INGOs). Research institutions such as the Nepal Academy of Science and Technology (NAST), universities, and other organisations can help to conduct phytochemical analysis to identify active colour compounds, optimise extraction, and develop methods to improve colourfastness. The University Grants Commission (UGC) and others can help in facilitating on-the-ground training, supporting the establishment of dye plant cultivation cooperatives that can help in product-to-trade and building a reliable and sustainable market (Paudel et al. 2020; Acharya et al. 2021).

Creating a validated and scalable supply chain for natural dyes is important not only in the harvesting of wild plant species but, more importantly, in the cultivation of dye-yielding plant species (Bhandari et al. 2020). This ensures that farmers benefit from a valuable cash crop. Thus, a synergistic integrated approach can help to foster a circular, green economy promoting environmental conditions in Nepal.

7.9 Recommendations

As there is a huge potential for natural dyes, here are key recommendations for their promotion and sustainability measures in Nepal. By implementing the below-mentioned recommendations, Nepal can transform its rich traditional and cultural heritage into a modern, sustainable, and economically important industry. The recommendations are as follows:

a. **Strengthen research, documentation, and their standardisation**

- Establishing institutional collaboration between scientific institutions and dyer communities (e.g., *Ranjitkars*, *Chippas*) can help to bridge the knowledge gap. The primary goal of collaboration should be systematically documenting indigenous knowledge and then there should be advancement in the use of scientific knowledge associated with dyeing to get maximum benefits.
- A *Nepal Natural Dye* campaign should be created with proper national standards to ensure product authenticity, quality, and sustainable sourcing. This will build trust with domestic and international consumers so that producers can get high prices for their products.

b. **Develop a sustainable and reliable supply chain**

- It is important to have programs that focus on transitioning from harvesting wild plants to the proper cultivation of highly demanded plant species. This will ensure proper yields and stabilise supply, improve quality control, reduce pressure on nature, and provide a reliable cash crop for farmers.
- There should be the establishment of centralised, efficient processing centres that are responsible for extraction, quality testing, and packaging in rural areas. This practice will reduce production costs, maintain consistent quality, and fulfil large demands.

c. **Enhance technical capacity and market accessibility**

- It is important to secure funding for research and development to overcome key technical limitations associated with natural dyes. This will make it possible to develop efficient, low-cost extraction technologies, increase dye yield, reduce processing time, and make natural dyes more affordable.
- There should be the development of robust platforms to interconnect natural dye producers to the market, eco-tourism and green manufacturing sectors, both within Nepal and at the international level.

d. **Implement supportive policies and incentives**

- Natural dye industries should be made aware of their advantage of having no-polluting effluents. At the same time, people associated with dyeing industries should be made aware of pollution and green campaigns, offering incentives for the sustainable approaches that are adopted.
- The government and funding bodies should prioritise and provide financial support to research projects and business ventures related to the natural dyeing sectors, so that others get inspired by such funding and align themselves with sustainable dyeing industries.

e. **Leverage cultural heritage for branding and tourism**

- The approach of developing *dye eco-tourism and cultural tourism* related to natural dyeing should be initiated. Such approaches can be implemented through workshops, visits to dye plant gardens, and demonstrations by master dyers to raise revenue and promote awareness at the international level.
- Rich indigenous knowledge of dyeing heritage in Nepal that has existed for centuries needs proper promotion using different approaches, such as storytelling, making documentaries, and publications, both in the form of popular articles or scientific research articles.

7.10 Conclusions

The tradition of dyeing is prevalent in Nepal, and it is an important part of the culture and has economic potential. This tradition in Nepal is still at a rudimentary stage. To establish a viable dye industry, different aspects related to standardising production and improving quality should be combined with novel scientific findings. For the improvement of the environment and the economy to be sustainable, it is important to establish proper supply chains and initiate dye plant cultivation rather than harvesting resources from the wild. This allows Nepal to develop a unique *Nepal Natural Dye* brand for different products that can be sold at high prices in the market.

References

Acharya KP, Phuyal S, Chand R, Kaphle K (2021) Current scenario of and future perspective for scientific research in Nepal. Heliyon 7:e05751. https://doi.org/10.1016/j.heliyon.2020.e05751

Alegbe EO, Uthman TO (2024) A review of history, properties, classification, applications and challenges of natural and synthetic dyes. Heliyon 10:e33646. https://doi.org/10.1016/j.heliyon.2024.e33646

Apau E, Tawiah B, Boadi MA (2025) Advancement in natural dyes extraction and application for textile finishing. In: Jiang S-X, Seidu RK, Tawiah B (eds) Advances in textile materials and processing techniques for sustainability. Springer Nature, Singapore, pp 77–131

AsiaInCH (2025) Natural dyeing with plant and mineral matter: cotton, silk, wool and leather. Glob InCH-Int J Intang Cult Herit. https://globalinch.org/craft/natural-dyeing-with-plant-mineral-matter-cotton-silk-wool-leather/. Accessed 22 Oct 2025

Bafana A, Devi SS, Chakrabarti T (2011) Azo dyes: past, present and the future. Environ Rev 19:350–371. https://doi.org/10.1139/a11-018

Bagyaraj DJ (2020) Microbes for restoration of degraded ecosystems, 1st ed. New India Publishing Agency (NIPA), New Delhi, India

Bhandari PKC, Bhusal P, Paudel G et al (2019) Importance of community forestry funds for rural development in Nepal. Resources 8:85. https://doi.org/10.3390/resources8020085

Bhandari NL, Pokhrel B, Bhandari U et al (2020) An overview of research on plant based natural dyes in Nepal: scope and challenges. J Agric Nat Resour 3:45–66. https://doi.org/10.3126/janr.v3i2.32328

Bhandari G (2021) An overview of research on plant-based natural dyes in Nepal: Scope and challenges. Lumbini Province (Province No. 5), Government of Province, Nepal
Casciani D, Chen M (2024) Conveying natural dyes in the fashion industry through a design-driven innovation. In: Cantista I, Ritch EL, Shearer L et al (eds) Fashion for the common good. Springer Nature Switzerland, Cham, pp 46–64
CBS (2023) National Census 2021 (Preliminary Report). Central Bureau of Statistics, Government of Nepal, Kathmandu, Nepal
Gautam R, Baral S, Herat S (2008) Opportunities and challenges in implementing pollution prevention strategies to help revive the ailing carpet manufacturing sector of Nepal. Resour Conserv Recycl 52:920–930. https://doi.org/10.1016/j.resconrec.2008.01.003
Ghimire SK, Sapkota IB, Oli BR, Parajuli-Rai R (2008) Non-timber forest products of Nepal himalaya: database of some important species found in the mountain protected areas and surrounding regions, 1st edn. WWF Nepal, Kathmandu, Nepal
Giri I, Ritika KC, Khadka UR (2022) Water quality status in Bagmati river of Kathmandu valley, Nepal. In: Madhav S, Kanhaiya S, Srivastav A et al (eds) Ecological significance of river ecosystems. Elsevier, pp 481–502
Gleba M, Vanden Berghe I, Aldenderfer M (2016) Textile technology in Nepal in the 5–7th centuries CE: the case of Samdzong. STAR Sci Technol Archaeol Res 2:25–35. https://doi.org/10.1080/20548923.2015.1110421
Green HM (2003) The effects of carpet dye on the Bagmati River. Thesis, Massachusetts Institute of Technology
Harriman A (2025) A perspective on Indigo: an iconic colorant. Colorants 4:12. https://doi.org/10.3390/colorants4020012
Hunger K (2007) Industrial dyes: chemistry, properties, applications. Wiley
Kachhipati CP (2022) Transferring traditional art and skills into contemporary craft as a sustainable business in Nepal. A case of Mithila art, Allo fibre and Dhaka weaving. In: Joubert L (ed) Craft shaping society: educating in the crafts—the global experience. Book One. Springer Nature, Singapore, pp 351–363
Khadka S (2020) Learning from river cleaning campaign: a case study of Bagmati river. Asian J Soc Sci Manag 1
Mabuhang TS, Khanal SK (2025) Ethnography of the traditional Dhaka makers: a case study of Terathum district. Adhyayan J 12:115–126. https://doi.org/10.3126/aj.v12i01.73547
Malla L (1993) The natural dyes of Nepal. Research Centre for Applied Science & Technology, Tribhuvan University, Kirtipur, Nepal
Manandhar NP (2002) Plants and people of Nepal. Timber Press, Portland, Oregon, USA
Miehe G, Pendry CA, Chaudhary R (eds) (2015) Nepal: an introduction to the natural history, ecology and human environment in the himalayas. Royal Botanic Garden Edinburgh, Edinburgh, United Kingdom
Modigh N (2018) True colours of Nepal : a minor field study on the Nepalease dyeing industry
Panda A, Maiti S, Madiwale P, Adivarekar R (2022) Natural dyes—a way forward. In: Textile dyes and pigments. Wiley, pp 323–343
Paudel PK, Giri B, Dhakal S (2020) Is research in peril in Nepal? Publication trend and research quality from projects funded by the University Grants Commission-Nepal. Account Res 27:444–456. https://doi.org/10.1080/08989621.2020.1768374
Pizzicato B, Pacifico S, Cayuela D et al (2023) Advancements in sustainable natural dyes for textile applications: a review. Molecules 28:5954. https://doi.org/10.3390/molecules28165954
Press JR, Shrestha KK, Sutton DA (2000) Annotated checklist of the flowering plants of Nepal. Natural History Museum, London
Rajbhandari KR (2001) Ethnobotany of Nepal. Ethnobotanical Society of Nepal, Kathmandu, Nepal
Rao NN, Rao SN (2022) Zero liquid discharge: water recycling in industries towards sustainability. In: Wasewar KL, Rao SN (eds) Sustainable engineering, energy, and the environment. Apple Academic Press, New York, USA

Repon MdR, Islam T, Paul TK et al (2024) Natural dyes in textile printing: parameters, methods, and performance. Environ Sci Pollut Res 31:47552–47583. https://doi.org/10.1007/s11356-024-34424-1

Salauddin MS, Mia R, Haque MA, Shamim AM (2021) Review on extraction and application of natural dyes. Text Leather Rev 4:218–233

Saxena S, Raja ASM (2014) Natural dyes: Sources, chemistry, application and sustainability Issues. In: Muthu SS (ed) Roadmap to sustainable textiles and clothing: eco-friendly raw materials, technologies, and processing methods. Springer, Singapore, pp 37–80

Shahid M, Shahid-ul-Islam MF (2013) Recent advancements in natural dye applications: a review. J Clean Prod 53:310–331. https://doi.org/10.1016/j.jclepro.2013.03.031

Sunkyoung L (2014) A study on the production and clinical evaluation of natural lip Balm using *Chamaecyparis obtusa* and *Lithospermum erythrorhizon*. J Fash Bus 18:29–44. https://doi.org/10.12940/JFB.2014.18.3.29

Wang X, Huang M, Wang H et al (2014) International collaboration activity Index: case study of dye-sensitized solar cells. J Informetr 8:854–862. https://doi.org/10.1016/j.joi.2014.08.004

Waring DR, Hallas G (2013) The chemistry and application of dyes. Springer Science & Business Media

Yameen M, Adeel S, Salman M et al (2024) Sustainable appraisal of lipstick tree seeds (*Bixa orellana*)-based bixin natural orange colorant for green mordanted silk fabrics and wool yarns. Sci Prog 107:00368504241242282. https://doi.org/10.1177/00368504241242282

Zahari NS, Sabran SF (2025) Bibliometric analysis of *Indigofera*-based natural dyes in the Asia–Pacific: research trends and future directions. Discov Sustain 6:562. https://doi.org/10.1007/s43621-025-01421-4

MIX
Papier aus verantwortungsvollen Quellen
Paper from responsible sources
FSC® C105338

If you have any concerns about our products,
you can contact us on
ProductSafety@springernature.com

In case Publisher is established outside the EU,
the EU authorized representative is:
Springer Nature Customer Service Center GmbH
Europaplatz 3, 69115 Heidelberg, Germany

Printed by Libri Plureos GmbH
in Hamburg, Germany